Monitoring Vertebrate Populations

Monitoring Vertebrate Populations

William L. Thompson
Department of Fishery and Wildlife Biology
Colorado State University
Fort Collins, CO 80523
Current Address: *USDA Forest Service, Rocky Mountain Research Station, Boise, ID 83702*

Gary C. White
Department of Fishery and Wildlife Biology
Colorado State University
Fort Collins, CO 80523

Charles Gowan
Department of Fishery and Wildlife Biology
Colorado State University
Fort Collins, CO 80523
Current Address: *Department of Biology, Randolf-Macon College, Ashland, VA 23005*

ACADEMIC PRESS, INC.
San Diego   London   Boston   New York   Sydney   Tokyo   Toronto

Academic Press
a division of Harcourt Brace & Company
525 B Street, Suite 1900, San Diego, California 92101-4495, USA
http://www.apnet.com

Academic Press
24-28 Oval Road, London NW1 7DX, UK
http://www.hbuk.co.uk/ap/

Library of Congress Catalog Card Number: 98-84423

International Standard Book Number: 0-12-688960-0

PRINTED IN THE UNITED STATES OF AMERICA
98 99 00 01 02 03 BB 9 8 7 6 5 4 3 2 1

Contents

Preface   xi

Chapter 1

Basic Concepts

1.1. Spatial Distribution, Abundance, and Density   1
1.2. Monitoring   3
    1.2.1. Baseline Research   3
    1.2.2. Population Monitoring   4
1.3. Characterizing a Species of Interest for Assessment   6
1.4. Obtaining Parameter Estimates   10
    1.4.1. Nonrandom Sampling   11
    1.4.2. Random Sampling   13
1.5. Usefulness of Parameter Estimates   19
    1.5.1. Precision   19
    1.5.2. Bias   38
    Literature Cited   41

Chapter 2

Sampling Designs and Related Topics

2.1. Plot Issues   44
    2.1.1. Plot Design   44
    2.1.2. Unequal Plot Sizes   48
2.2. Categorizing a Sampling Design   49

2.3. Sampling Designs for Moderately Abundant and Abundant Species   51
2.3.1. Simple Random Sampling   51
2.3.2. Stratified Random Sampling   52
2.3.3. Systematic Sampling with a Random Start   59
2.3.4. Simple Latin Square Sampling +1   63
2.3.5. Ranked Set Sampling   64
2.4. A Sampling Design for Rare and Clustered Species   67
Literature Cited   72

Chapter 3

Enumeration Methods

3.1. Complete Counts   75
3.2. Incomplete Counts   76
3.2.1. Indices   77
3.2.2. Adjusting For Incomplete Detectability   83
Literature Cited   117

Chapter 4

Community Surveys

4.1. Number of Species Versus Their Density   124
4.2. Assessment of Spatial Distribution   125
4.2.1. Presence–Absence   125
4.2.2. Other Methods for Assessing Distribution   134
4.3. Assessment of Abundance or Density   134
4.3.1. Presence–Absence   135
4.3.2. Index of Relative Abundance or Density   139
4.3.3. Abundance or Density Estimation   140
4.4. Recommendations   141
Literature Cited   142

Chapter 5

Detection of a Trend in Population Estimates

5.1. Types of Trends   146
5.2. Variance Components   149

5.3. Testing for a Trend   156
    5.3.1. Graphical Methods   156
    5.3.2. Regression Methods   157
    5.3.3. Randomization Methods   163
    5.3.4. Nonparametric Methods   165
5.4. General Comments   168
Literature Cited   168

Chapter 6

Guidelines for Planning Surveys

6.1. Step 1—Objectives   172
6.2. Step 2—Target Population and Sampling Frame   172
6.3. Step 3—Plot Design and Enumeration Method   173
6.4. Step 4—Variance among Plots   174
6.5. Step 5—Plot Selection, Plot Reselection, and Survey Frequency   176
    6.5.1. Plot Selection   176
    6.5.2. Plot Reselection   177
    6.5.3. Frequency of Surveys   179
6.6. Step 6—Computing Sample Sizes   179
6.7. Step 7—Power of a Test To Detect a Trend   180
6.8. Step 8—Iterate Previous Steps   186
6.9. Example   187
Literature Cited   188

Chapter 7

Fish

7.1. Sampling Design   192
    7.1.1. Overall Goals   192
    7.1.2. Suggestions   195
7.2. Fish Collection Methods   202
    7.2.1. Ichthyocides   202
    7.2.2. Underwater Observation   202
    7.2.3. Active Gear   204
    7.2.4. Passive Gear   204
    7.2.5. Electrofishing   205
    7.2.6. Hydroacoustics   207
    7.2.7. Comparisons among Gears   208

7.3. Estimating Populations   208
7.3.1. Complete Counts   208
7.3.2. Presence–Absence   212
7.3.3. Indices of Relative Abundance   213
7.3.4. Capture–Recapture and Removal Estimators   214
7.3.5. Distance Sampling   217
7.4. Example   218
7.4.1. Background   218
7.4.2. Pilot Study   220
7.4.3. Sample Size Calculations   222
7.5. Dichotomous Key to Enumeration Methods   223
Literature Cited   224

Chapter 8

Amphibians and Reptiles

8.1. Complete Counts   234
8.1.1. Entire Sampling Frame   234
8.1.2. Portion of Sampling Frame   234
8.2. Incomplete Counts   236
8.2.1. Index Methods   236
8.2.2. Adjusting for Incomplete Detectability   244
8.3. Recommendations   252
8.3.1. General Comments   253
8.3.2. Dichotomous Key to Enumeration Methods   253
Literature Cited   255

Chapter 9

Birds

9.1. Complete Counts   262
9.1.1. Entire Sampling Frame   262
9.1.2. Portion of Sampling Frame   262
9.2. Incomplete Counts   265
9.2.1. Index Methods   266
9.2.2. Adjusting for Incomplete Detectability   277
9.3. Recommendations   290
9.3.1. General Comments   290
9.3.2. Dichotomous Key to Enumeration Methods   291
Literature Cited   293

Chapter 10
Mammals

10.1. Complete Counts   301
    10.1.1. Entire Sampling Frame   301
    10.1.2. Portion of the Sampling Frame   302
10.2. Incomplete Counts   303
    10.2.1. Index Methods   303
    10.2.2. Adjusting for Incomplete Detectability   307
10.3. Recommendations   315
    10.3.1. General Comments   315
    10.3.2. Dichotomous Key to Enumeration Methods   315
    Literature Cited   317

Appendix A: Glossary of Terms   323
Appendix B: Glossary of Notation   331
Appendix C: Sampling Estimators   337
Appendix D: Common and Scientific Names of Cited
Vertebrates   355

Subject Index   359

Preface

Encroachment of human populations on fish and wildlife populations has caused changes in animal numbers and spatial distribution. Measuring these changes is the subject of this book. Some animal populations have declined drastically because of human encroachment—examples include most endangered species, such as chinook salmon, northern spotted owls, whooping cranes, black-footed ferrets, and grizzly bears.[1] With these species, few would question that human encroachment has caused problems. For other species, we suspect problems (e.g., desert tortoise, Mexican spotted owl, and neotropical migrants). And in a third category, we can put species whose status is uncertain, but that may experience problems in the future. Correctly judging the status of these species requires knowledge of both the species in question and the appropriateness of the methods needed to assess them.

With declining habitat and animal numbers has come increased public scrutiny of remaining habitats containing species of interest, particularly those occurring on public lands. Hence, recent years have seen an increase in the number of court actions brought against various state and federal agencies for their land use and natural resource policies. For instance, Murphy and Noon (1991) stated, "Environmental organizations have filed appeals of virtually every National Forest Plan that has been completed in the past several years." Thus, management plans and policies require the support of rigorously collected scientific data. The *Reference Manual on Scientific Evidence* (Federal Judicial Center, 1994) recounted a Supreme Court decision in *Daubert vs. Merrell Dow Pharmaceuticals, Inc.* that stated, "Evidentiary reliability will be based upon scientific validity." Further, "to be admissible as 'scientific knowledge', scientific testimony 'must be derived by the scientific method'." Management of our fish and wildlife resources

---

[1] Scientific names for species mentioned in the text are given in Appendix D.

can only benefit from an increased reliance on approaches based on good science. Even if a management policy is correct, it should be based on something other than opinion or sparse evidence; that is, it should be defensible in court. Otherwise, resource management decisions are in danger of being taken out of the hands of biologists and natural resource managers and put into the hands of court judges. In the end, it will be the fish and wildlife resources that could suffer.

We have written this book to serve as a general reference to biologists and resource managers who have been charged with monitoring vertebrate numbers within some area. Thus, we have focused on both basic concepts and practical applications. Our overall goal is to provide these professionals with the basic tools needed to assess feasibility of meeting project objectives within funding constraints. We discuss approaches to minimizing sampling error so that monitoring methods have a good chance of detecting trends in populations. By applying the methods described here, the investigator can have a reasonable idea a priori of whether the survey can detect a biologically important population trend, and hence make the decision to proceed with the survey or wait and try to obtain more resources to perform the survey correctly. The important message is that "no survey" results in the same default decision as a poorly performed survey, and "no survey" is cheaper. Worse still, a poorly performed survey means the inertia of its imprecise results must be overcome before a legitimate survey can be conducted. Conversely, detection of a nonexistent trend may lead to some remedial measure, restriction, or management action that could adversely affect both public and private interest. A poorly designed program based on biased estimates could lead to mistaken recognition of a decline in numbers. Hence, well-designed and properly conducted monitoring programs are essential to intelligently managing our fish and wildlife resources while maintaining public support.

We have attempted to combine classical finite population sampling designs (e.g., Cochran, 1977) with population enumeration procedures (e.g., Otis *et al.*, 1978; Buckland *et al.*, 1993) in a unified approach for obtaining abundance estimates. We then use these estimates in a test for trend. Inherent in this approach is the importance of proper design—one that realistically produces abundance estimates with minimum bias and maximum precision at a reasonable cost. Because this book deals mainly with abundance estimation, our focus is on survey design rather than on experimental design. That is, our focus is on methods for obtaining valid inferences from information on a portion of a population in order to detect a change in number over time, rather than ways of manipulating a population to evaluate possible causes for a change in numbers over time. However, a survey design can be applied within treatment and nontreatment groups to obtain abundance estimates. Authors such as Eberhardt and Thomas (1991); Hurlbert (1984); Manly (1992); Skalski and Robson (1992); and Underwood (1994, 1997) have discussed issues related to designing "field experiments" or "quasi-experiments", and we refer readers to these sources.

We have organized the major portion of this book into fundamentals (Chapters 1–6) and applications (Chapters 7–10). Chapter 1 covers basic statistical concepts and terminology as they apply to monitoring vertebrate populations. Particularly important are the ideas of variance and bias. Chapter 2 discusses finite population sampling designs useful in area sampling, and other related topics. We explain how these designs are implemented and make recommendations about which designs are best for decreasing variance. Further, we stress the need for gathering preliminary information for calculating sample sizes and survey costs required to obtain abundance estimates at a prespecified level of precision. Chapter 3 reviews methods for obtaining abundance estimates within selected units, or within an area as a whole (if an area is small enough to be completely surveyed), when complete counts are not realistic. In Chapter 4, the feasibility of the community survey approach is discussed. Chapter 5 deals with methods for detecting trends in populations, with discussion of how temporal and sampling variations affect our ability to detect a trend. Chapter 6 provides step-by-step guidelines for planning a survey for use in a monitoring program.

This book is written primarily for biologists and resource managers with relatively little statistical training. Hence, when possible, we have tried to present statistically related information from a practical perspective. Consequently, Chapters 7–10 represent "applied" chapters in that concepts discussed in earlier chapters are applied to specific vertebrate groups, namely, fish (Chapter 7), amphibians and reptiles (Chapter 8), birds (Chapter 9), and mammals (Chapter 10). These chapters are not meant to be exhaustive descriptions of every possible population enumeration technique. Such information is available from a host of other sources, a number of which are cited as references. Rather, we discuss and evaluate the more common population survey methods, list pertinent references, and provide a dichotomous key to methods of population enumeration at the end of each of these chapters as a general guide to aid readers in choosing the appropriate technique (assuming any is feasible). These keys are meant as a general guide only; they are not intended to provide a "cookbook" approach to designing surveys. Sampling vertebrate populations is much too complex and situation specific for simple answers. We stress that assumptions underlying proposed enumeration methods always should be evaluated for validity, especially from a biological perspective. The final four chapters also contain a number of hypothetical examples that apply some of the concepts discussed earlier in the book, particularly the use of pilot studies to assess the cost and feasibility of a proposed monitoring program. We have attempted to make these examples as realistic as possible, given that they are necessarily simplistic for illustrative purposes. Finally, readers may be tempted to focus only on the chapter that covers the vertebrate group of interest to them, while skipping the first six chapters. We strongly discourage this approach. The fundamentals discussed in earlier chapters of this book apply equally to all groups and must be understood if proper design decisions are to be made.

Because of the subject matter of this book, use of statistical terms, notation, and formulas is unavoidable. We have attempted to limit these technical aspects to make this book as readable as possible. We also have tried to clearly define statistical terminology. Consequently, a glossary of terms (Appendix A) is offered as a guide to readers. Unfortunately, providing a notation that is consistent with existing literature is considerably more challenging because of the often contradictory use of notation both within the fish and wildlife literature and between it and the statistical literature. Appendix B contains our attempt to use a consistent notation. Finally, for interested readers, Appendix C contains a list of selected sampling estimators and cost functions for a variety of one- and two-stage survey designs.

We cannot stress enough the need for a basic understanding of the principles involved in properly planning and executing a rigorous program for monitoring vertebrate populations. When setting up such a program, biologists and managers should seek advice from statisticians and quantitative biologists, and use this input in conjunction with their own (and others') expert knowledge of the species of interest. We do not wish to downplay the importance of this latter information because a good survey design *cannot* be implemented without proper knowledge of the ecology of the species of interest. Our hope is that this book will aid biologists and managers throughout planning and execution of a monitoring program so that monitoring goals can be achieved in a cost-efficient manner.

We gratefully acknowledge the Colorado Division of Wildlife (CDOW) for funding the work on this book. We especially thank CDOW biologists J. Sheppard, T. Nesler, and G. Skiba, and former CDOW biologist L. Carpenter, for their help in providing the funding and impetus for this work. We also thank K. Burnham, D. Anderson, G. Olson, A. Franklin, and R. Ryder for helpful discussions of various topics covered herein. Finally, we are indebted to various anonymous reviewers for their helpful comments on various aspects of this book.

LITERATURE CITED

Buckland, S. T., Anderson, D. R., Burnham, K. P., and Laake, J. L. (1993). "Distance Sampling: Estimating Abundance of Biological Populations" Chapman and Hall, New York.

Cochran, W. G. (1977). "Sampling Techniques," 3rd ed. John Wiley, New York.

Eberhardt, L. L., and Thomas, J. M. (1991). Designing environmental field studies. *Ecol. Monogr.* **61:** 53–73.

Federal Judicial Center. (1994). Reference manual on scientific evidence. Federal Judicial Center, U.S. G.P.O. 1994-384-831-814/20399, Washington, D.C.

Hurlbert, S. H. (1984). Pseudoreplication and the design of ecological field experiments. *Ecol. Monogr.* **54:** 187–211.

Manly, B. F. J. (1992). "The Design and Analysis of Research Studies," Cambridge Univ. Press, Cambridge, UK.

Murphy, D. D., and Noon, B. D. (1991). Coping with uncertainty in wildlife biology. *J. Wildl. Manage.* **55:** 773–782.

Otis, D. L., Burnham, K. P., White, G. C., and Anderson, D. R. (1978). Statistical inference from capture data on closed animal populations. *Wildl. Monogr.* **62:** 1–135.

Skalski, J. R., and Robson, D. S. (1992). Techniques for Wildlife Investigations: Design and Analysis of Capture Data, Academic Press, San Diego.

Underwood, A. J. (1994). Things environmental scientists (and statisticians) need to know to receive (and give) better statistical advice. In "Statistics in Ecology and Environmental Monitoring" (D. J. Fletcher and B. F. J. Manly, eds.), pp. 33–61. Otago Conf. Ser. No. 2, Univ. Otago Press, Dunedin, N. Zeal.

Underwood, A. J. (1997). "Experiments in Ecology: Logical Design and Interpretation Using Analysis of Variance," Cambridge Univ. Press, Cambridge, UK.

Chapter 1

Basic Concepts

1.1. **Spatial Distribution, Abundance, and Density**
1.2. **Monitoring**
  1.2.1. Baseline Research
  1.2.2. Population Monitoring
1.3. **Characterizing a Species of Interest for Assessment**
1.4. **Obtaining Parameter Estimates**
1.4.1. Nonrandom Sampling
1.4.2. Random Sampling
1.5. **Usefulness of Parameter Estimates**
  1.5.1. Precision
  1.5.2. Bias
  Literature Cited

Before describing how to assess animal populations, we must first present basic concepts and terminology associated with population monitoring. A fundamental understanding of the subject matter is needed to both correctly choose and appropriately apply a given approach. Moreover, relevant terminology is required when consulting a statistician about project design so that project objectives can be communicated in a concise and understandable manner, and subsequent recommendations can be understood. In this chapter, we address concepts and terminology associated with monitoring spatial distribution, abundance, and density of species in a given area during some period of time. We assume that biologists have a specific management goal or question that they are attempting to address before they begin designing an assessment protocol.

1.1. SPATIAL DISTRIBUTION, ABUNDANCE, AND DENSITY

The occurrence and spatial arrangement of a species within a defined area at a particular time are called its *spatial distribution*. That is, does a species occur in an area and, if so, where? The most basic distributional information may be obtained from previous records of trapped, harvested, sighted, or other form of documented occurrence of a given species. A more rigorous approach is to collect distributional data as part of a designed study, which we will discuss later in this chapter.

Assessing the status of a species within an area requires more than knowledge of its spatial occurrence. You also must know approximately how many individuals are present. For instance, 100 individuals of a species may occur in each of three sites within an area during year 1; 10 years later there may only be 10 individuals within each site but the species' spatial distribution is unchanged (i.e., all three sites still contain at least 1 individual). Thus, you would like to have a good idea of "how many" in addition to "where." Two terms commonly used to describe "how many" are *abundance* (number of individuals) and *density* (number of individuals per unit area), both of which are defined with respect to a specific area and time period. For example, suppose there are 100 deer in a 100-ha area during 1995. The quantity of deer may be presented in terms of abundance (100 deer) or density (100 deer/100 ha or 1 deer/ha).

Abundance sometimes is used in an unbounded sense, namely, as the number of animals within a site that has no well-defined boundary or area. Examples include the number of migrating waterfowl at a lake and animals coming to a bait site. When abundance is defined in this loose manner, there is no way to delineate a specific group of individuals. Thus, changes in numbers may be from immigration and/or emigration rather than a true change in abundance. That is, changes in numbers of animals recorded at a bait site could be due to changes in numbers of individuals drawn to the bait rather than an actual increase or decrease in their overall numbers. Therefore, we strongly favor explicitly defining each population of interest both spatially and temporally.

Validly assessing spatial distribution, abundance, or density requires either a *survey* or a *census*. A survey is a partial count of animals or objects within a defined area during some time interval, whereas a census refers to a complete count within a particular area and time period. These two terms are not synonymous, although they often are used incorrectly as such.

Spatial distribution, abundance, and density are *parameters*, i.e., they are fixed but unknown quantities within a defined area and time period. Obviously, the number and spatial distribution of animals will change over time and space; therefore, these parameters are "fixed" only over a short time within a defined space. Because biological populations are subject to processes of birth, death, immigration, and emigration, collection of assessment data within a study area over a certain time interval represents just a "snapshot" of a continually changing system. A biological population is considered *demographically closed* when the sampling period is short enough so that no births or deaths occur. A population is *geographically closed* when it is confined to a distinct area or space during the sampling period; hence, there are no movements by individuals across the study area boundary (i.e., no immigration or emigration). Hence, a *closed population* is a group of individuals that is fixed in number and composition during a specified time period. Conversely, an *open population* has one or more processes operating that affect the number and composition of its individuals, i.e., births, deaths, immigration, and emigration (Seber, 1982).

1.2. MONITORING

Monitoring, in its most general sense, implies a repeated assessment of status of some quantity, attribute, or task within a defined area over a specified time period. Implied in this definition is the goal of detecting important changes in status of the quantity, attribute, or task. What is considered important depends on the system being analyzed, and must be defined by investigators based on their expert knowledge. The term monitoring has been used in a variety of contexts in natural resource studies, ranging from collecting baseline ecological information to appraising effectiveness of an assessment program. McDonald *et al.* (1991) listed and defined several different types of monitoring; however, we will only use the term monitoring in conjunction with approaches using repeated measurements collected at a specified frequency over multiple time units (Fig. 1.1). We often will use year as the time unit, but any time period is acceptable as long as it is properly defined. Whatever meaning is employed, monitoring is a tool to be used for both assessing and achieving some management objective.

1.2.1. BASELINE RESEARCH

Gathering baseline information on a species' spatial distribution and abundance sometimes has been referred to as *baseline monitoring*, *inventory monitoring*, or *assessment monitoring* (MacDonald *et al.*, 1991). Although a number of counts may be used to produce an estimate for a particular year, simply collecting data during a single year does not constitute monitoring, i.e., when year is the time unit. A single year's estimate must be placed within a larger framework of estimates from multiple years. However, we have included baseline research in the monitoring section because it represents an initial step in setting up a monitoring program.

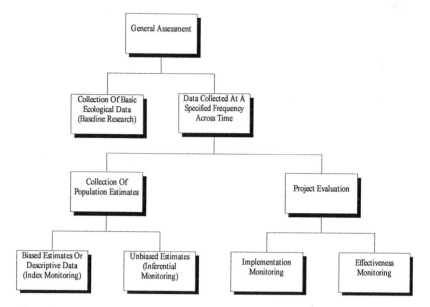

Figure 1.1 Relationship between baseline research and different types of monitoring.

Baseline research could be accomplished through a review of historical records of occurrence and abundance, interviews with experts familiar with the ecology and spatial distribution of the species of interest, actual collection of field data, or a combination of these. This approach may be the only option initially available if basic data are lacking within a particular area. That is, some basic ecological information is required to define the level and frequency of data collection in a monitoring program. The importance of this information should not be underestimated.

1.2.2. POPULATION MONITORING

Population monitoring refers to an assessment of spatial distribution, abundance, density, or other population attributes for one or more species of interest within a defined area over more than one time unit. How often these data are collected, and over what period of time, must be initially determined by the investigator. A goal of population monitoring is to detect an important change, in both magnitude and direction, in average number of animals over a defined time period (i.e., a *trend*).

There are a number of population attributes that can be monitored other than abundance, including reproductive rate, survival rate (e.g., Gutierrez *et al.*, 1996), spatial distribution, and density. This book focuses more on monitoring abundance because distributional data may be obtained as a by-product of abundance, density estimates can be easily converted into abundance estimates, and estimates of survival and reproductive rates generally require more intensive data collection than estimates of abundance. The fundamental concepts presented in this book apply equally well to obtaining valid information on a number of demographic parameters.

Population monitoring can be divided into two categories, *index monitoring* and *inferential monitoring*. These two approaches differ in the degree of potential bias in their population estimates and therefore the strength of the inferences that is possible from collected data. Designs that yield unbiased abundance estimates are usually both cost- and labor-intensive, and therefore may only be applied to a relatively few species.

Index monitoring refers to an assessment protocol that collects data that are at best a rough guess of population trend. A worst, this type of monitoring scheme could lead to incorrect conclusions regarding population trends. Examples of index monitoring include collecting nonrandom samples (see Section 1.4.1), index data, or descriptive data. For instance, a researcher may review current aerial photography for an area of interest once every 5 years to ensure that a particular habitat type is still present. If so, it may be assumed that a species associated with that type is still present. The problem is that factors other than habitat availability may be affecting a biological population. Numbers of breeding neotropical migratory birds, for instance, may be adversely affected by influences occurring on their wintering grounds that have nothing to do with their breeding grounds. Another example of an index monitoring approach is conducting road surveys. Whether presence–absence or relative abundance data are recorded, the results can only be applied to the area of detection around the surveyed road and cannot be validly applied to a larger area.

Inferential monitoring refers to an assessment protocol that uses unbiased or nearly unbiased estimators of spatial distribution and abundance that can be validly expanded to the entire area of interest for assessing trend. Target species could be those of special concern due to their restricted geographic distribution and/or low abundance, their attractiveness to the general public, agency mandates, or other reasons. The objectives of an inferential monitoring program can be stated more rigorously than those for an index-based program. For example, an objective could be to detect a 10% change in number of individuals of species A over a 10-year period about 90% of the time. Statistical techniques for analyzing trend data collected during such a monitoring program will be discussed in Chapter 5. The U.S. Environmental Protection Agency's EMAP (Environmental

Monitoring and Assessment Program; Overton *et al.,* 1990; Messer *et al.,* 1991) is, at least in part, an attempt at a large-scale inferential monitoring program.

Monitoring programs for assessing project-related objectives are called *implementation monitoring* and *effectiveness monitoring* (Fig. 1.1). These activities are primarily administrative in nature. Implementation monitoring has to do with evaluating whether a monitoring program was actually put into place, whereas effectiveness monitoring deals with judging the successfulness of a monitoring program in meeting its predetermined goals (MacDonald *et al.,* 1991). We will not discuss these monitoring approaches in this book, but point out that they are important to an overall assessment program.

The process of assessing population change should be viewed as part of an overall program that leads to some management action. Which and how many species are monitored may depend on various factors including legislative mandates, public opinion, commonness or rareness of a species, extinction potential of a species, and available funding, to name just a few. Funding constraints, in particular, force agencies to prioritize which species will be monitored. That is, available funding directly influences which species will be studied intensively enough to provide adequate data to detect important changes in average numbers and spatial distribution. Thus, the choice of species on which to spend available funds is an extremely important step.

1.3. CHARACTERIZING A SPECIES OF INTEREST FOR ASSESSMENT

Terms used in either a survey or a census are more or less hierarchical (Fig. 1.2), although certain ones may describe the same quantity depending on the situation. We will begin with the basic unit of interest. An *element* is an item on which some type of measurement is made or some type of information is recorded (Scheaffer *et al.,* 1990). This could be an individual animal, an object (such as a nest), or some other item of interest. How an

Figure 1.2 General hierarchy of classification terms, from most basic (element) to most general (target population), used in designing either a survey or a census. Certain terms may refer to the same quantity depending on the situation.

element is specifically defined may differ from one study to the next. We often will be using "animal" in place of "element" for descriptive purposes with the understanding that the definition of element is not so restrictive.

Our next step takes us from the most specific level to the most general. That is, we must define the collection of elements as a distinct and quantifiable entity, or "target," of our assessment. Hence, the *target population* represents all animals contained within some defined space and time interval (Cochran, 1977, p. 5). We use the term target population in a statistical sense; it could contain any part of, or all of, one or more *biological populations,* depending upon size of the area of interest. If you are only interested in assessing abundance of brown trout in a particular lake, then the target population would be defined as all individual brown trout occurring in that lake during the survey. If you are interested in a survey of pronghorn throughout Colorado, the target population would be all pronghorn occurring in the state during the time of assessment.

The next level of classification above an element is the *sampling unit.* A sampling unit is generally defined as a unique collection of elements (Scheaffer *et al.,* 1990). Note that sampling units do not necessarily have to contain any elements, such as in an area survey where the sampling unit is a plot of ground and the element is an animal or object. Further, elements and sampling units sometimes may represent the same quantity. A common example of this occurs in telephone surveys of licensed hunters in which respondents are chosen from a list of licensed hunters. Each hunter represents both the unit of measurement (i.e., each hunter is asked questions) and the unit of selection (i.e., each interviewed hunter had her/his name chosen from an entire list of licensed hunters).

A complete list of sampling units is called a *sampling frame.* If a sampling unit is a plot of ground, then the sampling frame will contain a numbered list or mapping of all plots of ground contained within the study area. One could construct a sampling frame by explicitly defining the size and geographical location of each plot of ground (Fig. 1.3). That is, some region of interest is divided into smaller, nonoverlapping sections, and each of these sections represents a sampling unit (e.g., plot, quadrat, strip transect, and so forth).

A more common approach to identifying the collection of elements to be sampled in an area survey is to simply outline the boundary of the target area while leaving the space within initially undelineated (Fig. 1.4). The surveyed plots are explicitly defined ("drawn out") only after they have been chosen; selection could be based on randomly generated x–y coordinates within the boundary, which then are treated as plot centers (e.g., see Fig. 1.10). In this case, the sampling frame is undefined in the sense that there is not a single, unique, prespecified list of sampling units or plots.

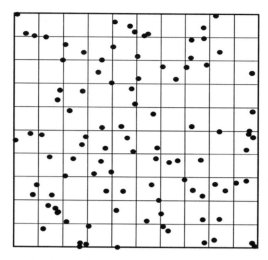

Figure 1.3 An example of a well-defined sampling frame composed of plots superimposed over randomly distributed elements (i.e., black dots). In reality, an area of interest is rarely as perfectly symmetrical as this; however, the concepts are the same regardless of the area configuration.

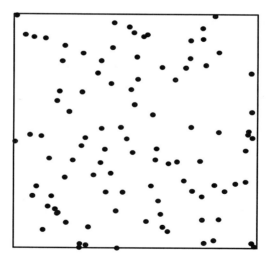

Figure 1.4 An example of an "undefined" sampling frame (i.e., represented only by a boundary with plots not delineated) superimposed over randomly distributed elements (i.e., black dots).

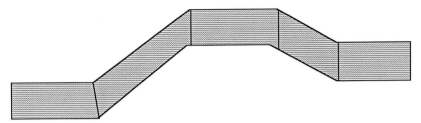

Figure 1.5 A short stretch of river delineated with five plots that may correspond to pools, riffles, or runs.

However, the collection of elements still is defined by the area within the boundary. An extension of this scenario is to conduct an incomplete count of individuals over the entire area (e.g., a mark–resight aerial survey). Thus, the defined area is not partitioned into discrete units in any way. This can be viewed either as a sampling frame consisting of a single unit (the entire area) or as a collection of sampling units (animals) of unknown number, i.e., as an undefined sampling frame. The latter interpretation is the one commonly taken in capture–recapture studies, and the survey process is referred to as *encounter sampling* (Manly, 1992).

The concept of a sampling frame may be readily applied to aquatic environments as well. The target area in rivers would obviously be linear. Sampling units could be delineated pools, riffles, runs, or some other defined stretch of stream (Fig. 1.5). Lakes may be divided into long strips, perhaps by depth, depending on the enumeration technique and species of interest (Fig. 1.6). Conversely, an undefined sampling frame could be used. One example is to randomly choose a starting point and direction of travel on a lake when conducting a trawl survey.

Quite often there are fewer elements within a sampling frame than are contained in the entire area of interest, which contains the target population.

Figure 1.6 A lake delineated with five plots, or strip transects (3-dimensional), such as may be used for a net haul or similar sampling method. These strips may be set at specific depths.

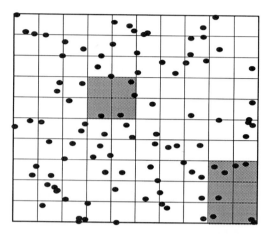

Figure 1.7 The sampling frame is composed of plots with no shading, whereas shaded areas are inaccessible for surveying. The target population includes black dots within all 100 squares, 10 of which are in shaded plots. Conversely, the sampled population only includes black dots in the 90 unshaded squares.

For instance, certain private lands, and the elements therein, may be contained within an area of interest, but may not be accessible for surveying (Fig. 1.7). Therefore, we use *sampled population* to refer to only those elements contained in a sampling frame (Cochran, 1977). Any inferences drawn from count data are applicable only to that part of the target population that has a chance of being surveyed, i.e., only applicable to the area contained within a sampling frame. Consequently, we must carefully consider just how broadly we can apply our survey information. If we are only interested in a particular portion of the state, then that is all that should be included in our target population. However, if we limit our sample to only the northwest portion of the state of Colorado, then we cannot make valid inferences about the other three-fourths of the state.

1.4. OBTAINING PARAMETER ESTIMATES

After the target population has been identified and sampling frame constructed, the next step is to obtain estimates for the parameter of interest. The general procedure is the same whether the parameter of interest is spatial distribution, abundance, or density. Generally, a subset of plots (called a *sample*) is selected, counts are conducted on each selected plot, and count results are inserted into a formula (*estimator*) to calculate a numerical value (*estimate*) for the parameter of interest. Only a subset of plots is chosen, rather than all plots, because costs are almost always too prohibitive for conducting counts on all plots.

The reason for conducting counts on a sample of plots is to make an inference to the sampled population and, hopefully, the target population. The strength of this inference depends on how similar, on average, the sample-based abundance estimate is to the true abundance. One factor affecting this inference is how the sample is chosen. There are two basic approaches to choosing a sample of plots for a survey. One approach, *nonrandom sampling,* relies on a subjective choice of plots to form samples, whereas the other, *random sampling,* chooses plots based on some probability-based scheme. Both approaches have implications regarding how comfortable we can be about the validity of our estimates of spatial distribution, abundance, or density.

1.4.1. NONRANDOM SAMPLING

Nonrandom sampling is the subjective choice of sampling units based on prior information, experience, convenience, or related criteria. General categories of nonrandom sampling include, but are not restricted to, purposive, haphazard, and convenience sampling (Cochran, 1977; Levy and Lemeshow, 1991).

A *purposive sample* contains sampling units chosen because they appear to be typical of the whole sampling frame (Levy and Lemeshow, 1991). For instance, a researcher may subjectively choose plots he or she considers representative habitat for the target species. Or, in assessing potential impacts of a management action, a comparison unit may be chosen based on its apparent similarity to a managed area. Representativeness is not something that can be accurately assessed subjectively. Moreover, animals could easily be focusing on different attributes than those used by the investigator to select a representative sample.

Choosing sampling units based on haphazard contact or unconsious planning is called a *haphazard sample* (Cochran, 1977). Repeatedly dropping a coin on a map overlay with delineated sampling units, and selecting the unit that the coin rests on after each drop, would be a haphazard sample. Another example could be choosing plots based on where a person, while traveling through a study area, encountered some predetermined number of predefined habitat features (e.g., sampling for terrestrial salamanders at the first 10 fallen logs that are encountered). In sampling terminology, haphazard is not synonymous with random.

A *convenience sample* contains sampling units chosen because they are easily accessible, such as those on or adjacent to a road or trail (Cochran, 1977). Examples include bird surveys conducted on roads, searches for animal sign on game trails, and fish surveys restricted to stretches of stream near bridges or road crossings. Roads and trails are usually placed where they are for a reason, and therefore adjacent habitats may be quite different from surrounding areas. Roads often follow watercourses through valleys or through level areas in general. Trails also may be placed for ease of travel or simply because of the scenic value of surrounding habitats. In addition, the rate of change in habitat composition and structure along roads and trails may be quite different from that of surrounding areas.

The problem with nonrandom sampling techniques is an inferential one. In a statistical sense, parameter estimates based on counts from nonrandomly chosen plots cannot be expanded to a larger area (i.e., unsampled plots). In other words, a misleading parameter estimate will result from nonrandom sampling because the selected plots are not truly representative of the unchosen plots (this is called *selection bias*). Consider a bird survey along a riparian area, which contains a variety of shrub and tree cover, surrounded by grassland. Would it be sensible to apply birds counts obtained within this habitat type to the surrounding grassland? The answer is obviously no. The two habitats contain very different species assemblages. Now consider a count of deer feeding at dusk in a roadside meadow surrounded by mature forest. Is it reasonable to calculate a density of deer within the meadow and then expand this to the surrounding forest? Again, the answer is obvious. Although these are extreme examples, the idea is still the same for less obvious examples. Attempting to generalize over a heterogeneous environment without the proper use of inferential statistics can lead to very misleading results.

All nonrandom samples share the common trait that one cannot assign a probability or chance of selection to each plot contained within the area of interest. If, as in a convenience sample, some plots have no chance of selection, then they are not part of the sampling frame. In this case, the sampling frame is only composed of sampled plots; hence, inferences are limited to animals within these plots. Further, to assess the representativeness of an estimate obtained from a nonrandom sample requires comparing it either with the true parameter of interest or with an unbiased estimate, which would require some type of random sample. The true parameter value is unknown, and one may just as well have obtained a random sample in the first place. Therefore, this book will concentrate on random sampling procedures because they, on average, yield unbiased results, as well as allow us to assign a known level of uncertainty to our parameter estimates.

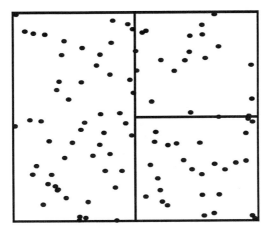

Figure 1.8 Three plots in which the two smaller ones are each half the size of the larger one. If chance of selecting a plot was based on size, the larger plot would be assigned twice the probability of selection as the smaller two.

We do not wish to imply that every aspect of a research study is, or should be, based on some type of random sampling. For example, choice of the area and time period of study could be dictated by funding agencies, political or public mandates, accessibility, and so forth. One does not randomly choose a 5-year interval from all possible intervals during some time period. Nonetheless, concepts of proper inference still apply, i.e., we can only make valid inferences to our parameter of interest within the selected study area during the selected time interval.

1.4.2. RANDOM SAMPLING

A procedure that chooses sampling units based on some known chance of selection is called random sampling (Stuart, 1984). To obtain a random sample, a researcher would assign some positive probability to each sampling unit and randomly select a subset of them based on those probabilities.[1] When sampling units are plots of ground, the probability assigned to each could be based on relative size, i.e., smaller plots could be assigned lower probabilities of selection than larger plots (Fig. 1.8). However, larger

[1] This approach is problematic when the sampling unit is an animal or some other item of unknown quantity and occurrence. We suggest using an area-based random sampling scheme in conjunction with encounter sampling, when feasible. For instance, when conducting a capture–recapture study, one could use some random sampling design to determine trap or grid location to help ensure a random sample of animals.

plots do not necessarily translate into more animals. Unless otherwise noted, estimators for the selection procedures discussed in this chapter are based on equal-sized units.

The simplest and most basic method for randomly selecting plots is called, appropriately enough, *simple random sampling*. Simple random sampling is a process by which a subset of a given number of plots is selected from all plots in a sampling frame in such a way that every subset of a given size has an equal chance of being chosen (Scheaffer *et al.*, 1990). Simple random sampling "without replacement" is a procedure in which every plot, and hence every possible subset of plots of a given size, may only be selected once. Conversely, a "with replacement" procedure allows for repeatedly drawing the same plots; however, this results in biased estimates at smaller sample sizes, whereas formulas based on without-replacement procedures yield unbiased estimates. Consequently, the without-replacement method is the preferred technique in most assessments of biological populations. Unless stated otherwise, when we use the phrase "simple random sampling" we are referring to simple random sampling without replacement. More complex plot selection methods will be discussed in Chapter 2.

In area sampling with a well-delineated sampling frame, a common way to draw a simple random sample is to assign each plot a unique number, and then use either a random number table or a computer to generate a set of random numbers to indicate which plots to include in the survey. For instance, we could construct a sampling frame of 100 equal-sized plots and assign each plot a unique number between 1 and 100 (Fig. 1.9). To obtain a simple random sample of 5 plots, we would randomly select 5 different numbers between 1 and 100, and survey the plots associated with these numbers. If the same number is picked on different draws, the subsequent draw is discarded, and another number is drawn until 5 unique numbers between 1 and 100 are obtained.

Choosing a simple random sample from an area without its units delineated (e.g., Fig. 1.4) requires a different procedure than the one described above. In this case, the bottom border of the defined area is treated as an x axis and the left border as a y axis. Each axis then is numbered appropriately and random numbers are chosen to generate x–y coordinates. For instance, randomly selected coordinates could be used as center points of plots, with the restriction that plots cannot overlap with each other and must fall entirely within the area boundary (Fig. 1.10). The advantage of choosing plots based on this method is the savings in time and effort from not having to delineate each plot. These savings can be significant in situations in which there are many plots, particularly small ones. There also may be some situations where plots are small enough to make their delineation difficult.

1	2	3	4	5	6	7	8	9	10
11	12	13	14	15	16	17	18	19	20
21	22	23	24	25	26	27	28	29	30
31	32	33	34	35	36	37	38	39	40
41	42	43	44	45	46	47	48	49	50
51	52	53	54	55	56	57	58	59	60
61	62	63	64	65	66	67	68	69	70
71	72	73	74	75	76	77	78	79	80
81	82	83	84	85	86	87	88	89	90
91	92	93	94	95	96	97	98	99	100

Figure 1.9 A sampling frame of 100 uniquely numbered plots to be used for selecting a simple random sample.

A potential source of bias with the approach shown in Fig. 1.10 could occur because some elements could have zero chance of being included in a given sample if they are located either between two selected plots or between a selected plot and the boundary. This statement applies only if the distance or "gap" is less than the width of a plot. In theory, every element should have equal probability of being included in a sample (i.e., each should be within one of the sampling units if all units were selected) or abundance estimates will be biased. However, computer simulations have shown this bias to be negligible. To limit the possibility of any bias, we still recommend constructing a well-defined sampling frame when feasible, but using an undefined frame when it is not.

The procedure for selecting a simple random sample from a defined sampling frame applies to estimation of spatial distribution, abundance, or density. The only difference is the amount of data collected within chosen plots. Estimating spatial distribution only requires recording whether a species is present or absent on each selected plot, whereas abundance and

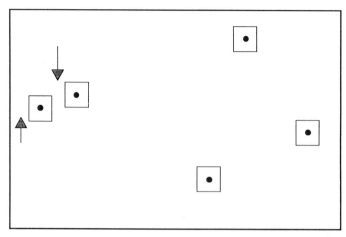

Figure 1.10 A simple random sample of five plots selected by means of randomly chosen *x–y* coordinates, which are represented by the black dots. The two arrows indicate "gaps" where placement of subsequent plots cannot occur.

density, or an index to these two, require a count of individuals. Techniques for acquiring these counts will be discussed in Chapter 3. At present, we will assume that both a species' occurrence and numbers on a plot can be assessed without error. Now we will go through an example in which estimates of spatial distribution, abundance, and density are obtained via a simple random sample. There are other random sampling schemes available; a number of these will be discussed in Chapter 2.

A simple random sample of 10 plots was chosen from a frame of 100 plots (Fig. 1.11) by picking 10 unique random numbers (i.e., 7, 16, 29, 39, 41, 51, 71, 86, 89, and 91), and matching them to plots numbered like those in Fig. 1.9. For estimating spatial distribution, we record either a "1" for present or "0" for absent for the *i*th selected plot. These values are contained in Y_i. We then sum these values across all sampled plots, which compose the estimator $\hat{Y}$. We can represent this as a formula or estimator (Cochran, 1977, p. 50),

$$\hat{Y} = \sum_{i=1}^{u} Y_i, \tag{1.1}$$

where $\sum_{i=1}^{u} Y_i$ means that there are *u* number of Y_i values added together, and *u* refers to sample size (i.e., number of selected plots). Inserting informa-

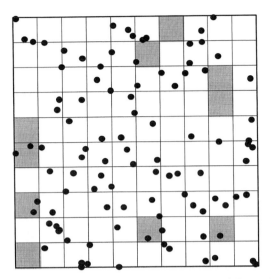

Figure 1.11 A simple random sample of 10 plots (shaded squares) from a frame containing 100 plots and 100 elements (i.e., black dots or "animals").

tion from Fig. 1.11 into Eq. (1.1) yields

$$\hat{Y} = \sum_{i=1}^{10} Y_i = 0 + 1 + 0 + 0 + 0 + 1 + 1 + 1 + 1 + 0 = 5.$$

Note that Y_i does not have a caret (^) because we assume that presence or absence of a species within the ith plot is known with certainty. Conversely, $\hat{Y}$ does have a caret because it is based on only a portion of the available sampling units, which represent only a portion of the total information.

In order to obtain an estimate of spatial distribution, we then use $\hat{Y} = 5$ in an estimator ($\hat{p}$) of the true proportion (p) of plots containing a species of interest (Cochran, 1977, p. 51),

$$\hat{p} = \frac{\hat{Y}}{u}. \tag{1.2}$$

Inserting the appropriate numbers into Eq. (1.2) yields

$$\hat{p} = \frac{5}{10} = 0.5.$$

Thus, we estimate that half (i.e., 0.5 or 50%) of the plots in the frame contain the species of interest.

Obtaining estimates of abundance and density requires a different set of formulas and additional notation. The number of individuals within the ith selected plot (N_i) must be obtained, an arithmetic average ($\overline{N}$) of these N_i computed, and this average used in an abundance estimator ($\hat{N}$) for the entire sampling frame by multiplying it by the total number of plots,[2] U. The relevant formulas (Cochran, 1977, pp. 20–21) to do this are

$$\overline{N} = \frac{\sum\limits_{i=1}^{u} N_i}{u} \tag{1.3}$$

and

$$\hat{N} = U \times \overline{N}. \tag{1.4}$$

Again, note that we assume for this example that we can conduct a complete count within each of the i plots, and hence there is no caret on N_i. However, there is uncertainty associated with the sample mean of the plots counts, $\overline{N}$, and correspondingly, with the abundance estimator, $\hat{N}$.

Based on information gathered from the sampling frame in Fig. 1.11, we calculate an abundance estimate as follows,

$$\overline{N} = \frac{\sum\limits_{i=1}^{u} N_i}{u} = \frac{0 + 1 + 0 + 0 + 0 + 2 + 2 + 1 + 1 + 0}{10} = 0.7$$

and

$$\hat{N} = U \times \overline{N} = (100)(0.7) = 70.$$

We can easily calculate a density estimate from the abundance estimate using the density estimator ($\hat{D}$),

$$\hat{D} = \frac{\hat{N}}{A},$$

where A is the area contained within the sampling frame (note that $\hat{D}$ is an estimator of average density). Both $\hat{p} = 0.5$ and $\hat{N} = 70$ are called *point estimates* (i.e., single values calculated to provide estimates for the parameters of interest). We will use the terms point estimate and parameter

---

[2] Unfortunately, sampling notation is not consistent between statistical texts and fish and wildlife literature. Statistical texts often use N to denote total number of sampling units, whereas fish and wildlife papers often use N to represent total number of animals (in the present example, elements within sampling units). Therefore, we have decided to remain consistent with fish and wildlife literature and keep N for number of animals. Further, we will use U (for "unit") to denote total number of sampling units in a sampling frame, and u to represent the number of sampling units in a given sample.

estimate interchangeably. From Fig. 1.11, we can see that $p = 0.65$ [i.e., 65 squares contain at least 1 black dot (or animal)] and $N = 100$ [i.e., there are 100 black dots (animals) in the sampling frame]. However, in real situations we obviously would not have this information. How, then, do we assess the quality of our point estimates when we do not know the true parameter values? We address this question in the next section.

1.5. USEFULNESS OF PARAMETER ESTIMATES

How well a parameter estimate represents the true spatial distribution, abundance, or density of the species of interest depends on the amount of *error* associated with this estimate. We are using the term error to refer to both variation (i.e., a measure of numerical spread of observations) and mistakes or bias introduced somewhere in the sample design and/or collection process. Error may come into play at both sampling frame and plot levels.

1.5.1. PRECISION

The amount of variation among parameter estimates over repeated samples is referred to as *precision*. For instance, suppose we obtain abundance estimates (i.e., complete plot counts) from surveys conducted once a year in two different areas over a 3-year period. The values for Area A are 19, 20, and 21 animals, and values for Area B are 10, 20, and 30 animals. Both areas have a mean of 20 animals for the 3 years combined, which are our two overall abundance estimates. However, we feel much more comfortable that the overall estimate from Area A is closer to the true abundance because its three values are closely centered around 20. Conversely, there is much more spread in Area B's values, indicating that the true abundance may fall somewhere within a larger range of values. Consequently, the abundance estimate from Area A is more precise than the one from Area B. Before we discuss formulas used in quantifying precision, we will describe different components of variation that affect overall precision of estimates.

1.5.1.1. Components of Variation

The general sources of variation commonly affecting population assessment data are *temporal variation, spatial variation,* and *sampling variation.* Sampling variation can be further divided into *among-unit variation* and

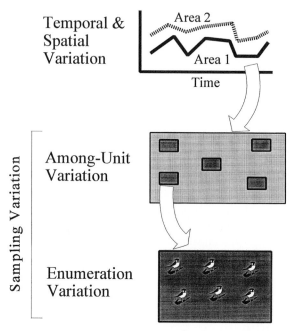

Figure 1.12 Sources of variation involved in estimating the size of an animal population. Populations vary across time and space (temporal and spatial variation), must be sampled from only a portion of plots on a sampling frame (among-unit variation), and often must be assessed from incomplete counts within plots (enumeration variation). Among-unit variation and enumeration variation together compose sampling variation.

enumeration variation (Fig. 1.12). Temporal and spatial variation are listed together because both are associated with environmental and demographic processes affecting animal numbers and spatial distribution. Because both temporal and spatial variation are process-based rather than selection-based (e.g., sampling variation), they sometimes are collectively referred to as *process variation.*

Changes in animal distribution and/or numbers within a sampling frame across time leads to temporal variation. That is, parameter estimates will usually vary from one time period to the next because biological populations are subject to environmental and demographic stochasticity. It is very unlikely that a species will occur in exactly the same numbers in a given area over a period of years. Temporal variation is caused by factors such as weather, other animal populations (predators, prey, competitors), plant community succession, fire, and humans. Note that length and spacing of time periods will affect how temporal variation is defined and perceived.

Differences in spatial distribution and/or numbers of individuals between/among different sampling frames or study areas during a specific time period leads to spatial variation. Spatial variation is usually influenced by habitat quality and heterogeneity, both of which may differ across time. Higher quality habitats may contain more individuals (but see Van Horne 1983), and habitat composition is usually correlated with environmental conditions such as soil type, aspect, slope, and precipitation. For example, one area may receive larger amounts of spring precipitation than another. Increased precipitation could result in increased forage availability for deer, improved doe condition (with corresponding gains in fawn survival), and therefore increased number of animals compared to an area with lower precipitation. In addition, spatial variation may be generated by habitat changes from succession and perturbations like fire and weather.

The net result of all these complicated interactions is that animal populations fluctuate, seemingly randomly in some cases. Classic examples documenting fluctuations in animal populations include the lynx and hare cycles (Keith, 1963), and the Kaibab mule deer population (Caughley, 1970). Thus, to monitor a biological population through time and across the landscape, we are aiming at a moving target. The population likely will not be fixed at some constant value, so that the decision of whether the population is increasing or decreasing will be made difficult by this innate variation.

Sampling variation stems from both random selection of sampling units and incomplete counts of animals within those selected units. The first subcomponent of sampling variation, called among-unit variation, arises from sampling only a portion of available plots. Hence, there are a range of possible values for the many different combinations of plots or samples that can be chosen from a larger number of available plots. Conducting counts within every plot on a sampling frame is frequently impossible. In fact, depending on the number of plots on a frame, only a small proportion of available plots may be sampled.

To better understand how variability in estimates arises from different samples of a given size, we have to know something about how many of these samples are chosen. Suppose our sampling frame is composed of 4 plots labeled A, B, C, and D. How many different simple random samples of size 2 are contained in this frame? We simply have to list all possible without-replacement combinations of 2 plots, of which there are 6 (Example 1.1a). This method for determining the total number of possible samples is fine for very small frames, but quickly becomes tedious, and impossible without a computer or calculator, as frames get larger. Luckily, there is a formula based on counting rules from probability to compute the number of simple random samples of size u from U plots,

$$\binom{U}{u} = \frac{U!}{(U-u)!u!}.$$

The term on the left side of the equation is read as "U choose u" and the term on the right side is its mathematical equivalent for computing the actual value; ! is the factorial sign and is equivalent to multiplying a given number by all positive integers (i.e., whole numbers above 0) less than that number (Note: $0! = 1$). For example, 4!, read as "4 factorial," is equal to 4 times 3 times 2 times 1, or 24. Thus, for a sample of 2 from 4 plots there are

$$\binom{4}{2} = \frac{4!}{(4-2)!\, 2!} = \frac{4 \times 3 \times 2 \times 1}{(2 \times 1)\, 2 \times 1} = \frac{24}{4} = 6 \text{ possible samples,}$$

where adjacent numbers with no symbol separating them are multiplied together as in the denominator above. There would be approximately 1.73×10^{13} or 17.3 trillion possible samples of size 10 from a frame of 100 plots! As noted earlier, every simple random sample has the same probability of being chosen. In the 4-plot frame example, this probability, $P(s)$, would be $1/\binom{U}{u}$ or 1/6. We are using such a small sampling frame for illustrative purposes only; frames in actual studies would normally be considerably larger.

How does the number of possible samples relate to among-unit variation? Because each possible sample generates an abundance estimate, some of which may or may not be the same, there is a range of possible values that could be picked by random chance alone. This spread of possible values is the source of among-unit variation, whereas the values themselves, and the frequency at which they occur, represent the *sampling distribution* of abundance estimates. For instance, the 6 $\hat{N}$ values listed for Example 1.1a represent the complete sampling distribution (each value occurs 1 out of 6 times) for a sample size of 2 from that particular frame.

Example 1.1. Among-Unit Variation

a. A simple random sample of size 2 is selected from a sampling frame composed of 4 plots, which are labeled A, B, C, and D. Plot A has 3 animals, plot B has 13 animals, plot C has 5 animals, and plot D has 12 animals. Therefore, the true abundance, N, is equal to 33. Information for all possible samples is as follows,

Plot pair	Plot counts (N_i)	$\overline{N}$	$\hat{N}$	Prob. of selection [$P(s)$]
(A, B)	(3, 13)	8	32	1/6
(A, C)	(3, 5)	4	16	1/6

(A, D)	(3, 12)	7.5	30	1/6
(B, C)	(13, 5)	9	36	1/6
(B, D)	(13, 12)	12.5	50	1/6
(C, D)	(5, 12)	8.5	34	1/6

Equation (1.3) was used to compute $\overline{N}$ and Eq. (1.4) was used for $\hat{N}$. The $\hat{N}$ selected is determined entirely from random chance, and in this example can take any 1 of 6 values. This is the source of among-unit variation. Note that multiplying each $\hat{N}$ by its probability of selection, $P(s)$, and then summing the resulting values, will produce an overall abundance estimate.

b. The same scenario as in (a) except all possible simple random samples of size 3 are chosen, with the following results.

Plot pair	Plot counts (N_i)	$\overline{N}$	$\hat{N}$	Prob. of selection $[P(s)]$
(A, B, C)	(3, 13, 5)	7	28	1/4
(A, B, D)	(3, 13, 12)	9.33	37.33	1/4
(A, C, D)	(3, 5, 12)	6.67	26.67	1/4
(B, C, D)	(13, 5, 12)	10	40	1/4

We can see that the number of possible samples is lower than in part (a), as is the numerical spread in the $\hat{N}$s (i.e., lower among-unit variation).

The larger the sample size, the more plot information is in each sample. This translates into generally more precise abundance estimates over all samples, and fewer possible samples (and abundance estimates) to choose from. This can be seen when comparing abundance estimates generated by sample sizes of $u = 2$ in Example 1.1a and $u = 3$ in Example 1.1b. Samples of size 2 have abundance estimates ranging from 16 to 50, whereas those for samples of size 3 only range from 26.67 to 40. The extreme case for this trend is choosing a sample size equal to the number of plots in the sampling frame, which would give us only one sample to choose from and all the count values for each plot; hence, there would be no among-unit variation. Another case where there would be no among-unit variation is if every plot contained exactly the same number of individuals. This rarely, if ever, occurs in real situations because of environmental heterogeneity and its effect on spatial distribution of organisms.

Another point to keep in mind regarding sample size and precision is that it is the absolute number of sampling units that is important, not the relative number (Hahn, 1979; Hahn and Meeker, 1991). That is, more precise results will be obtained from a sample size of 50 from a frame of 1000 units (5% of frame) than from a sample size of 10 from a frame of 100 units (10% of frame). Similarly, a sample of size 10 will provide, on average, the same amount of information whether it is drawn from a sampling frame of 100 units or 1000 units.

The second subcomponent of sampling variation, called enumeration variation, arises from incomplete counts of animals on selected plots, which is the usual situation in population studies of fish and wildlife species. That is, there is a range of possible counts that can be produced from repeated surveys on a given plot even if the absolute number of animals does not change. Enumeration variation may occur either by conducting complete counts on only randomly chosen subplots within a plot, or by conducting an incomplete count of individuals over an entire plot. Complete counts within subplots are much more common in vegetational studies than in studies of animal populations. Unbiased estimation techniques, like capture–recapture methods (in certain circumstances), or biased techniques, like index methods, are more the norm in obtaining incomplete plots counts for animal species.

1.5.1.2. Rigorously Quantifying Precision

To this point, we have only used the range or numerical spread of possible sample estimates to roughly quantify precision. Now we will progress to actually calculating a single numerical value that will represent the precision of a parameter estimate. The estimators we present in this section are all unbiased, a subject that will be discussed later.

One common measure of precision, called *variance,* is generally defined as the average of the squared differences between a set of values and the mean of the probability distribution of those values. The variance of N_i, or true counts of the i plots, may be written as (Cochran, 1977, p. 23),

$$S_{N_i}^2 = \frac{\sum_{i=1}^{U} (N_i - \overline{N}_{\text{true}})^2}{U}, \tag{1.5}$$

where $\overline{N}_{\text{true}}$ is a parameter representing the true mean of counts across all plots. The subscript "true" distinguishes this quantity, the population mean, from the mean of counts obtained from a sample of plots, $\overline{N}$, the sample mean. Equation (1.5) represents the true variance of the N_i across plots, or *population variance.*[3]

Let us use the data from Example 1.1a to illustrate this concept. We have a sampling frame divided into 4 plots that contain 3, 13, 5, and 12 animals; these values represent the true distribution or *population distribu-*

---

[3] Statistical textbooks often use μ to denote the population mean and σ^2 to denote the population variance.

tion of counts within a single frame and time period (i.e., no process or enumeration variation). To obtain the true abundance, N, we simply add up the counts across the 4 plots. This gives us $N = 33$ animals. To obtain the true mean of counts, we simply compute the arithmetic average of the 4 counts,

$$\overline{N}_{\text{true}} = \frac{3 + 13 + 5 + 12}{4} = 8.25.$$

This value and the values for the 4 counts are used to compute the population variance in Eq. (1.5),

$$S^2_{N_i} = \frac{[(3 - 8.25)^2 + (13 - 8.25)^2 + (5 - 8.25)^2 + (12 - 8.25)^2]}{4} = 18.69.$$

Unfortunately, we rarely, if ever, have the luxury of conducting complete counts across all plots in realistic situations. Therefore, we must use the among-unit variance of $\overline{N}$, denoted $\text{Var}(\overline{N})$, which is based on a simple random sample from only a portion of available plots. However, this estimator includes the population variance, $S^2_{N_i}$, which we normally do not know. Consequently, we also need an estimator for $S^2_{N_i}$, which is (Cochran, 1977, p. 26)

$$\hat{S}^2_{N_i} = \frac{\sum_{i=1}^{u} (N_i - \overline{N})^2}{u - 1}. \tag{1.6}$$

(Note that the subscripted N_i in $\hat{S}^2_{N_i}$ denotes a variance estimator based on complete plot counts; in later chapters, this will be contrasted by the variance estimator $\hat{S}^2_{\hat{N}_i}$, with a subscripted $\hat{N}_i$, which denotes a variance estimator based on incomplete plot counts.) The expression in Eq. (1.6) then is inserted into the estimator for the variance of $\overline{N}$ (Cochran, 1977, p. 26),[4]

$$\hat{\text{Var}}(\overline{N}) = \left(1 - \frac{u}{U}\right) \frac{\hat{S}^2_{N_i}}{u}, \tag{1.7}$$

where $\left(1 - \dfrac{u}{U}\right)$ is called the *finite population correction* (fpc). The fpc is used to adjust the variance estimator because the sample was taken from a finite or countable population of units rather than an infinite one. As a general rule of thumb, the fpc can be ignored when sample size, u, is 5% or less of the total number of plots, U (Scheaffer *et al.*, 1990). The reason

---

[4] Statistical textbooks commonly represent the "infinite population" (i.e., no fpc) form of this estimator as $s^2_{\overline{x}} = s^2/n$.

for this is that when the *sampling fraction,* $\frac{u}{U}$, is low, the effect of the fpc on the variance estimator is negligible. For instance, a sample of size 5 selected from 1000 plots has an fpc (i.e., 0.995) that is nearly equal to 1. This would have a trivial effect on the variance estimator. Conversely, a sample of size 200 would have an fpc (i.e., 0.8) that is appreciably lower than 1. In the latter case, the fpc would lower the variance estimate by 20%.

When estimating abundance, we are interested in its variance estimator, $\hat{V}ar(\hat{N})$, rather than the one for the mean. The variance estimator for the abundance estimate is obtained by squaring the total number of plots and multiplying this quantity by $\hat{V}ar(\bar{N})$ (Cochran, 1977, p. 26),

$$\hat{V}ar(\hat{N}) = U^2 \left[\left(1 - \frac{u}{U}\right) \frac{\hat{S}^2_{N_i}}{u} \right]. \tag{1.8}$$

This formula can be used to obtain the variance estimator for density by

$$\hat{V}ar(\hat{D}) = \frac{\hat{V}ar(\hat{N})}{A^2}. \tag{1.9}$$

Now that we know how to compute an abundance estimate [Eqs. (1.3) and (1.4)] and its variance estimate [Eqs. (1.6) and (1.8)] from a simple random sample, what does this tell us? We can see from Example 1.1 that there is a range of abundance estimates possible from a simple random sample, some of which may be quite different from the true abundance, N. And if we are taking a sample, we will not know what the true abundance is. How do we judge how close our estimate is to the true abundance? The answer to this is related to the nature of the sampling distribution of abundance estimates.

According to the Central Limit Theorem, a sampling distribution of abundance estimates will follow the form of a normal probability distribution for sufficiently large sample sizes, u. This is true regardless of the form of the underlying distribution of abundance values. We will discuss later how large a sample must be to invoke the Central Limit Theorem. For now, we will assume our sample is sufficiently large.

There is, in fact, a family of normal distributions, with each distribution being more or less bell-shaped in appearance (Fig. 1.13). Moreover, each distribution is defined and identified by its population mean (location) and population standard deviation (shape), and all are symmetrical about their centers or means. Normal distributions with larger standard deviations are wider and less "peaked" than those with smaller standard deviations. With many different normal distributions, how do we know which one to use as an approximation for our sampling distribution? What we can do is

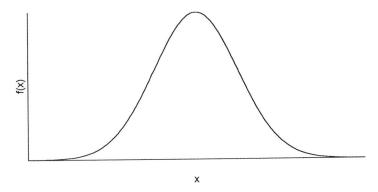

Figure 1.13 An example of a normal probability distribution.

standardize each so that they all follow a single distribution. Specifically, for a given item or measurement (e.g., abundance value, N_i) from some normal distribution, we subtract this item from its population mean (e.g., $\overline{N}_{\text{true}}$), and divide this difference by the population standard deviation (e.g., S_{N_i}). The resulting quantity is commonly referred to as a *z score, z value,* or *normal deviate.*[5] The distribution of all possible z values follows a particular normal distribution, one with a population mean of 0 and population standard deviation of 1. This distribution is called, appropriately enough, a *standard normal distribution.*

Now that we have a way of standardizing any value from any normal distribution, we can apply this approach to a sampling distribution of abundance estimates as we did with the known distribution of abundance values above. Recall that the theoretical sampling distribution contains abundance estimates from all possible samples of size u from a sampling frame (i.e., population distribution). The standard deviation of a sampling distribution is referred to as the *standard error* (or *population standard error* for the theoretical sampling distribution). For the theoretical sampling distribution of abundance estimates, the standard error, $\text{SE}(\hat{N})$, is based on all possible samples.[6] We obviously cannot collect all possible samples, but we can obtain an estimate of the population standard error from a single sample. The sample standard error, $\hat{\text{SE}}(\hat{N})$, is defined as the square root of the variance estimator in Eq. (1.8) or

$$\hat{\text{SE}}(\hat{N}) = \sqrt{\hat{\text{Var}}(\hat{N})},\tag{1.10}$$

[5] Statistical textbooks often represent this as $z = (x - \mu)/\sigma$.

[6] Conversely, an approximated sampling distribution is based on fewer than all possible samples (e.g., Figs. 1.16 and 1.19), and is nearly always, if not always, generated by a computer.

and is usually just referred to as the standard error. Note that this estimator is generated, in part, by the sample standard deviation, $\hat{S}_{N_i}$, which is usually just referred to as the standard deviation.

Another useful measure of precision is called the *coefficient of variation.* This is the standard error of an estimator divided by the estimator. For an abundance estimator this would be

$$\hat{\text{CV}}(\hat{N}) = \frac{\hat{\text{SE}}(\hat{N})}{\hat{N}}. \tag{1.11}$$

The coefficient of variation is used as a measure of relative precision and, more importantly, in determining adequate sample size during project design. This estimator is commonly multiplied by 100 and given as a percentage. We will be using the coefficient of variation in calculating sample sizes in some examples later in this book.

Let us review what we have learned to this point about the sampling distribution of abundance estimates. First of all, assuming an adequate sample size, the sampling distribution approximates a normal distribution regardless of the underlying spatial distribution of animals. Second, if we convert the values of the sampling distribution into z values, these values will approximate a standard normal distribution with mean 0 and standard deviation 1. Now, how can we use this information to quantify the precision of abundance estimate from a single sample? The answer relates to a property shared by all normal distributions. Specifically, about 68% of the values within any normal distribution fall within 1 standard deviation from its center or mean; about 95% fall within 2 standard deviations, and just under 100% fall within 3 standard deviations. For a sampling distribution of abundance estimates, these would be described as 1, 2, and 3 standard errors from the center. And, for a standard normal distribution, the preceding would be in terms of z values from the center. For instance, approximately 95% of the values in the distribution would be within $z = 2$ of the center (Fig. 1.14).

We now can use the properties of our standardized sampling distribution to generate a symmetrical, two-sided interval around each parameter estimate that will give us a measure of confidence regarding how close our estimate is to the true parameter. The two-sided *confidence interval* (CI) for an abundance estimate is defined as

$$C[L_L \leq N \leq U_L] = 1 - \alpha, \tag{1.12}$$

where C stands for confidence, N is the true abundance, L_L is the lower limit of the interval, U_L is the upper limit of the interval, and $1 - \alpha$ is the *confidence level* (Graybill and Iyer, 1994). In addition, L_L is equal to $\hat{N}$ −

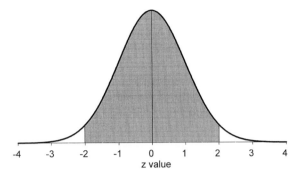

z value

Figure 1.14 A standard normal distribution, where the gray area represents about 95% of the area under the curve, and hence contains about 95% of the items in the distribution.

[z value $\times \hat{SE}(\hat{N})$] and U_L is equal to $\hat{N} + $ [z value $\times \hat{SE}(\hat{N})$] when we assume a standard normal. Thus, the upper and lower limits are based on the abundance estimate derived from the sample. How is this interval interpreted? We say that we are $100(1 - \alpha)\%$ confident that the true parameter will occur somewhere between the lower and upper limits. What do we mean by "confident" in this context? Recall that a theoretical sampling distribution is composed of all possible samples of size u randomly selected from a sampling frame containing U sampling units. Each of these samples has its own abundance estimate and associated confidence interval. Out of all these possible confidence intervals, $100(1 - \alpha)\%$ will contain the true abundance. A similar statement can be made for fewer than all possible samples. Namely, over many repeated samples, $100(1 - \alpha)\%$ of the computed confidence intervals, on average, will contain the true abundance. The key point is that $100(1 - \alpha)\%$ relates to the procedure for interval construction rather than a specific interval (Hahn and Meeker, 1991). That is, there is not a $100(1 - \alpha)\%$ probability that a given interval contains the true abundance because the true abundance is a constant and therefore is either within the interval or is not.[7] A fixed number cannot be 95% inside an interval and 5% outside. Therefore, we use the term confidence from a procedural standpoint as described above.

Confidence level is directly related to the z value. Suppose, for instance, that we would like to generate a 95% confidence interval around our abundance estimate. Assuming the standard normal distribution is a valid approximation for our sampling distribution, we can look in a table of standard normal deviates (located in the back of most introductory statistics text-

---

[7] An exception to this is a confidence interval based on Bayesian inference, an approach that we do not address in this book.

books) to find the z value that corresponds to 95% of the area under the standard normal distribution. Let us see how this is done. The α in the confidence level represents the proportion of area in both tails of the sampling distribution, and is set by the investigator. In this case, we have set $\alpha = 0.05$, i.e., $100(1 - 0.05) = 95\%$. Therefore, because the normal distribution is symmetrical, each tail contains $0.05/2 = 0.025$ or 2.5% of the area. Now, we find the z value corresponding to 0.025, which is designated as either $z_{1-\alpha/2}$ or $z_{\alpha/2}$. Depending on how a particular table of normal deviates is constructed, we would usually look for either 0.025 or 0.4750 in the main body of the table, and then find the corresponding z value along the axes. In either case, the z value would be 1.96. (The 0.4750 is the proportion of the area between 0 and $z = 1.96$; doubling this value, because of symmetry, yields 0.95). A common approximation for this is $z = 2$, which actually corresponds to 95.44% of the area. Using the exact z value, our 95% confidence interval for an abundance estimate is $\hat{N} \pm 1.96 \times \hat{SE}(\hat{N})$.

Assuming unbiased abundance estimates (see next section), the validity of the previous confidence interval depends on how closely the sampling distribution of standardized deviates approximates a standard normal. Recall that knowledge of the true standard error is required to compute z values of abundance estimates,[8] i.e.,

$$z = \frac{\hat{N} - N}{SE(\hat{N})}.$$

In reality, the true standard error is unknown and its estimator must be used. However, when $\hat{SE}(\hat{N})$ is substituted for $SE(\hat{N})$, the resulting standardized deviates (called t values) no longer follow a standard normal. Assuming the population being sampled is at least approximately normally distributed, t values will follow a t distribution that, like the standard normal, is symmetrical about 0. An important difference between the two is that the shape of the t distribution depends on the sample size, or more specifically, the *degrees of freedom* (df; equal to the sample size minus 1 or $u - 1$), whereas the standard normal retains the same shape (i.e., z values are constant) because it assumes the standard error is known and constant. For example, consider the abundance estimate calculated from a simple random sample of size 10 from 100 sampling units in Fig. 1.11. The estimated stan-

---

[8] z values are usually calculated based on the mean, i.e., $z = (\bar{N} - \bar{N}_{true})/SE(\bar{N})$ (our notation) or $z = (\bar{x} - \mu)/s_{\bar{x}}$ (common notation). However, a z value is the same whether it is based on abundance (i.e., total) or a mean because the two differ only by a constant, U. That is, $\hat{N} = U \times \bar{N}$, $N = U \times \bar{N}_{true}$, and $SE(\hat{N}) = U \times SE(\bar{N})$, and all U terms cancel out.

dard error would be [combining Eqs. (1.8) and (1.10)],

$$\hat{SE}(\hat{N}) = \sqrt{\hat{Var}(\hat{N})} = \sqrt{(100)^2 \left[\left(1 - \frac{10}{100} \right) \frac{0.823}{10} \right]} = 27.2.$$

A 95% confidence interval, assuming a standard normal, would be 70 ± 1.96 × 27.2 or (16.7, 123.3). When computed assuming a t distribution, this interval becomes 70 ± 2.262 × 27.2 or (8.5, 131.5), and is based on 9 df. (To obtain the proper t value from the usual t table, find the appropriate degrees of freedom in the left column, and then locate the specified α across the top row, which in this case would be 0.025. The correct t value is at the cross-section of this row–column combination in the main body of the table.) The differences in width (i.e., precision) between these two intervals is evident. The increased width is due to the added uncertainty of estimating the standard error in calculating a t value, whereas it is assumed to be known for the z value.

How do we decide when to use a z value or t value? At large sample sizes, a t distribution adequately approximates a standard normal distribution of z values. But how large is "large"? The answer depends on the shape of the underlying distribution of items from which a sample is obtained. Samples from an underlying normal distribution will produce a sampling distribution that is normal even at very low sample sizes. Conversely, underlying populations that are highly skewed may require sample sizes of 100 or more before their sampling distributions can be considered approximately normal (Snedecor and Cochran, 1989). A general rule of thumb is that sample sizes should be at least 20–30 to use a z value, assuming the underlying population is not overly skewed (Hahn and Meeker, 1991). For sample sizes less than this, a t distribution usually provides an adequate approximation except for highly skewed populations. For example, Fig. 1.15 is a bar

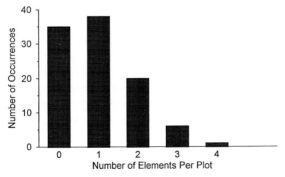

Figure 1.15 A bar graph showing the frequency of occurrence of plots containing 0, 1, 2, 3, and 4 elements in Fig. 1.11.

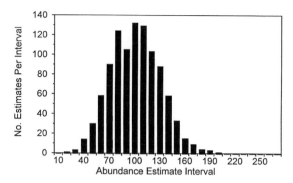

Figure 1.16 A histogram of abundance estimates obtained from 1000 simple random samples of size 10 selected from a frame with an underlying random distribution of points shown in Fig. 1.11. Abundance estimates are grouped into intervals; e.g., "100" represents the interval of values greater than 90 and less than or equal to 100.

graph corresponding to the underlying random spatial distribution of points as seen in Fig. 1.11. Bar heights correspond to the number of plots containing a given number of points (e.g., there are 35 plots that contain no points). We can see that this distribution of points is asymmetrical or skewed. Now consider Fig. 1.16, which is the estimated sampling distribution based on 1000 samples of size 10 from a sampling frame like the one in Fig. 1.11. Despite being based on relatively small samples from a skewed distribution, the estimated sampling distribution approximates a bell shape or normal distribution.[9] Figure 1.17 depicts a sampling frame with an even more skewed underlying distribution of points (as seen in Fig. 1.18). But even in this case, an estimated sampling distribution based on samples sizes of 10 more or less approximates a bell shape, although there is a definite tail to the right (Fig. 1.19). Thus, a *t* distribution would appear to provide at least a rough approximation for constructing confidence intervals with a sample size of 10 from a frame with a very skewed underlying population. However, assuming a *t* distribution when a sampling distribution is skewed, such as in Fig. 1.19, will adversely affect the nominal coverage of the computed confidence intervals. In other words, a 95% confidence interval may only contain, on average, the true abundance, say, 80% of the time. This is obviously not a good property. A possible remedy for this is to use a bootstrapping approach to generate confidence intervals, which we will discuss next.

[9] Note that Fig. 1.13 is a theoretical distribution based on an infinite number of values (i.e., a continuous distribution characterized by a solid line), whereas Fig. 1.16 is a histogram based on a finite number of values (i.e., finite or discrete distributions). However, we can use the properties of the theoretical distribution as an approximation.

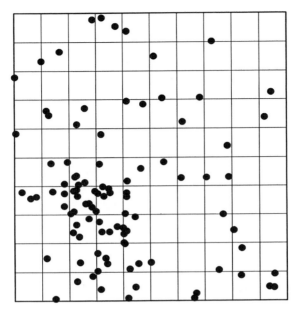

Figure 1.17 A sampling frame of 100 plots with an underlying clumped distribution of elements.

To this point, we have used properties of *external* or *reference distributions* to produce measures of how close our sample-based abundance estimate is to the true abundance. Specifically, distributional properties of the standard normal and *t* distributions were assumed to adequately describe sampling distributions of abundance estimates (actually, their standardized

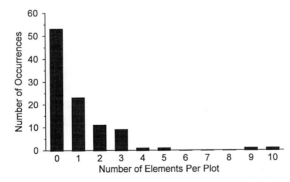

Figure 1.18 A bar graph showing the frequency of occurrence of plots containing 0–10 elements shown in Fig. 1.17.

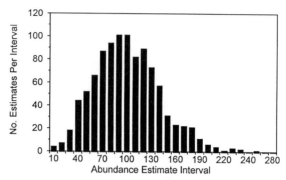

Figure 1.19 A histogram of abundance estimates generated from 1000 simple random samples of size 10 selected from a frame with an underlying clumped distribution of points (sampling frame in Fig. 1.17 and corresponding bar graph in Fig. 1.18).

deviates) collected from samples drawn from a sampling frame. The validity of these properties depends upon how closely these reference distributions approximate the sampling distributions. As we have just mentioned, poor approximations can lead to misleading measures of precision. An alternative approach is to use observed data to generate an *internal* or *empirical distribution* to serve as the estimated sampling distribution. For example, suppose we have drawn a random sample of size u from a sampling frame. Rather than assuming that the sampling distribution of abundance estimates follows the form of some reference distribution, we instead could take some number of random resamples of size u with replacement from our empirical data, compute an abundance estimate for each resample, and use this collection of estimates to produce the estimated sampling distribution. The "with replacement" specification means that a particular data value in the sample could be included in any given resample more than once. This computer-intensive procedure, called *bootstrapping,* has become fairly easy to run with the advent of powerful computers. Hence, it has only been within the last 10–15 years that bootstrapping has become the topic of extensive investigation and review. Because this is a continuing topic of research, we provide only a brief overview here, and refer interested readers to Mooney and Duval (1993) for a clear and concise introduction to bootstrapping; more technical reviews are given by DiCiccio and Efron (1996), Efron and Tibshirani (1993), and Hall (1988).

    We will briefly discuss two of a number of possible techniques for computing bootstrapped confidence intervals. One method assumes that the bootstrapped-generated sampling distribution adequately approximates the true sampling distribution. In the *percentile method,* one orders from low to high the abundance estimates obtained from the resamples and then uses the abundance estimates occurring at $100(\alpha/2)\%$ from each end of the list for the lower and upper limits of the $100(1 - \alpha)\%$ confidence intervals. Suppose we obtained 1000 bootstrap resamples from a randomly selected sample and wanted a 95% confidence interval about our abundance estimate. Using the percentile method, we would order the 1000 resample estimates from low to high, and then use the 25th estimate on the list as the lower limit and the 975th estimate as the upper limit (Mooney and Duval, 1993). This interval could be asymmetrical about the sample-based abundance estimate, which is a good property if the true sampling distribution is skewed and the center of the bootstrapped interval is correctly located. However, a fundamental assumption of bootstrapping is that the original sample adequately represents the range of values in the underlying population that is contained in the sampling frame (Mooney and Duval, 1993). If the sample contains a limited range compared to the true population, then the bootstrapped sampling distribution will be of improper width and incorrectly centered, which would produce biased (see next section) intervals of improper width. Unfortunately, an unrepresentative original sample is more likely to occur at smaller sample sizes (Schenker, 1985). Thus, the problem we faced in Fig. 1.19 would probably not be corrected by the percentile method.

A bootstrapping approach that has initially shown at least some promise for producing confidence intervals with adequate coverage for small, skewed samples is the *percentile–t method* (Hall, 1988; Mooney and Duval, 1993; Singh, 1986). This approach is analogous to constructing a confidence interval based on *t* values from a *t* distribution. In this case, however, we would generate an estimated sampling distribution composed of bootstrapped *t* values, which we will denote t_b^*. In essence, we are constructing a *t* table based on our 1000 or more t_b^* values. Details of this procedure are given in Mooney and Duval (1993). Once our t_b^* distribution has been constructed and ordered from low to high, we simply substitute the $100(\alpha/2)\%$ and $100(1 - \alpha/2)\%$ values from the ordered list into a standard confidence interval such as in Eq. (1.12), where the estimated standard error is generated from bootstrap resampling. Unfortunately, only limited software is available to perform bootstrap operations like the percentile–t method.[10] Further, the properties of different methods for generating bootstrap confidence intervals have not been thoroughly investigated; the best method will probably depend on the situation. We stress that bootstrap methods

[10] SAS code (SAS Institute, Inc., 1990) for various methods for computing bootstrap confidence intervals is available via the Internet at www.sas.com/techsup/download/stat/jack-boot.sas. Efron and Tibshirani (1993, Appendix) provided S-Plus code (Becker *et al.*, 1988) for similar procedures.

may not perform well with very small samples and when the underlying population is highly skewed.

Another type of confidence interval that may be helpful when a sampling distribution is skewed is the profile likelihood confidence interval. We discuss this interval in some detail in Chapter 3.

One aspect of confidence interval construction that has not been addressed is the choice of α. Although $\alpha = 0.05$ is the conventional choice, there is nothing magical about this number. The level of confidence depends on the objectives of the study. Moreover, there is a trade-off between precision and confidence. For example, suppose $\hat{N} = 100$, $\hat{SE}(\hat{N}) = 35$, and we could assume a standard normal sampling distribution. If we wanted to be extremely confident, say 99%, that the true abundance will be within our interval, the resulting confidence interval would be (9.7, 190.3). Although this interval, on average, would contain the true abundance 99 out of 100 times, it is so wide (i.e., imprecise) that it is essentially useless. Conversely, what if we wanted a precise interval and were willing to accept a confidence level of 50%? The 50% confidence interval would be much more precise, (76.4, 123.6), but would only contain the true abundance, on average, half of the time. One may as well flip a coin. Again, there is no straightforward answer; choice of α must be done on a case-by-case basis.

Before moving on, let us consider one more aspect of confidence intervals. Figure 1.20 shows one possible group of 95% confidence intervals obtained from 20 simple random samples of size 10 from the sampling frame in Fig. 1.11. Although, on average, 19 of 20 intervals should contain

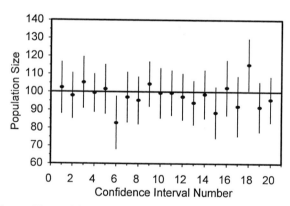

Figure 1.20   Twenty 95% confidence intervals for abundance estimates from 20 simple random samples of size 10 drawn from the sampling frame in Fig. 1.11. The true population size, $N = 100$, is denoted by the dark horizontal line. The black dots represent abundance estimates centered on the confidence intervals.

the true parameter, only 18 do so (i.e., cross the black line). In fact, it is possible that anywhere from 0 to 20 intervals may include the true abundance in this example, although there is a decreasing probability of occurrence the farther a number is from 19. Now, suppose that these 20 abundance estimates represented 20 years of survey data with associated measures of precision. Further suppose that a biologist wished to inspect the data for trend, but was provided only with point estimates from years 11 to 13. Note that these estimates show, by random chance, a steady decrease from year 11 to year 13. If one only considered the three point estimates without their associated confidence intervals, he/she may conclude there was a decreasing trend in animal numbers over this particular 3-year period. If so, the conclusion would be incorrect. These three confidence intervals broadly overlap, so one cannot validly conclude there is a trend. Given the width and overlap of these three intervals, their point estimates could have just as easily shown an increase or no pattern at all. This indicates the difficulty in concentrating too much on point estimates while ignoring precision and related measures.

1.5.1.3. Spatial Distribution and Precision

Many species are spatially arranged in clusters or clumps (Cole, 1946). This is probably due mainly to habitat heterogeneity (i.e., composition, structure, and/or quality), although some species are gregarious by nature. Let us investigate how spatial distribution of a species of interest can influence precision of abundance estimates.

If a species is randomly distributed, then the mean of the number of individuals per plot will be about the same as the variance of the number of individuals per plot. This relationship exists because, for a randomly distributed species, the number of individuals per plot follows a *Poisson distribution*. Given that most species are clumped, we expect to see the variance of individuals per plot greater than the mean of individuals per plot. The ratio of the variance to the mean of individuals per plot often is used as an index of the spatial aggregation of species. For the example in Fig. 1.11, the mean is 1 and the variance is 0.89, giving a variance to mean ratio of 0.89. This is close to the expected value of 1 because the data points are randomly distributed. Conversely, the variance to mean ratio for Fig. 1.17 is 1.72, which is much greater than 1 because points follow a clumped distribution. The *negative binomial distribution* is a useful statistical distribution for modeling this process (e.g., White and Eberhardt, 1980; White and Bennetts, 1996). For the negative binomial distribution, the clumping parameter, k, provides another measure of aggregation. Typically, these measures of aggregation are species specific. Aggregation or clumping also is specific to the size of plot being sampled.

Estimates obtained from sampling frames with larger numbers of plots with zero counts (e.g., underlying clumped distribution) will likely be less precise than those obtained from frames in which animals are more evenly distributed. For example, Figs. 1.11 and 1.17 both contain 100 plots and 100 elements, where elements represent animals. The frame with a random distribution of animals has 35 empty plots, whereas the one with a clumped distribution has 50 empty plots. We know that the frame with the random distribution of elements has a variance of 0.89, whereas the one with the clumped distribution has a variance of 1.72.

1.5.2. BIAS

The second aspect of error is *bias,* which sometimes is referred to as *nonsampling error,* and is systematic in nature (i.e., exhibiting either a consistent positive or a consistent negative influence on results). This is distinct from *sampling error,* which arises from random or variable sources such as the random selection of sampling units.[11] Before we define bias, let us return to Example 1.1, where we saw that estimates fluctuated from sample to sample. Note, however, that the arithmetic average of all estimates, either $\hat{N}$ or $\overline{N}$, in the sampling distribution is exactly equal to the true parameter. This is what determines an unbiased estimator. The arithmetic average of all possible sample estimates is called the *expected value.* Therefore, bias is the difference between the expected value of an estimator and the true value of the parameter of interest. For instance, $E(\hat{N}) = N = 33$ in Example 1.1 so that the estimator based on simple random sampling is unbiased.

Bias can impact the plot selection method the counting technique or both. Data collected under a biased selection scheme still will yield biased results even if inserted into an estimator that is otherwise considered unbiased. Conversely, properly collected count data could be inserted into a biased estimator and still produce biased results. We will initially discuss sources of bias at the plot selection or frame level, and then describe sources of bias associated with counting elements (i.e., unit or plot level).

Frame level sources of bias are those that may occur among units. These errors are called selection bias (discussed previously) and *coverage error.* Coverage error arises from imperfect sampling frames and results from

---

[11] Kish (1965) presented a more complex model of survey errors that included sampling biases and variable nonsampling errors. However, for simplicity, we use a more general and sharper distinction between sampling and nonsampling errors.

excluding elements that are part of the target population and/or including those that are not or those that have already been counted (Lessler and Kalsbeek, 1992). Imperfect sampling frames may be caused by inaccessibility of part of the target population (e.g., limited access to private lands; Fig. 1.7) or poor delineation of sampling units so that gaps containing animals occur among them. Selecting a random sample from an undefined sampling frame (Fig. 1.10) is a milder example of this. The overall result is that animals in the target population may not be included in the sampled population. Plots that include area outside of the sampling frame or overlapping plots that lead to double-counting of individuals are also examples of coverage error.

Sources of bias that may occur during the course of a count are termed *response error* and *nonresponse error*. Response error is produced by mismeasuring or misrecording information associated with an animal that has been observed, contacted, or otherwise detected during a count. This may be due to incorrectly recording or reading a number on a data sheet, incorrectly reading a value from some type of measurement device, or incorrectly identifying and then recording the wrong species during a count within a sampling unit. Examples include misidentifying/misrecording one species for another that looks similar (e.g., white crappie and black crappie), sounds similar (e.g., chipping sparrow and dark-eyed junco), or leaves similar sign (e.g., moose and elk tracks). This could lead to both failing to record a species that is present and incorrectly recording a species that is in fact absent. Training observers may lessen the potential for misrecording data, but probably will not eliminate it.

Nonresponse error arises from failing to detect animals included in a sample. Although animals within selected plots are theoretically included in the sample, all animals are not detected and therefore do not contribute to count totals. *Detectability* or *detection probability* is the chance of confirming the occurrence of an animal within some defined space and time period. Factors adversely affecting detectability of animals within selected sampling units are diverse and may differ depending on degree of physical structure and cover, weather, behavior of individuals, age of individuals (as related to size), type of counting method, and so forth. Detectability is an often overlooked or ignored source of bias that commonly afflicts population surveys. This is especially true for relative index techniques, which assume that a constant proportion of animals are detected in surveys across space and time. That is, these techniques assume that any changes or differences in index results are proportional to true changes or differences in the population of interest. However, it is very possible that a difference among uncorrected counts may only be due to a difference in detectability of animals between areas or within the same area across time. A common example of this is comparing uncorrected counts of songbirds in cut and uncut forest stands. One would expect a difference in detectability, and hence numbers of counted birds, because of major differences in vegetational structure alone. We will be considering these problems and other possible effects of errors on estimates of spatial distribution and abundance when we discuss and evaluate specific survey methods and counting procedures later in this book.

Unlike sampling errors, errors due to bias will not decrease with increasing sample size. Further, magnitude of bias can be assessed directly only through comparisons either between the sample estimate and the true value of the parameter or between the biased sample estimate and an unbiased sample estimate (Yates, 1981). A census would be required to obtain the true value of the parameter, which is very unlikely. One may as well have obtained an unbiased estimate to begin with rather than obtaining one for comparative purposes. Uncorrected bias could have serious consequences regarding validity and usefulness of parameter estimates. For instance, suppose the detectability of animals was only 75%, rather than 100%, within plots used to generate the 20 abundance estimates (with confidence intervals) displayed in Fig. 1.20. That is, what if animals within each selected plot were systematically undercounted by 25%? Figure 1.21 shows the effect of this systematic undercounting; only 1 of 20 95% confidence intervals contain the true abundance, whereas 18 of 20 contained the true abundance in Fig. 1.20. In reality, detectability would likely vary among plots and across time, but the effect of biased abundance estimates is obvious.

Although we may strive for unbiased estimates, slightly biased estimates

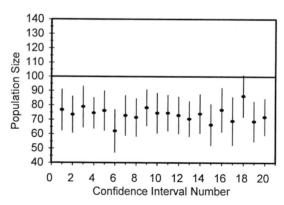

Figure 1.21   The same collection of confidence intervals shown in Fig. 1.20 except the detection probability is 0.75 rather than 1; hence, only 75% of the animals on the plots were detected rather than 100%. Only 1 of 20 95% confidence intervals include the true abundance, $N = 100$.

may provide more information. That is, there is a trade-off between bias and precision. A slightly biased but precise estimate is better than an unbiased and imprecise one. But how biased is "slightly biased"? The effect of bias on an estimate can be ignored if the ratio of bias to the standard error is less than 10% (Cochran, 1977). Whereas a complete count would be best to assess the magnitude of bias, comparing a corrected count (i.e., a count properly corrected for incomplete detectability; see Chapter 3) to an uncorrected count should provide, on average, a decent estimate of bias.

LITERATURE CITED

Becker, R., Chambers, J., and Wilks, A. (1988). *The S Language.* Wadsworth, Belmont, CA.
Caughley, G. (1970). Eruption of ungulate populations, with emphasis on Himalayan thar in New Zealand. *Ecology* **51:** 53–72.
Cochran, W. G. (1977). *Sampling Techniques,* third ed. Wiley, New York.
Cole, L. C. (1946). A study of the cryptozoa of an Illinois woodland. *Ecol. Monogr.* **16:** 50–86.
DiCiccio, T. J., and Efron, B. (1996). Bootstrap confidence intervals. *Stat. Sci.* **11:** 189–228.
Efron, B., and Tibshirani, R. J. (1993). *An Introduction to the Bootstrap.* Chapman and Hall, New York.
Graybill, F. A., and Iyer, H. K. (1994). *Regression Analysis: Concepts and Applications.* Duxbury, Belmont, CA.
Gutierrez, R. J., Forsman, E. D., Franklin, A. B., and Meslow, E. C. (1996). History of demographic studies in the management of the northern spotted owl. *Stud. Avian Biol.* **17:** 6–11.
Hahn, G. J. (1979). Sample size determines precision. *Chemtech* **4:** 16–18.
Hahn, G. J., and Meeker, W. Q. (1991). *Statistical Intervals: A Guide for Practitioners.* Wiley, New York.
Hall, P. (1988). Theoretical comparison of bootstrap confidence intervals (with discussion). *Ann. Stat.* **16:** 927–985.
Keith, L. B. (1963). *Wildlife's Ten-Year Cycle.* University of Wisconsin Press, Madison, WI.
Kish, L. (1965). *Survey Sampling.* Wiley, New York.
Lessler, J. T., and Kalsbeek, W. D. (1992). *Nonsampling Errors in Surveys.* Wiley, New York.
Levy, P. S., and Lemeshow, S. (1991). *Sampling of Populations: Methods and Applications.* Wiley, New York.
MacDonald, L. H., Smart, A. W., and Wissmar, R. C. (1991). *Monitoring Guidelines to Evaluate Effects of Forestry Activities on Streams in the Pacific Northwest and Alaska.* U.S. Environ. Protect. Agency Water Div. Seattle, WA. [EPA/910/9-91-001].
Manly, B. F. J. (1992). *The Design and Analysis of Research Studies.* Cambridge Univ. Press, Cambridge, UK.
Messer, J. J., Linthrust, R. A., and Overton, W. S. (1991). An EPA program for monitoring ecological status and trends. *Environ. Monit. Assess.* **17:** 67–78.
Mooney, C. Z., and Duval, R. D. (1993). *Bootstrapping: A Nonparametric Approach to Statistical Inference.* Sage, Newbury Park, CA.
Overton, W. S., White, D., and Stevens, D. L., Jr. (1990). *Design Report for EMAP.* U.S. Environ. Prot. Agen., Corvallis, OR. [EPA/600/3-91/053].

SAS Institute (1990). *SAS Language: Reference, Version 6,* first ed. SAS Institute, Cary, NC.

Scheaffer, R. L., Mendenhall, W., and Ott, L. (1990). *Elementary Survey Sampling,* fourth ed. PWS-Kent, Boston.

Schenker, N. (1985). Qualms about bootstrap confidence intervals. *J. Am. Stat. Assoc.* **66:** 360–361.

Seber, G. A. F. (1982). Estimation of Animal Abundance, Second ed. Griffin, London.

Singh, K. (1986). Discussion of Wu. *Ann. Stat.* **14:** 1328–1330.

Snedecor, G. W., and Cochran, W. G. (1989). *Statistical Methods,* eighth ed. Iowa State University Press, Ames, IA.

Stuart, A. (1984). *The Ideas of Sampling.* Oxford Univ. Press, New York.

Van Horne, B. (1983). Density as a misleading indicator of habitat quality. *J. Wildl. Mgmt.* **47:** 893–901.

White, G. C., and Bennetts, R. E. (1996). A model selection approach to the analysis of frequency count data using the negative binomial distribution. *Ecology* **77:** 2549–2557.

White, G. C., and Eberhardt, L. E. (1980). Statistical analysis of deer and elk pellet group data. *J. Wildl. Mgmt.* **44:** 121–131.

Yates, F. (1981). *Sampling Methods for Censuses and Surveys,* fourth ed. MacMillan, New York.

Chapter 2

Sampling Designs and Related Topics

2.1. Plot Issues
    2.1.1. Plot Design
    2.1.2. Unequal Plot Sizes
2.2. Categorizing a Sampling Design
2.3. Sampling Designs for Moderately Abundant and Abundant Species
    2.3.1. Simple Random Sampling
    2.3.2. Stratified Random Sampling

2.3.3. Systematic Sampling with a Random Start
2.3.4. Simple Latin Square Sampling +1
2.3.5. Ranked Set Sampling
2.4. A Sampling Design for Rare and Clustered Species
Literature Cited

In the previous chapter, we discussed some initial steps in designing a population survey, such as specifying a target population and constructing a sampling frame. One aspect of frame construction that was not addressed was plot design; therefore, we will initially address this issue in this chapter. Next, we will discuss how to obtain abundance estimates once the target population has been identified and sampling frame constructed. On which plots should we conduct counts? How many of these plots should we select? How do we use collected data to obtain distribution, abundance, or density estimates? These questions relate to *sampling design,* which is the protocol for obtaining parameter estimates for a sampled population. This protocol includes a *sampling plan* for selecting plots and estimators that use collected data for calculating parameter estimates (Levy and Lemeshow, 1991). Specific enumeration methods will be discussed in the next chapter.

A preferred sampling plan is one that, within cost constraints, produces estimates that are both unbiased and precise. The plans discussed in this chapter are the ones used more commonly for moderately abundant to abundant species. Because the more common sampling designs often are inefficient for surveying rare species, we also describe a specialized design that could be useful for obtaining estimates for species that are rare and spatially arranged in clusters or clumps.

2.1. PLOT ISSUES

Constructing a sampling frame requires selection of an appropriate shape and size for plots. One also must determine if plots will be the same size or different sizes. Both of these aspects of plot design may affect precision of estimators, which in turn affects the ability to detect a population trend.

2.1.1. PLOT DESIGN

A number of factors may influence the choice of shape and size of plots, and some of these may suggest entirely different optimal designs. We first will discuss these various factors and then offer some general recommendations for designing plots.

2.1.1.1. Factors Affecting Plot Shape

Choice of plot shape is influenced by the ratio of a plot's perimeter to its inside area (Krebs, 1989), the method for obtaining plot counts, the detectability of individuals (Thompson, 1992), and underlying arrangement of individuals (Krebs, 1989). All four factors may be related to one another as well as to plot shape.

Perimeter to area ratio is a measure of *edge effect*. A plot with a large perimeter to area ratio (e.g., strip quadrat) is more susceptible to having individuals (or objects) wrongly included or excluded in counts than the same size plot with a small ratio (e.g., circle; Fig. 2.1). A plot with a longer perimeter and smaller core area tends to have more individuals occurring either at the perimeter of a plot or crossing plot boundaries than a plot with a shorter perimeter and larger core area. This may not be a problem in riverine systems if physical barriers are used to restrict individual movements from sampled stretches of a stream. However, unrestricted plots in lakes are susceptible to edge effect. Counting errors due to edge effect

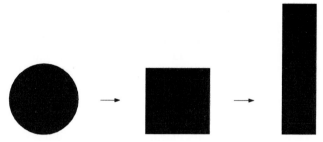

Figure 2.1 Three plot shapes containing the same area but differing in perimeter to area ratios, which is a measure of edge effect. The circular plot has a ratio of 2.36:1, the square plot 2.67:1, and the strip plot 3.33:1. A lower ratio signifies a lower potential for edge effect.

could add a significant source of bias to overall estimates depending on the situation.

The method used to count mobile individuals could have a significant bearing on the magnitude of edge effect and hence plot shape. A complete count within a small time interval can circumvent the problem of movements across plot boundaries. Even if movement occurs, it may not occur during the count so that numbers of animals may be treated as fixed in space. Unfortunately, complete plot counts are uncommon or rare in fish and wildlife studies. Further, methods for adjusting partial counts for incomplete detectability (e.g., capture–recapture) could produce biased abundance estimates because of edge effect. Therefore, plot shapes that have the lowest perimeter/area ratio would be best for methods most susceptible to edge effect.

Detectability of individuals varies according to shape of the surveyed plot. Long and narrow plots are more efficient for detecting individuals than either square or circular plots (Thompson, 1992). For example, consider a person conducting a count along the centerline of a plot. There is a shorter distance from the line to the plot boundary in a long and narrow plot than in a square or circular plot, and therefore the observer is less likely to overlook individuals because they are closer to the centerline.

The spatial arrangement of individuals also may dictate which plot shape is best for a given situation. This is essentially a question of variability. As discussed in the previous chapter, abundance estimates obtained from a sampling frame with an underlying clumped distribution of animals are usually less precise than those obtained from one with a random distribution. And most animal populations are spatially clumped because of habitat heterogeneity, social behavior, or other factors. Consequently, we would like to use a plot shape that would increase precision of resulting estimates, i.e., one that would minimize the numbers of plots with zero counts or unusually high counts. Long and narrow plots accomplish this because they are more likely to intersect a "clump" or animal group than square plots (Clapham, 1932; Bormann, 1953; Greig-Smith, 1964), and therefore are less likely to have extreme counts (either zeros or high numbers). One exception to this is if some sampled plots are oriented along topographic contours. Because habitats are usually more homogenous along a contour than adjacent areas, a long and narrow plot may intersect fewer clumps than a square plot that covers more area away from the contour (Bormann, 1953). This could lead to less precise estimates if densities of animals are at all correlated with habitat.

2.1.1.2. Factors Affecting Plot Size

How large plots should be depends on spatial scale and species of interest being sampled, underlying spatial distribution of individuals, potential magnitude of edge effect, enumeration method, and cost of data collection (Krebs, 1989). These factors overlap among themselves and with those associated with plot shape.

Spatial scale is an obvious place to begin when determining size of plots, especially with regard to size of the species of interest. Larger animals tend to have larger home ranges or daily movements (McNab, 1963), so plot sizes must be larger. Stream units containing 50 longnose dace would likely be smaller than those containing 50 rainbow trout.

As with shape, plot size is influenced by the underlying spatial arrangement of individuals within the sampling frame. Again, the objective is to increase precision of estimates. This may be accomplished by having plots large enough so that few have zero counts and counts are less extreme (Fig. 2.2). The closer the distribution of individuals is to random, the smaller the plots will likely have to be to satisfy this requirement. For instance, a clumped distribution of points means that, on average, some plots will contain more points than others. Conversely, if the 100 points within Fig. 2.2 were randomly distributed, each of the 100 plots, on average, would contain the same number of points (i.e., 1) if many such frames were constructed. Thus, a frame with a clumped distribution of animals would likely have more empty plots than one with a random distribution, and thus require more area to be covered by each unit to reduce the number of empty plots.

Edge effect is associated with plot size in addition to shape. Given a constant shape, ratio of perimeter length to plot area decreases with increasing plot size. For example, a 1×1-m square plot has a ratio of $4:1$, whereas

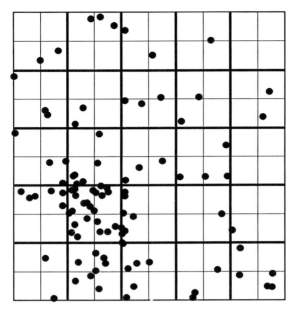

Figure 2.2 An underlying clumped distribution of 100 animals within a sampling frame of 100 plots (light lines, ignore dark lines) or one with 25 plots (dark lines, ignore light lines). The frame with 100 plots has 50 empty plots, whereas the one with 25 plots has 2. The coefficient of variation based on the 100-plot frame is larger ($CV(\hat{N})$ = 1.31 or 131%) than that ($CV(\hat{N})$ = 1.11 or 111%) obtained via the 25-plot frame.

a 100 × 100-m unit has a ratio of 0.04:1 or 1:25. The magnitude of edge effect depends on the mobility and movement patterns of the species of interest. A 10 × 10-m plot used to sample a species with an average home range of 200 m² is obviously subject to considerable edge effect because the plot area (100 m²) is not even large enough to contain the entire home range of a single individual.

Size of a sampling unit also is influenced by the enumeration method used to gather abundance information. For complete plot counts, a unit must be small enough to conduct a complete count, but large enough to minimize the number of plots with no individuals. Moreover, it would probably be more costly and less efficient to travel among many small plots than fewer large ones. Plot sizes used for incomplete plot counts are dictated by minimum area required to contain sufficient numbers of individuals for accurately adjusting counts for detectability (e.g., capture–recapture methods) or otherwise obtaining unbiased estimates. Plot sizes for unadjusted incomplete counts (i.e., index methods) may largely depend on maximizing the number of individuals within a given plot that can be realistically surveyed.

Funding availability is a limiting factor in many survey situations, and plot design is no different. Cost constraints may limit plots to sizes smaller than could be realistically surveyed if maximum time and manpower were available. In addition, cost of the enumeration method may dictate how large a plot can be as well as how many plots of a given size are possible.

2.1.1.3. Choosing a Plot Design

An optimum plot design is one that produces unbiased and precise estimates under cost constraints. However, choosing an appropriate plot design is not a straightforward decision because of the different factors affecting shape and size of sampling units. Further, some of these factors may lead to conflicting shapes and sizes. For instance, long and narrow plots may lead to more precise abundance estimates, but square plots offer less potential for edge effect. Therefore, there is no single plot design that is universally applicable; plot design is study-specific.

The best way to determine which plot design is best is by trying several in a pilot study. Previous studies of the species of interest, if available, may offer some ideas for initial designs. A pilot study allows a researcher to assess how different shapes work with a given counting technique, provides estimates of precision for different plot shapes and sizes, and provides cost estimates for different plot designs. Wiegart (1962) offered a method for computing optimum plot shape or size based on relative cost and relative variance. Krebs (1989, p. 70) provided an example of Wiegart's method and has written program QUADRAT[1] to calculate optimum plot shapes and sizes.

2.1.2. UNEQUAL PLOT SIZES

In area sampling, selection probabilities for plots may be based on the size of each plot. When units are the same or nearly the same size, the selection procedure is simplified, i.e., each plot initially has an equal chance of being included in a sample. However, if plots differ much in size, with a corresponding increase or decrease in numbers of animals, treating them as equal-sized in calculating estimates will usually result in an increase in variance estimates. This would only be the case if plot size was strongly correlated with number of animals.

[1] Program QUADRAT can be obtained via the Internet at: nhsbig.inhs.uiuc.edu/krebs/.

Plots may not be of equal size because of inherent physical attributes of the underlying substrate or habitat feature. In streams, for example, sampling units may be defined to coincide exactly with stream habitat features (e.g., riffles, pools, and runs) that will vary in size (i.e., length). At a larger spatial scale, one may wish to select a sample from various-sized wetlands, ponds, lakes, or streams in an area.

There are a few options if plots vary significantly in size. First of all, one could either assume or show via survey results that size of sampling units is not highly correlated with number of animals. Habitat quality and composition within a plot likely have more to do with the number of animals it contains than its size. However, wrongly assuming this may lead to imprecise results, whereas investigating it adds additional costs and time. Second, estimators that account for unequal plot sizes could be used. These estimators, like the Horvitz–Thompson estimator (Horvitz and Thompson, 1952, Appendix C), are more complicated and their calculation requires specialized computer programs. Third, units could be stratified by their size (Cochran, 1977). Carefully stratifying plots by size will probably result in a minimal loss of precision in many situations. Determination of which of the three options to use will have to occur on a study-by-study basis. In this chapter, we will assume that plot sizes are equal.

2.2. CATEGORIZING A SAMPLING DESIGN

Sampling designs may be placed within a general framework based on how many levels of selection-based and/or assessment-related uncertainty they include (Fig. 2.3). These levels of uncertainty are termed *stages* within a design context. This step is necessary for assessing how many levels of sampling are needed as well as which estimators are required.

A *one-stage design* is one that has either among-unit variation or enumeration variation as a source of uncertainty. Such a design could be based on a sample of plots within which complete counts are conducted (i.e., among-unit variation) or an incomplete count over the entire sampling frame or defined area of interest (i.e., enumeration variation). In this latter case, the sampling frame sometimes may be thought of as one large plot, where parameter estimates are obtained from estimators based on the data collection method employed (e.g., aerial mark–resight, discussed in Chapter 3). Also, note that in these two cases sampling variation may refer to either among-unit variation or enumeration variation.

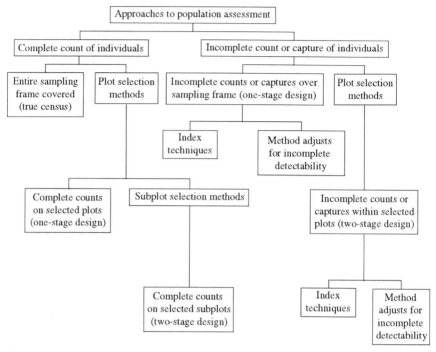

Figure 2.3 Relationships among different approaches to assessing population abundance within a defined area during a specific time interval. Assume plot selection methods have associated estimators for calculating parameter estimates and sample sizes. Note that "captures" may be sightings of marked/identifiable animals or objects.

Designs containing both among-unit variation and enumeration variation are called *two-stage designs*. A two-stage design could be based on a random selection of plots within which a random selection of subplots is chosen. Complete counts must be possible on subplots if unbiased estimates are to be obtained. This type of selection procedure sometimes is referred to as a *two-stage cluster sample*. Two-stage cluster sampling is not commonly used for obtaining unbiased abundance estimates for animal populations because complete enumeration of subplots often is not realistic for animals in many terrestrial environments, and is not very suitable for aquatic environments.

Another example of a two-stage design is one in which a sample of plots is randomly chosen and then an incomplete count, using some type of method that accounts for incomplete detectability (e.g., mark–recapture), is conducted within each chosen plot.[2] This is a much more common scenario in fish and wildlife studies. Note that sampling variation in two-stage designs refers to a combination of among-unit variation and enumeration variation. We could increase the number of stages to three or more, depending on the situation, but for simplicity this book discusses mainly one- and two-stage designs. Three-stage sampling designs are reviewed by Cochran (1977) and Gilbert (1987).

2.3. SAMPLING DESIGNS FOR MODERATELY ABUNDANT AND ABUNDANT SPECIES

This section lists and discusses several sampling designs for obtaining information on numbers and spatial distributions of animals that are moderately abundant to abundant. We will focus mostly on abundance because distributional data may be obtained as a by-product of abundance data and density estimates are easily converted from abundance estimates. Further, we will offer general examples of some of the more common sampling plans and estimators. Interested readers may refer to Appendix C for a more complete listing of estimators.

2.3.1. SIMPLE RANDOM SAMPLING

The procedure for choosing a simple random sample has already been covered in the previous chapter so we will not repeat it here. However, we wish to reiterate that the ability of a simple random sampling procedure to produce precise estimates of abundance depends a large part on the underlying distribution of animals. Another characteristic of simple random sampling is that it may or may not provide a sample with good area coverage. Drawing a sample of units clustered within one part of a study area is just as likely as one evenly spread throughout. As stated above, this is not a problem if individuals are more or less uniformly arranged throughout an area, but could result in inflated estimates of variation and poor abundance estimates over only one or a few samples.

---

[2] Our broad definitions of one- and two-stage designs follow those of Hankin (1984) and Skalski (1994), which were adapted for fish and wildlife applications. Conversely, statistical sampling texts strictly define a one-stage design as one based on a random selection of plots that have complete counts conducted on them, and a two-stage design as one based on a two-stage cluster sample.

Because most biological populations are not randomly distributed across the landscape, simple random sampling is not a particularly good design to obtain abundance estimates with low variation. Thus, we will discuss other sampling designs that may yield estimates with higher precision, and therefore are better for detecting a trend in animal numbers.

2.3.2. STRATIFIED RANDOM SAMPLING

A way to increase precision of estimates when individuals are spatially clumped is by stratification of the sampling frame. A stratified random sampling plan groups similar and/or adjacent sampling units together and draws a random sample separately from each group or stratum. The simplest stratified design is the stratified simple random sample in which a simple random sample is chosen from each stratum. The abundance and variance estimators would be the same as those already given for simple random sampling except that each would be applied separately to each stratum, and then all stratum estimates would be added together to calculate estimates for the entire sampling frame. For example, if a stream was stratified into pools, riffles, and runs, then abundance and variance estimates would be calculated for each of the three strata and added together. Estimators for a one-stage design with complete plot counts (Cochran, 1977, pp. 93, 95) would be

$$\hat{N} = \sum_{h=1}^{H} U_h \overline{N}_h = \sum_{h=1}^{H} \hat{N}_h \tag{2.1}$$

and

$$\mathrm{Var}(\hat{N}) = \sum_{h=1}^{H} U_h^2 \left[\left(1 - \frac{u_h}{U_h} \right) \frac{\hat{S}_{N_{hi}}^2}{u_h} \right] = \sum_{h=1}^{H} \mathrm{Var}(\hat{N}_h), \tag{2.2}$$

where h is the identifier for a particular stratum, H is the number of strata, and other terms are defined as in a previous chapter (except now are defined in terms of each stratum h). Note that the fpc is now based on individual strata. That is, if the number of selected plots within each stratum (u_h) is 5% or less of the total plots within each stratum (U_h), then the fpc term can be removed from Eq. (2.2). Also, the sum of the selected plots across all strata (u_h) is equal to the total sample size, u, just as the sum of all U_h is equal to the total number of plots, U, in the sampling frame. Summing over the stratum estimates of abundance, $\hat{N}_h$, and variance, $\mathrm{Var}(\hat{N}_h)$, to obtain overall estimates is applicable for any stratified random sampling design.

Two-stage estimators require additional notation to account for incomplete counts within selected plots. However, as above, estimates are calculated for each strata and then summed over all strata to obtain estimates for the entire sampling frame. Interested readers should refer to Appendix C for relevant estimators.

Prior information is required to group sampling units with respect to some auxiliary variable known to influence animal numbers and spatial distribution. Common criteria used in creating strata include grouping units by similar environmental characteristics, such as habitat; this information can usually be obtained via maps or aerial photographs. Sampling units do not have to be adjacent to one another to be contained within the same stratum. A specific habitat type may be lumped together into one stratum even if its component areas occur in separate areas within a sampling frame. Examples of this include stream habitat units. A researcher may wish to group all riffles, pools, and runs into each of three strata, but the three types will obviously be spread out along a stream.

Stratification, when applied intelligently, nearly always leads to increases (sometimes large increases) in precision (Cochran, 1977). Stratification will usually produce estimates with no less precision than simple random sampling even when based on little prior information (Levy and Lemeshow, 1991). In fact, because of the clumped distributions of most animal populations, simply dividing a sampling frame into a number of equal-sized strata, and choosing the same number of plots from each (Fig. 2.4), will nearly always improve sample coverage over the frame, and very often produce estimates with higher precision than those from simple random sampling (Example 2.1). The more clumped the distribution of animals, the greater the potential for large improvement in precision of estimates through dividing a frame into some number of equal-sized strata. Therefore, stratification is a good approach even when little prior information is available.

Example 2.1. One-Stage Stratified Simple Random Sample (Equal-Sized Strata–No Prior Information about Strata)

A sampling frame was divided into 4 equal-sized strata consisting of 25 plots each. A simple random sample of 4 plots was chosen from each stratum (Fig. 2.4), and complete counts were conducted on each one. Results of counts were: 0, 1, 1, and 1 for stratum I (top left of frame); 0, 0, 0, and 1 for stratum II (top right); 0, 4, 4, and 1 for stratum III (bottom left); and 0, 1, 1, and 0 for stratum IV (bottom right). Using estimators for simple random sampling, we calculated the within-stratum estimates for abundance and variance. Estimates for stratum I were:

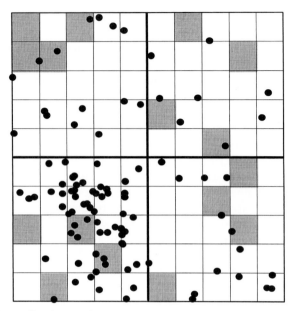

Figure 2.4 A sampling frame, with an underlying clumped spatial arrangement of animals, is divided into four strata each containing 25 plots. A simple random sample of four plots (gray fill) is selected from each stratum.

$$\overline{N}_1 = \frac{0 + 1 + 1 + 1}{4} = 0.75$$

$$\hat{N}_1 = (25)(0.75) = 18.75$$

$$\hat{S}^2_{N_{1i}} = \frac{[(0 - 0.75)^2 + (1 - 0.75)^2 + (1 - 0.75)^2 + (1 - 0.75)^2]}{4 - 1} = 0.25$$

$$\hat{V}\text{ar}(\hat{N}_1) = (25)^2 \left(1 - \frac{4}{25}\right) \frac{0.25}{4} = 32.81.$$

Estimates for the other 3 strata were: $\hat{N}_2 = 6.25$, $\hat{V}\text{ar}(\hat{N}_2) = 32.81$; $\hat{N}_3 = 56.25$, $\hat{V}\text{ar}(\hat{N}_3) = 557.81$; and $\hat{N}_4 = 12.50$, $\hat{V}\text{ar}(\hat{N}_4) = 43.75$. Therefore, the estimated abundance was $\hat{N} = 18.75 + 6.25 + 56.25 + 12.50 = 93.75$ and its estimated variance was $\hat{V}\text{ar}(\hat{N}) = 32.81 + 32.81 + 557.81 + 43.75 = 667.18$. The coefficient of variation was $\hat{C}\text{V}(\hat{N}) = \sqrt{667.18}/93.75 = 0.276$ or 27.6%. Now, compare the variance estimate to one obtained from a simple random sample of the same 16 plots from a nonstratified sampling frame. The estimated abundance still would be 93.75, but its estimated variance would be 872.81. Consequently, in this particular example, simply subdividing a frame into 4 equal-sized strata and randomly selecting the same number of plots from each yielded almost a 25% gain in precision (i.e., $1 - [667.18 \div 872.81] \times 100\% = 23.6\%$) over a simple random sample of the same plots taken from the entire sampling frame.

How to best distribute plots among strata depends on sampling costs, variability, and size of each stratum (Cochran, 1977). Each stratum should contain at least two selected plots to obtain a variance estimate. The simplest way to determine how many plots should be chosen from each stratum is to base the number of selected plots solely on the relative size of each stratum. For instance, a stratum that is three times larger than another would have three times as many plots selected within it (Fig. 2.5). This approach is called *proportional allocation* and assumes that sampling costs and variation are the same across strata (Scheaffer *et al.*, 1990). Figure 2.4 is a special case of proportional allocation in which all strata are the same size and hence contain the same number of selected plots.

Proportional allocation produces estimates that are usually less precise than other allocation techniques because it does not account for variation across strata. This is especially true when variability differs among strata. For instance, the coefficient of variation from Example 2.2a is almost twice as large as the one in Example 2.2b even though both were obtained from the same sampling frame. The difference was that the more variable stratum in Example 2.2a was undersampled (i.e., did not have an adequate number

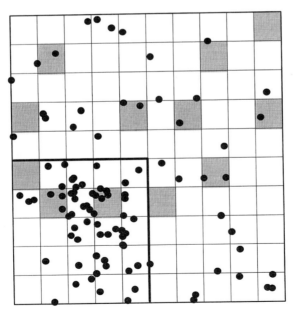

Figure 2.5   A sampling frame divided into two strata, the smaller of which ($U_1 = 25$) is one-third the size of the larger ($U_2 = 75$). Three selected plots are proportionally allocated to stratum I and nine to stratum II.

of selected plots to obtain a precise stratum estimate), whereas the more variable stratum in Example 2.2b was sampled with more effort than the less variable stratum. Thus, strata with higher variability should contain more selected plots than those with less variability (Cochran, 1977).

Example 2.2. One-Stage Stratified Simple Random Sample (Unequal-Sized Strata)

A sampling frame was divided into 2 unequal-sized strata based on habitat (Fig. 2.5). Stratum I contained 25 plots and stratum II had 75 plots. A sample of 12 plots was chosen. Complete counts were conducted on each selected plot. The same formulas used in Example 2.1 were used to calculate estimates for abundance and variance.

a. *Proportional Allocation.* The sample of plots was proportionally allocated to each stratum. A simple random sample was selected from each stratum: 3 plots from stratum I and 9 from stratum II. Counts recorded for stratum I were 0, 3, and 5, and for stratum II were 0, 1, 0, 0, 2, 1, 1, 2, and 0. The resulting estimates for each stratum were $\hat{N}_1 = 66.67$, $\hat{S}^2_{N_{1i}} = 6.33$, $\text{Var}(\hat{N}_1) = 1160.50$, $\hat{N}_2 = 58.33$, $\hat{S}^2_{N_{2i}} = 0.69$, and $\text{Var}(\hat{N}_2) = 379.5$. Thus, overall estimates were $\hat{N} = 66.67 + 58.33 = 125$, $\text{Var}(\hat{N}) = 1160.50 + 379.5 = 1540.00$, and $\text{C}\hat{\text{V}}(\hat{N}) = \sqrt{1540.00} \div 125 = 0.314$ or 31.4%.

b. *Neyman Allocation.* Based on strata variance estimates from part (a), the 12 plots are divided between the 2 strata according to the Neyman allocation [Eq. (2.3)]. The number of selected plots for stratum I would be

$$u_1 = (12) \left[\frac{(25)\sqrt{6.33}}{(25)\sqrt{6.33} + (75)\sqrt{0.69}} \right] = 6.03 \doteq 7,$$

where the calculated sample size is always rounded up to the next integer. If $u_1 = 7$ then $u_2 = 12 - 7 = 5$. Counts for stratum I were 4, 2, 3, 4, 4, 0, and 2, and those for stratum II were 1, 1, 1, 0, and 2. Overall estimates were $\hat{N} = 142.86$, $\text{Var}(\hat{N}) = 668.88$, and $\text{C}\hat{\text{V}}(\hat{N}) = 0.181$ or 18.1%. Note that the coefficient of variation is considerably less than the one in part (a) and also less than the one in Example 2.1.

c. *Optimum Allocation.* Assume sampling costs for stratum I are twice those of stratum II. We use this information along with variance estimates from part (a) to compute an optimum allocation for the 12 plots [Eq. (2.4)]. The resulting allocation is $u_1 = 5$ and $u_2 = 7$. A new simple random sample is taken from each stratum based on the new stratum sample sizes. The counts for stratum I were 4, 3, 3, 0, and 2, and for stratum II were 0, 1, 2, 1, 1, 0, and 0. Overall estimates were $\hat{N} = 113.57$, $\text{Var}(\hat{N}) = 646.33$, and $\text{C}\hat{\text{V}}(\hat{N}) = 0.224$ or 22.4%. This coefficient of variation is greater than the one computed under Neyman allocation because of the trade-off between costs and stratum variability.

Because variability among strata is such an important part of allocating samples, we would like to be able to incorporate this into our allocation scheme. The allocation method that accounts for variability among strata, but assumes costs of collecting data are equal, is known as *Neyman allocation* (Cochran, 1977; Scheaffer *et al.*, 1990). The formula for a one-stage stratified design (after Cochran, 1977, p. 98) is

$$u_h = (u) \left(\frac{U_h S_{N_{hi}}}{\sum\limits_{h=1}^{H} U_h S_{N_{hi}}} \right), \tag{2.3}$$

where $\hat{S}_{N_{hi}}$ is used in place of $S_{N_{hi}}$. Note in Example 2.2 the large gains in precision that can be obtained by accounting for variation among strata when allocating sampling effort.

Sampling costs may affect how many plots can be selected in a given stratum. For example, a stratum that is located in a more remote area requiring more travel time to and from sites will cost more to sample than a stratum with more accessible sites. This could be an important consideration because of limited availability of funding. Thus, we would like to balance the cost of collecting data within a particular stratum against its variation. This approach is called *optimum allocation* (Cochran, 1977). The allocation formula (after Cochran, 1977, p. 98) is the same as the one for Neyman allocation except for an additional cost term,

$$u_h = (u) \left(\frac{U_h S_{N_{hi}} / \sqrt{c_h}}{\sum\limits_{h=1}^{H} U_h S_{N_{hi}} / \sqrt{c_h}} \right), \tag{2.4}$$

where $\hat{S}_{N_{hi}}$ is used in place of $S_{N_{hi}}$ and c_h is the sampling cost associated with the hth stratum. Sampling costs can be represented either in absolute units (e.g., $c_1 = \$100$ and $c_2 = \$50$) or relative units (e.g., $c_1 = 2$ and $c_2 = 1$). Optimum allocation represents a trade-off between stratum costs and variances; therefore, stratum estimates may not be as precise as those produced under Neyman allocation (Example 2.2). How much less precise will depend on the differences in costs among strata, which strata are most costly to sample, and the stratum sizes.

In order to allocate sampling effort among strata, one must first set an overall sample size. We set our sample size equal to 12 in the previous example for illustrative purposes, but we would normally have to compute this value. This requires an estimate of variance, which is usually obtained through a pilot study. We will discuss this further in the chapter on guidelines for planning surveys. For now, we will assume that we have a variance estimate. For a one-stage stratified random sample, we need an $\hat{S}_{N_i}^2$ for each stratum (written $\hat{S}_{N_{hi}}^2$, where h is a subscript used to denote to which stratum this estimator applies). In addition, we can set the precision of our estimate to be within some percentage of the true parameter, which is

called *relative error* (*e*) and is equal to $\dfrac{\hat{N} - N}{N} \times 100$ in its percentage
form. We can further stipulate this in terms of α so that we can be confident that our estimate is within a certain percentage of the true parameter
about $100(1 - \alpha)\%$ of the time. All of this information is used to calculate
sample size. The formula for calculating sample size under a presumed
optimum allocation (for a fixed u; Cochran, 1977, p. 106) in estimating
abundance is

$$u = \frac{\left(\sum\limits_{h=1}^{H} U_h S_{N_{hi}}\right)^2}{\left(\dfrac{eN}{t}\right)^2 + \sum\limits_{h=1}^{H} U_h S_{N_{hi}}^2}, \tag{2.5}$$

where e is relative error and t is the value obtained from a t table correspond-
ing to $1 - \alpha/2$ and $u - 1$ degrees of freedom. Also, $\hat{N}$, $\hat{S}_{N_{hi}}$, and $\hat{S}_{N_{hi}}^2$ are
used in place of their associated parameters. We have to solve iteratively
for t because we obviously do not know the value of u to obtain the proper
degrees of freedom. This is done by inserting the z value associated with
the same $1 - \alpha/2$ and solving for u. Then, this u value is used to determine
the appropriate t value, which is inserted into the formula. This process
continues until a single u value is reached. Estimates from Example 2.2a
will be used for illustrative purposes. Suppose we wished to calculate a
sample that, on average, would put us within $\pm10\%$ of the true abundance
about 95% ($\alpha = 0.05$) of the time. Our initial computation would be

$$u = \frac{[(25)\sqrt{6.33} + (75)\sqrt{0.69}]^2}{\left[\dfrac{(0.10)(125)}{1.96}\right]^2 + [(25)(6.33) + (75)(0.69)]} = 62.53 \doteq 63.$$

The t value corresponding to $1 - (0.05/2)$ and 62 degrees of freedom is
1.999. Substituting this value in for 1.96 again yields a sample size of 63;
therefore, we would use some allocation method to divide the 63 plots
between the 2 strata.

Now that we know how to calculate sample sizes and allocate plots
among strata, how do we now if we can afford to sample that many plots?
Overall cost (C) can be computed using the estimated sampling costs per
strata (c_h), number of selected plots per strata (u_h), and overhead costs of
a project (c_0) into the following cost function (Cochran, 1977, p. 96),

$$C = c_0 + \sum\limits_{h=1}^{H} u_h c_h. \tag{2.6}$$

In this way, an estimated total cost can be computed, which allows adjustments/decisions to be made if estimated costs exceed available funds.

2.3.3. Systematic Sampling with a Random Start

A sampling procedure designed to simplify the selection process and select plots distributed across the sampling frame is called *systematic sampling* (Schaeffer *et al.*, 1990). A systematic sampling procedure randomly chooses a plot from the first k_s plots and then includes every k_sth plot after (Scheaffer *et al.*, 1990). Therefore, systematic sampling as we use it implies a random starting point. Choice of k_s depends on desired sample size; if there are 500 sampling units and desired sample size is 20, then k_s would be $500/20 = 25$.

If k_s does not divide evenly into the total number of plots, then biased estimates will result from drawing a 1-in-k_s sample. The reason for this is that not all systematic samples will contain the same number of plots. For instance, suppose we wish to take a 1-in-4 systematic sample from a sampling frame of 10 plots. There are four possible systematic samples: (1, 5, 9), (2, 6, 10), (3, 7), and (4, 8). Thus, two of the possible samples are based on 3 plots, whereas the other two are based on 2 plots. This results in biased estimates. However, Cochran (1977) suggested that the magnitude of this bias is trivial when sample sizes exceed 50 and is likely insignificant even when sample sizes are small. Therefore, we will ignore this possible source of bias in subsequent discussions. We refer interested readers to Cochran (1977, p. 206) and Levy and Lemeshow (1991, pp. 82–84) for descriptions of two modifications of the systematic sampling method that always yield unbiased estimates of means, totals, and proportions. Further, the Horvitz–Thompson estimator (Horvitz and Thompson, 1952) listed in Appendix C may be used to achieve the same result (Overton and Stehman, 1995).

Systematic sampling does not require a well-defined sampling frame. If distance or dimension data are available for an area of interest, then a 1-in-k_s sample may be drawn based on some distance. For instance, a biologist wishes to select five sampling points using a 1-in-10 systematic sampling design along 50 km of a riparian area. She randomly chooses a point within the first 10 km (e.g., 2.5 km) to use as her starting point. She then would have sampling points at 2.5, 12.5, 22.5, 32.5, and 42.5 km.

Although a single systematic sample improves sample coverage over a frame, it is based only on a single random point so its variance estimates must be approximated. Just as a minimum of 2 plots must be randomly selected from each stratum to directly compute variance estimates, so too must at least 2 random points be selected for systematic sampling. The variance approximation for a single systematic sample is based on the variance estimator for a simple random sample and is unbiased only if animals are randomly distributed across a frame. The magnitude of this bias will depend on the underlying distribution of animals.

Drawing a single systematic sample from a frame with either periodically arranged or spatially autocorrelated individuals will likely produce biased estimates of abundance and variance. Applying the Horvitz–Thompson estimator (Horvitz and Thompson, 1952) will yield unbiased abundance estimates regardless of the underlying arrangement of animals (Overton and Stehman, 1995); however, variance estimates still would be underestimated. Spatial autocorrelation can occur due to habitat. That is, two adjacent units occurring in habitat supporting a large number of animals are much more likely to both contain high numbers than two nonadjacent units occurring in different habitats with different animal densities. An example of periodicity could be the pool–riffle–run sequence in rivers. However, because these stream units vary in size, a systematic sample based on distances is unlikely to coincide exactly with only one or mostly one stream habitat type. Periodic variation is unlikely to occur in most situations because of the clumped distributions of many biological populations (Milne, 1959; Krebs, 1989). Further, periodicity will probably not occur at the scale at which most fish and wildlife populations are sampled.

Taking repeated systematic samples will yield unbiased estimates of abundance and variance in any situation. Instead of randomly selecting 1 point or plot within the first k_s, we could randomly select more than 1 and use the resulting counts from each to calculate an overall estimate of abundance and variance (Fig. 2.6). Or, multiple starting points could be selected across the sampling frame (Scheaffer et al., 1990). Unfortunately, a repeated systematic sample on average, will yield a less precise estimate than a single one when based on the same number of plots. For instance, a single sample of 10 plots will produce a more precise estimate than two samples of 5 plots each. Larger repeated samples would produce more precise estimates, but also would increase overall sample size. Estimators for repeated systematic sampling are similar to those for simple random sampling except they are based on multiple samples rather than simply multiple observations within a sample; hence, notation must be modified (Scheaffer et al., 1990, pp. 221–222) to

$$\overline{N}* = \frac{\sum_{j=1}^{n_s} \overline{N}_j}{n_s}, \tag{2.7}$$

$$\hat{N} = U \times \overline{N}* \tag{2.8}$$

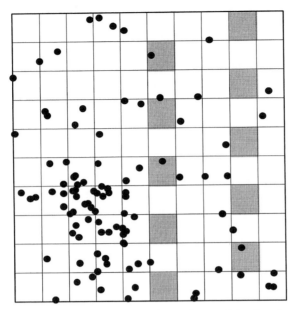

Figure 2.6 A repeated systematic sample, each sample containing five plots, from a sampling frame.

and

$$\text{V\^ar}(\hat{N}) = U^2 \left(1 - \frac{u}{U}\right) \frac{\sum\limits_{j=1}^{n_s} (\overline{N}_j - \overline{N}*)^2}{n_s(n_s - 1)}, \qquad (2.9)$$

where $\overline{N}_j$ is the arithmetic mean of the jth systematic sample, n_s is the number of systematic samples chosen, and $\overline{N}*$ is the overall mean across all systematic samples chosen. All other terms are defined as before.

Selecting a repeated systematic sample begins with setting its sampling interval, k_s^*, which is obtained by multiplying the number of repeated systematic samples chosen by the sampling interval for a single systematic sample ($k_s = U \div u$), or $k_s^* = n_s k_s$. Then, n_s random starting points are selected from the first k_s^* plots. Finally, the k_s^* value is added as a constant to each random starting value until u/n_s numbers are picked between each starting point and U. For example, say we wanted $n_s = 2$ systematic samples totaling $u = 10$ plots from a sampling frame of $U = 100$ plots. Our sampling interval for a single systematic sample would be $k_s = 100 \div 10 = 10$ or 1-in-10; hence, the sampling interval for 2 systematic samples would be $k_s^* = 2 \times 10 = 20$ or 1-in-20. We then randomly choose two numbers without replacement between 1 and 20, say, 3 and 19. Next, we generate our string of 10 ÷ 2 = 5 plot values for each systematic sample by consecutively adding 20 to each starting point. Thus, our first systematic sample would contain plot numbers 3, 23, 43, 63, and 83, whereas our second would contain 19, 39, 59, 79, and 99. We further illustrate this procedure in Example 2.3, but may require a larger sample size (also see Scheaffer *et al.*, 1990, pp. 222–223). The sample size formula for repeated systematic sampling is in Appendix C.

Even though systematic sampling usually provides better coverage than simple random sampling, we can see from Fig. 2.6 and Example 2.3 that this is not always the case. That is, even a repeated systematic sample missed the majority of the clumped distribution of "animals" in Fig. 2.6 (i.e., a spatially autocorrelated population). A remedy for this is to stratify the frame and then take repeated systematic samples within each strata. This would pretty well ensure good coverage of the frame and avoid problems that occurred in Example 2.3.

Example 2.3. One-Stage Repeated Systematic Sample

Two systematic samples were taken from a sampling frame containing 100 plots (Fig. 2.6). The sampling interval for each, k_s, was set at 10 plots to correspond to an overall sample size of $100/10 = 10$; therefore the overall sampling interval was $k_s^* = (2)(10) = 20$. Two random numbers were chosen between 1 and 20 and used as the starting points for each sample. The first systematic sample had counts of 0, 0, 0, 0, and 1; the second had counts of 1, 0, 1, 0, and 0. Therefore, the mean of the first sample, $\overline{N}_1$, was $(0 + 0 + 0 + 0 + 1)/5$ or 0.2, whereas the mean for the second, $\overline{N}_1$, was $(1 + 0 + 1 + 0 + 0)/5$ or 0.4. The overall mean across both samples was:

$$\overline{N}^* = \frac{0.2 + 0.4}{2} = 0.3.$$

The estimates for abundance and variance were

$$\hat{N} = (100)(0.3) = 30$$

and

$$\hat{Var}(\hat{N}) = (100)^2 \left(1 - \frac{10}{100}\right) \frac{(0.2 - 0.3)^2 + (0.4 - 0.3)^2}{2(2 - 1)} = 90.0.$$

Although the above variance estimate is considerably smaller than those in Example 2.2, the abundance estimate is not even close to the true abundance of 100. A better approach would be to stratify the sampling frame and then take a repeated systematic sample within each stratum.

Single systematic samples are commonly used in fish and wildlife studies when a systematic scheme is employed. As stated previously, periodicity in animal distributions is probably not a problem but spatial autocorrelation likely is (e.g., Fig. 2.6). Stratifying a frame by habitat may alleviate this difficulty, but without knowing the exact distribution of animals, stratification alone is probably not the answer. Although a repeated systematic sample within each stratum does not offer the precision of a single systematic sample, it will produce an unbiased estimate of variance; data can be collected from two systematic samples at the same time as long as plot information from each sample is recorded separately, so there should not be a loss in efficiency when conducting a repeated sample. Therefore, we recommend a repeated systematic sample when a systematic design is used.

2.3.4. SIMPLE LATIN SQUARE SAMPLING +1

Munholland and Borkowski (1996) have recently developed a sampling design that attempts to ensure good coverage of plots across a sampling frame while providing unbiased estimates of precision. *Simple Latin square sampling +1* is a design in which 1 plot is initially chosen randomly from each unique row and column combination of a square sampling frame, and then a single plot is randomly selected from the remaining group of unselected plots irrespective of location. This process is best described via example. Figure 2.7A is a 3 × 3-square sampling frame consisting of 9 plots

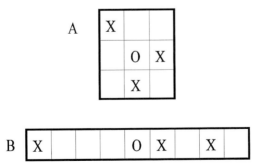

Figure 2.7 The top figure (A) is a simple Latin square sample +1 of size 4 drawn from a square sampling frame of nine plots. Plots marked with an "X" were initially chosen, whereas the plot marked with an "O" was the "+1" unit selected from the remaining six plots. The bottom figure (B) is figure A "stretched out" to be applied to a linear habitat such as a stream or riparian zone.

from which we will draw a simple Latin square sample +1 of size 4 (equal to $n + 1$, where n is equal to the number of rows or columns). We begin by randomly selecting a plot (marked with an X) from the first row of 3 units. Then, the second plot is randomly chosen from the 2 units not sharing the same column as the first plot. The third plot is the remaining unit in the third row that is not associated with the columns occupied by plots in the first 2 rows. Finally, the +1 plot is randomly chosen from the group of 6 units not previously selected, and is marked with an O.

Areas do not have to be shaped as squares for a simple Latin square sample +1 to be applied. Figure 2.7B shows a stretched out form of Fig. 2.7A that could be superimposed over a linear habitat such as a stream or riparian area (Munholland and Borkowski, 1996). Moreover, a group of smaller squares could be made to approximately fit an irregularly shaped sampling frame.

Simple Latin square sampling +1 generally results in more precise estimates than either simple random sampling or systematic sampling when the underlying distribution of animals exhibit spatial autocorrelation (Munholland and Borkowski, 1996). Also, unlike single systematic sampling, unbiased variance estimates can be calculated because there are 2 plots that are randomly selected. However, variance estimators under this design are complex, and their calculation requires a computer program. Nonetheless, simple Latin square sampling +1 shows some promise as a more precise alternative to other sampling designs.

2.3.5. RANKED SET SAMPLING

As stated previously, an ideal sampling procedure is one that produces unbiased and precise estimates at low effort and cost; much of the cost and effort from sampling derive from the large number of plots necessary to meet precision requirements. Therefore, an ideal sampling design would provide unbiased and reasonably precise estimates based on a relatively low number of measured plots. A design that may potentially accomplish this is called *ranked set sampling* (McIntyre, 1952; Johnson *et al.*, 1996). This design is based on an initial random selection of plots from which a smaller subset of plots is intensively surveyed. The subset consists of a collection of plots thought to best represent the sampled population. For example, a subset of 3 plots containing low, medium, and high densities of animals would probably be more representative than a subset containing low densities of animals, all else being equal.

Ranked set sampling has been applied mostly to vegetational studies (Halls and Dell, 1966; Muttlak and McDonald, 1992), but has potential for broader applications. For instance, this design could be used in estimating animal abundance. The procedure for choosing a ranked set sample for estimating abundance would be: (1) randomly choose a sample of n_s^2 plots from a sampling frame, (2) randomly allocate selected plots into n_s subsets of size n_s, (3) rank each plot within a particular subset based on perceived number of animals contained within it relative to other plots within that particular subset, (4) choose a plot with the smallest perceived number of animals from the first subset, the second smallest from the second subset, and so on until 1 plot is chosen from each subset to compose the actual sample that will be used in computing estimates, and (5) repeat the previous four steps m times until $u = n_s m$ plots are obtained (Johnson *et al.*, 1996). The ranking procedure could be based on habitat classifications (e.g., from GIS information, cover maps, and related sources; Fig. 2.8). That is, if abundance or density of a particular species can be correlated with habitat structure and composition, then subsets of plots could be ranked using this information. This would be much less costly than on-site visits. In fact, ranked set sampling may be viewed as a form of *double sampling* in which a "cheap" method is used to collect general information from a large sample of plots, and an "expensive" method is used to acquire detailed data from a smaller subset of plots.

We will use the example shown in Fig. 2.8 to illustrate how abundance and variance estimates are calculated from a ranked set sample that is

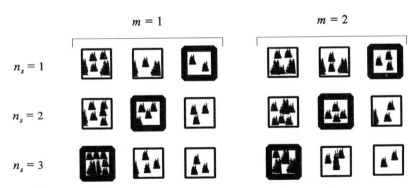

Figure 2.8 An example of a ranked set sample with ranking based on vegetational cover and composition within 2 repetitions ($m = 2$) of simple random samples of $n_s^2 = 9$ plots from a sampling frame superimposed on an aerial photograph. Each set of 9 is randomly partitioned into 3 subsets of 3 plots. Each subset then is ranked by degree of vegetational cover (high, medium, and low), which is assumed to be correlated with animal density. Only 6 plots ($u = n_s m = 3 \times 2 = 6$), heavily outlined in black, are included in the actual sample; that is, a low-cover plot was taken from the first subset, a medium from the second, and a high from the third for each repetition.

chosen using a simple random sampling scheme. Suppose complete counts on the 3 plots from the first sample (i.e., $m = 1$) were 4, 16, and 20, and from the second were 12, 14, and 18. Further, assume the sampling frame contains 500 plots. We use the following formula (Stokes, 1986, p. 585) to calculate an abundance estimate,

$$\overline{N} = \frac{\sum_{i=1}^{m} \sum_{j=1}^{n_s} N_{[j]i}}{mn_s}, \tag{2.10}$$

where $N_{[j]i}$ is the count within the jth ranked plot selected in the ith repetition. Inserting the 6 counts into this formula gives us

$$\overline{N} = \frac{4 + 16 + 20 + 12 + 14 + 18}{(2)(3)} = 14.0.$$

Estimated abundance is calculated as usual, i.e., $\hat{N} = U \times \overline{N} = (500)$ $(14.0) = 7000$. The formula for sample variance (Stokes, 1986, p. 587) is

$$\hat{S}^2_{N_{[j]i}} = \frac{\sum_{i=1}^{m} \sum_{j=1}^{n_s} (N_{[j]i} - \overline{N})^2}{mn_s - 1}. \tag{2.11}$$

Inserting the appropriate numbers in the previous formula gives a sample variance of

$$\hat{S}^2_{N_{[j]i}} = \frac{(4 - 14.0)^2 + (16 - 14.0)^2 + \cdots + (18 - 14.0)^2}{(2)(3) - 1} = 32.0$$

and the estimated variance of $\hat{N}$ (ignoring the fpc) is the usual

$$\hat{V}ar(\hat{N}) = U^2 \left(\frac{\hat{S}^2_{N_{[j]i}}}{u} \right) = (500^2) \left(\frac{32.0}{6} \right) = 1,333,333.$$

There are certain conditions that must be met for ranked set sampling to be both usable and useful. First of all, a cheap method of collecting data must be available that produces information (i.e., a ranking criterion) that is strongly correlated with the parameters of interest. Second, sets should be randomly partitioned into subsets of size n_s before ranking is conducted. Nonrandom partitioning will lead to selection bias. Third, ranking should be reasonably accurate for optimum precision. However, under equal allocation of intensively sampled plots to ranking categories (as in Fig. 2.8), variance estimates obtained from a ranked set sample will never be worse than those from an equal-sized simple random sample even if ranking errors occur (Stokes, 1980). Last, the underlying spatial arrangement of individuals can negatively affect the performance of ranked set sampling (Patil *et al.*,

1994). Given that animal populations tend to be spatially clustered, some ranking categories will be more variable than others. If so, one should allocate more of the intensively sampled plots to the more variable categories to increase overall precision. For example, suppose we had 3 categories of rank, where the high category had 2.5 times the variability of the other 2 categories. For a final sample of size 9 (e.g., $m = 3$, $n_s = 3$), instead of equally allocating 3 plots to each category, we could allocate 5 plots to the high rank and 2 to each of the others. That is, 5 of the 9 subsets would have a plot selected in the "high" ranking category. Unequal allocation should be done carefully, however. Incorrectly allocating samples could lead to low precision, even lower than that of a simple random sample (Johnson *et al.*, 1996).

A stratified sampling scheme also may be used in conjunction with rank set sampling. In fact, ranked set sampling is itself a form of stratified sampling—samples are stratified according to a judgement criterion. The procedure using a stratified frame would be the same as described above, but would be conducted within each stratum. Overall abundance and variance estimates would be obtained by summing across strata. Moreover, ranked set sampling could be used in a two-stage scenario in which a method for correcting estimates for incomplete detectability (e.g., capture–recapture) could be used to obtain plot abundance estimates. A second component of variation would be included in the overall variance estimate.

How do we determine an appropriate sample size for a ranked set sample? That is, how do we set values for m and n_s? The optimum value for number of repetitions (m) is 1 (Takahasi and Wakimoto, 1968). Unfortunately, reducing m to 1 means increasing n_s to at least twice its size, depending on the original number of repetitions, and thereby greatly increasing the potential for ranking error; choice of n_s depends on how many plots can be easily sampled using a cheap method, accuracy of the associated ranking procedure (Patil *et al.*, 1994), and how many plots can be intensively sampled under available funding. Johnson *et al.* (1996) recommended an n_s of 3–5 to reduce potential for ranking error. In any event, a pilot study should be conducted to obtain estimates of precision and cost for use in determining both the best way to allocate plots to different ranking categories and the optimum values of m and n_s.

2.4. A SAMPLING DESIGN FOR RARE AND CLUSTERED SPECIES

Sampling designs discussed to this point are most useful in obtaining information for moderately common to abundant species; they are neither cost nor labor efficient when applied to assessing rare populations. A large percentage of the sampling frame may have to be included for resulting estimates to have adequate precision. That is, many zero counts in a sample will likely decrease the precision of resulting estimates even when nonzero counts are not that much larger than zero. Moreover, spatial clustering tends to be an attribute of rare species as well as abundant ones. A sampling procedure specifically designed to yield more precise estimates than commonly used designs when individuals are rare and clustered is called *adaptive cluster sampling*. There are other sampling designs that may be applied specifically to rare populations (e.g., see reviews by Sudman *et al.,* 1988, and Schreuder *et al.,* 1993), but adaptive cluster sampling is probably the one best suited for surveying animal populations.

Adaptive cluster sampling is a design that allows for increased counting effort adjacent to a selected plot containing an individual or variable of interest (Thompson, 1992; Thompson and Seber, 1996). For instance, a simple random sample of plots could be initially selected. If a plot contains an individual or item of interest, then surrounding plots to the top, bottom, right, and left are included in the sample (fewer if the selected plot happens to be adjacent to the edge of the sampling frame). If any of the surrounding plots contain animals, then counts are conducted on plots to the top, bottom, right, and left of these as well. Note that counts may have already been performed on one or more of these surrounding plots during the course of plot selection. This process continues until animals are no longer encountered within additionally selected plots. The collection of initially chosen and additionally counted plots is called a *network*. Initially selected plots that do not contain animals can be considered networks of 1 plot (Thompson, 1992). Thompson (1992) provided estimators for adaptive cluster sampling schemes in which initial selection of plots was done via simple random sampling, repeated systematic sampling, and stratified sampling. Munholland and Borkowski (1993) derived estimators for adaptive simple Latin square sampling +1 designs, which were presented in Thompson and Seber (1996).

Adaptive cluster sampling is probably best explained through an example. We will use a stratified adaptive cluster design because stratified designs quite often provide a more precise estimator than other designs. Also, for illustrative purposes, we will assume a complete count is feasible for each plot.

Suppose a sampling frame was divided into 2 strata, where stratification was based on perceived density of a rare species (Fig. 2.9). An initial simple random sample of 5 plots was chosen from each stratum. One initial plot from each stratum contained at least one animal and therefore adjacent plots also were included in the sample as described above. The other 8

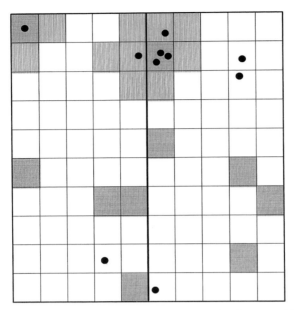

Figure 2.9 A stratified adaptive cluster sample from a sampling frame partitioned into two equal-sized strata ($U_1 = U_2 = 50$) by perceived density of a rare species of interest. Five initial plots (gray fill) were chosen via simple random sampling; adjacent plots selected are shown with hatching.

initially chosen plots did not contain animals. Note that in Fig. 2.9 the chosen plot containing an animal in stratum 1 was located in a corner so that only 2 adjacent plots were available for additional counts. Neither of these additional plots contained animals; hence, no additional plots were part of this network. Conversely, an initial plot chosen in the second stratum had 2 adjacent plots that contained animals, one of which was located in stratum 1. Therefore, plots adjacent to these 2 also were counted. This network was composed of 9 plots.

Abundance and variance estimators for stratified adaptive cluster designs are based on networks as well as plots. Count data for the network containing the ith initially selected plot in the hth stratum are inserted into an estimator, w_{hi}. The formula (Thompson, 1992, p. 306) is

$$w_{hi} = \frac{\left(\dfrac{u_h}{U_h}\right) \sum\limits_{j=1}^{H} \tau_{jhi}}{\sum\limits_{j=1}^{H} \dfrac{u_j}{U_j} (m_{jhi})}, \tag{2.12}$$

where u_h is the number of initially selected plots in stratum h, U_h is the total number of plots in stratum h, τ_{jhi} is the total count within stratum j for the network containing the ith initially selected plot in stratum h (strata are additionally referenced with a j in case 1 or more plots with animals within the ith network are inside a different stratum than the hth stratum containing the initially selected plot), u_j is number of initially selected plots in the jth stratum, U_j is the total number of plots in stratum j, and m_{jhi} is the number of plots that contribute to the τ_{jhi}.

We now will apply sample results from Fig. 2.9 to Eq. (2.12) to clarify its notation and use. The first plots initially selected from both strata were the only initially selected plots that contained animals; therefore, $w_{12} = w_{13} = w_{14} = w_{15} = 0$ and $w_{22} = w_{23} = w_{24} = w_{25} = 0$. None of the plots adjacent to plot 1 in stratum 1 contained animals, so w_{11} was computed as

$$w_{11} = \frac{(5/50)(1)}{(5/50)(1)} = 1.$$

There is only a single term in the denominator because the network containing plot 1 was entirely within stratum 1. Conversely, the network containing plot 1 in stratum 2 occurred within both strata, which must be accounted for in the computation of w_{21},

$$w_{21} = \frac{(5/50)(3 + 1 + 1)}{(5/50)(2) + (5/50)(1)} = 1.67,$$

where the 2 in left term in the denominator refers to the fact that 2 plots surveyed in this network in stratum 2 contained animals. The 1 in the right term represents the 1 plot containing animals in the same network that was located in stratum 1. Both strata in this example were the same size and had the same number of initially selected plots, so the u_j / U_j are the same.

Now that we have computed w_{hi}, how do we use these data to obtain estimates of abundance and variance? The estimator of total number of animals within each stratum (Thompson, 1992, p. 307) is

$$\hat{N}_h = \frac{U_h}{u_h} \left(\sum_{i=1}^{u_h} w_{hi} \right), \tag{2.13}$$

and the estimated overall abundance is simply the sum of the $\hat{N}_h$ as in other stratified designs. Consequently, our stratum estimates were

$$\hat{N}_1 = \frac{50}{5}(1 + 0 + 0 + 0 + 0) = 10$$

and

$$\hat{N}_2 = \frac{50}{5}(1.67 + 0 + 0 + 0 + 0) = 16.7$$

Thus, the estimate of overall abundance would be $\hat{N} = 10 + 16.7 = 26.7$. The variance estimator for $\hat{N}$ is the same as in Eq. (2.2) except that $\hat{S}^2_{N_{hi}}$ now is defined in terms of w_{hi} (Thompson, 1992, p. 307)

$$\hat{S}^2_{w_{hi}} = \frac{\displaystyle\sum_{i=1}^{u_h} (w_{hi} - \overline{w}_h)^2}{u_h - 1}, \tag{2.14}$$

where $\overline{w}_h$ is the sum of the w_{hi} divided by u_h. For our example, $\overline{w}_1 = (1 + 0 + 0 + 0 + 0) \div 5 = 0.2$ and $\overline{w}_2 = (1.67 + 0 + 0 + 0 + 0) \div 5 = 0.334$. The sample variance estimates for each stratum were

$$\hat{S}^2_{w_{1i}} = \frac{(1 - 0.2)^2 + (0 - 0.2)^2 + \cdots + (0 - 0.2)^2}{5 - 1} = 0.2$$

and

$$\hat{S}^2_{w_{2i}} = \frac{(1.67 - 0.334)^2 + (0 - 0.334)^2 + \cdots + (0 - 0.334)^2}{5 - 1} = 0.558.$$

The overall variance estimates for each stratum were

$$\hat{Var}(\hat{N}_1) = (50)^2 \left(1 - \frac{5}{50}\right)\frac{0.2}{5} = 90$$

and

$$\hat{Var}(\hat{N}_2) = (50)^2 \left(1 - \frac{5}{50}\right)\frac{0.558}{5} = 251.1.$$

Therefore, our overall estimate of variance for the entire sampling frame was $\hat{Var}(\hat{N}) = 90 + 251.1 = 341.1$. Putting this in terms of the coefficient of variation ($[\sqrt{341.1/26.7}] \times 100 = 69.2\%$), we can see that our estimate is imprecise. We could improve precision by increasing sample size, but how many more plots should be included? Unfortunately, the nature of this design does not allow us to compute sample sizes as in a number of other designs. The reason is that we can only control how many initial samples are chosen; the number of adjacent plots that are included in the overall sample cannot be predetermined. For instance, 10 plots were initially chosen in Fig. 2.9, but the total sample ended up being 21 plots. One approach to estimating sample size for an adaptive cluster sampling design is to use results from a previous survey and formulate a best guess based on number of adjacent plots included with the initially selected plots and resulting variance estimates. Also, note that applying an adaptive cluster sampling scheme to a moderately common to abundant population could result in a substantial number of plots being counted, which could be quite costly and inefficient. Therefore, care should be exercised in the use of adaptive cluster sampling. Thompson and Seber (1996) discussed a number of issues related to sample size and efficiency in adaptive cluster sampling. Smith *et al.* (1995) discussed efficiency of adaptive cluster designs for estimating density of waterfowl, and offered recommendations for wildlife surveys in general.

Rarely can we assume a complete count of animals at any scale. Hence, Thompson and Seber (1994, 1996) presented estimators for instances when animals are less than completely detectable. However, incomplete detectability of animals can be especially difficult to handle when sampling rare populations. There may not be enough individuals within a reasonably sized plot to be able to obtain valid estimates from techniques like capture–recapture (but see Rosenberg *et al.,* 1995) and distance sampling. In this case, unbiased and precise abundance estimates may not be possible under current technology. Perhaps the only time adaptive cluster sampling would be feasible in these circumstances is when the item of interest is a collection of individuals, such as a roost. Then, a capture–recapture method may be feasible within located roosts, and would correspond to a two-stage design. In any event, sampling rare populations will likely be a very costly endeavor regardless of how it is performed.

LITERATURE CITED

Bormann, F. H. (1953). The statistical efficiency of sample plot size and shape in forest ecology. *Ecology* **34:** 474–487.

Clapham, A. R. (1932). The form of the observational unit in quantitative ecology. *J. Ecol.* **20:** 192–197.

Cochran, W. G. (1977). *Sampling Techniques,* third ed. Wiley, New York.

Gilbert, R. O. (1987). *Statistical Methods for Environmental Pollution Monitoring.* Van Rostrand Reinhold, New York.

Greig-Smith, P. (1964). *Quantitative Plant Ecology,* second ed. Butterworths, London.

Halls, L. K., and Dell, T. R. (1966). Trial of ranked-set sampling for forage yields. *For. Sci.* **12:** 22–26.

Hankin, D. G. (1984). Multistage sampling designs in fisheries research: Applications in small streams. *Can. J. Fish. Aquat. Sci.* **41:** 1575–1591.

Horvitz, D. G., and Thompson, D. J. (1952). A generalization of sampling without replacement from a finite universe. *J. Am. Stat. Assoc.* **47:** 663–685.

Johnson, G. D., Nussbaum, B. D., and Patil, G. P. (1996). Designing cost-effective environmental sampling using concomitant information. *Chance* **9:** 4–11.

Krebs, C. J. (1989). *Ecological Methodology.* Harper Collins, New York.

Levy, P. S., and Lemeshow, S. (1991). *Sampling of Populations: Methods and Applications.* Wiley, New York.

McIntyre, G. A. (1952). A method of unbiased selective sampling, using ranked sets. *Aust. Res.* **3:** 385–390.

McNab, B. K. (1963). Bioenergetics and the determination of home range size. *Am. Natur.* **97:** 133–140.

Milne, A. (1959). The centric systematic area-sample treated as a random sample. *Biometrics* **15:** 270–297.

Munholland, P. L., and Borkowski, J. J. (1993). *Adaptive Latin Square Sampling +1 Designs.* Department of Mathematical Sciences, Montana State University, Bozeman. [Tech Rep. No. 3-23-93].

Munholland, P. L., and Borkowski, J. J. (1996). Simple Latin square sampling +1: A spatial design using quadrats. *Biometrics* **52:** 125–136.

Muttlak, H. A., and McDonald, L. L. (1992). Ranked set sampling and the line intercept method: A more efficient procedure. *Biom. J.* **34:** 329–346.

Overton, W. S., and Stehman, S. V. (1995). The Horvitz-Thompson theorem as a unifying perspective for probability samping: With examples from natural resource sampling. *Am. Stat.* **49:** 261–268.

Patil, G. P., Sinha, A. K., and Taillie, C. (1994). Ranked set sampling. In *Environmental Statistics* (G. P. Patil and C. R. Rao, Eds.), pp. 167–200. North Holland/Elsevier Science, Amsterdam.

Rosenberg, D. K., Overton, W. S., and Anthony, R. G. (1995). Estimation of animal abundance when capture probabilities are low and heterogeneous. *J. Wildl. Mgmt.* **59:** 252–261.

Scheaffer, R. L., Mendenhall, W., and Ott, L. (1990). *Elementary Survey Sampling,* fourth ed. PWS-Kent, Boston.

Schreuder, H. T., Gregoire, T. G., and Wood, G. B. (1993). *Sampling Methods for Multiresource Forest Inventory.* Wiley, New York.

Skalski, J. R. (1994). Estimating wildlife populations based on incomplete area surveys. *Wildl. Soc. Bull.* **22:** 192–203.

Smith, D. R., Conroy, M. J., and Brakhage, D. H. (1995). Efficiency of adaptive cluster sampling for estimating density of wintering waterfowl. *Biometrics* **51:** 777–788.

Stokes, S. L. (1980). Estimation of variance using judgement ordered ranked set samples. *Biometrics* **36:** 35–42.

Stokes, S. L. (1986). Ranked set sampling. In *Encyclopedia of Statistical Sciences* (S. Kotz and N. L. Johnson, Eds.) Vol. 7, pp. 585–588. Wiley, New York.

Sudman, S., Sirken, M. G., and Cowan, C. D. (1988). Sampling rare and elusive populations. *Science* **240:** 991–996.

Takahasi, K., and Wakimoto, K. (1968). On unbiased estimates of the population mean based on the sample stratified by means of ordering. *Ann. Inst. Stat. Math.* **20:** 1–31.

Thompson, S. K. (1992). *Sampling.* Wiley, New York.

Thompson, S. K., and Seber, G. A. F. (1994). Detectability in conventional and adaptive sampling. *Biometrics* **50:** 712–724.

Thompson, S. K., and Seber, G. A. F. (1996). *Adaptive Sampling.* Wiley, New York.

Wiegart, R. G. (1962). The selection of optimum quadrat size for sampling the standing crop of grasses and forbs. *Ecology* **43:** 125–129.

Chapter 3

Enumeration Methods

3.1. Complete Counts
3.2. Incomplete Counts
   3.2.1. Indices

3.2.2. Adjusting for
     Incomplete Detectability
Literature Cited

The sampling methods described in Chapter 2 provide for a valid sample of the population of interest. However, nothing in Chapter 2 pertains to the problem of estimating the actual number of animals on the sampling unit. In this chapter, we describe how each of the sampling units is assessed to obtain either the actual abundance or an estimate of the abundance on each sampling unit. We discuss complete counts (censuses), indices, and enumeration methods that adjust for incomplete detectability of individuals. A review of these methods also is provided by Lancia *et al.* (1994) and van Hensbergen and White (1995).

Efficient population estimators are needed to minimize the variance of the estimated population size of each sampling unit. Inefficient estimators will result in high sampling variation because there is a lot of noise associated with each sampling unit. That is, the population estimates all have large standard errors, and as a result, the variation among sampling units is large. As discussed in Chapter 1, the enumeration variance contributes to the overall variance of the survey, so we want to keep the enumeration variance as small as possible.

3.1. COMPLETE COUNTS

Methods for assessing the numbers of individuals within a sampling unit can be categorized as complete and partial counts. Complete counts are a complete enumeration (census) of individuals within a sampling unit. Thus, a random sample of quadrats might be drawn, and all the individuals counted on each of the quadrats. Such counts are rarely possible in studies of animal populations. The assumption is that if the sampling unit could be counted repeatedly, exactly the same count would be made for each replicate. Thus, there is no variance associated with the count.

An example of a survey that is treated as a complete count is quadrat counts of mule deer (Kufeld *et al.*, 1980; Bartmann *et al.*, 1986) and moose (Gasaway *et al.*, 1986). For mule deer surveys, the Colorado Division of Wildlife divides the winter range area to be surveyed into 1 mile^2 or $\frac{1}{4}$ mile^2 quadrats based on the land survey section corners. Then, a random sample from this list of quadrats is drawn to be surveyed by helicopter. The count of animals on each quadrat is treated as a complete count of the deer on the quadrat. In Alaska, the moose range to be surveyed is divided into subareas based on topography and vegetation types. Then, a random sample of these subareas is selected to be surveyed by helicopter. The count of moose on each subarea is treated as a complete count of the subarea, and a weighted mean and variance are computed from the surveyed subareas based on the size of each of the surveyed subareas.

3.2. INCOMPLETE COUNTS

Incomplete or partial counts mean that not all individuals are counted on the sampling unit or, possibly, over the sampling frame (e.g., aerial survey). All counts of animals (or other items of interest) may be represented by a simple relationship between the observed count of animals (O), the probability of detection (p), and the true number of animals (N) in some defined area and time period. That is, our observed count is equal to the detectability multiplied by the total number of animals (Lancia *et al.*, 1994). For plot counts, we write this relationship as $O_i = p_i \times N_i$, where the subscript i refers to the ith plot. Detecting 100% of animals in the ith plot means that $p_i = 1$ and $O_i = N_i$, i.e., the observed count is equal to the true number of animals. This is rarely the case. Instead, we need an estimator for detectability, $\hat{p}_i$, to be used in conjunction with our observed count, O_i, to estimate abundance such as

$$\hat{N}_i = \frac{O_i}{\hat{p}_i}.$$

For instance, if 50% of the animals were included in an observed count of 50, our abundance estimate would be 50 ÷ 0.5 or 100 animals. It is clear from this that our ability to reliably estimate abundance revolves around

$\hat{p}_i$ (Nichols, 1992; Lancia *et al.*, 1994). There are two basic approaches to dealing with this issue: (1) assume that the proportion of animals detected is constant across all plots and time intervals (i.e., index methods); and (2) compute $\hat{p}_i$ from the sampling process and use it to adjust abundance estimates accordingly. In this latter case, the method for calculating $\hat{p}_i$ can be based on statistical theory, or computed *ad hoc*. *Ad hoc* methods are an improvisation based on properties of empirical data; there are no known statistical theories underlying these methods so they cannot be rigorously assessed for statistical validity. Therefore, we will concentrate on methods that were founded on statistical theory because, under the correct conditions, they have known statistical properties. The key in this latter case is the biological validity of underlying assumptions. That is, methods that have known statistical properties still will produce invalid abundance estimates if the assumptions underlying these methods are not biologically realistic. Thus, assumptions underlying any method of population enumeration must be tested for validity before the method is used as part of a monitoring program.

3.2.1. INDICES

An index is a statistic assumed to be correlated to the true parameter of interest in some way. Unadjusted partial counts of animals is a common example. We know that not all animals or items are detected, but there is no mechanism for providing a valid estimate of how many are missed, i.e., there is no adjustment for incomplete detectability of individuals. If a constant proportion of animals is counted across all sampling units and across time, then there is a reasonable chance to detect a true trend in abundance estimates (although the trend statistic is still biased; see Barker and Sauer, 1992). The popularity of indices is due to their ease of measurement as well as the belief that attempts at standardizing counting protocols satisfies the constant proportionality assumption.

We will discuss two types of indices: the *presence–absence index* and the *index of relative abundance*. The presence–absence or frequency index is based on the proportion of plots that contain at least 1 individual, sign, or other variable of interest within a sampling frame (Scattergood, 1954). This approach may be used for indexing spatial distribution or both spatial distribution and abundance (or density). The term "presence–absence" is actually a misnomer. We only know for certain that an animal is present when it is detected; absence may indicate either true absence or undetected presence.

An index of relative abundance is an incomplete count that is assumed to be proportional to the true abundance. In other words, an average change in an index value is assumed to reflect the true average change in numbers. An index of relative abundance also may be stated in terms of density because abundance is easily converted into density. Consequently, we will be using relative abundance and relative density interchangeably throughout this book.

3.2.1.1. Presence–Absence

A common approach to evaluating the spatial distribution of a species is to perform a presence–absence survey (e.g., Jackson and Harvey, 1989; Owen, 1990; Debinski and Brussard, 1994). A set of sampling units (plots) is drawn from the sampling frame, and each sampling unit is surveyed to determine if the species is present. The percentage of the sampling units containing the species then is taken as an index of the species' spatial distribution, abundance, or both. Valid inferences require a linear relationship between mean density of animals per plot and proportion of occupied plots. We will discuss the validity of this assumption in the following paragraphs.

The nonlinearity of the presence–absence index is demonstrated in Fig. 3.1. In this graph, we have assumed that the individuals are randomly distributed in the area surveyed, and that a sample of plots is taken to assess presence–absence. The mean density per plot is the expected number of individuals to be encountered on each plot. The proportion of plots that

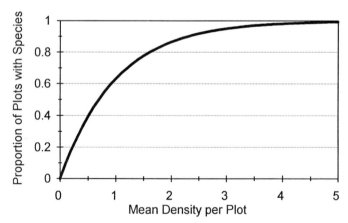

Figure 3.1 Relationship between proportion of plots that contain 1 or more individuals of a species as a function of mean density per plot for a species that is randomly distributed across a sampling frame.

contain ≥1 individual (i.e., the line in Fig. 3.1) is described as a function
of the mean density for a randomly distributed species by the equation

$$\text{Proportion of plots with} \geq 1 \text{ individual} = 1 - e^{-(\text{mean density})}.$$

Nonlinearity of the presence–absence index is a function of the plot size
because larger plots have a larger mean density. As the mean density per
plot rises above 1 in the graph in Fig. 3.1, the index becomes less sensitive
to increases in density. That is, as density goes from 1 to 2 animals per
plot, the index goes from 63 to 86% (23 units) of the plots occupied. But
as density goes from 2 animals per plot to 4 animals per plot, the index
only changes from 86 to 98% (12 units). Consequently, even under ideal
conditions, comparisons across the range of densities shown in the figure
must be made with some concern for this nonlinearity. Further, the relationship between degree of nonlinearity of this index to plot size essentially
precludes its use in comparisons in which plot sizes vary. An example could
be in riverine systems where unequal-sized stream habitat units are used
as sampling units. Moreover, collecting information for multiple species
that occur at different spatial scales basically requires different sampling
frames with different plot sizes, which would make comparisons based on
index-derived density estimates problematic at best. Finally, the relationship in Fig. 3.1 assumes that the individuals are randomly distributed across
the environment, which is hardly likely with real populations. Clumping of
individuals would further increase the nonlinearity of the function, whereas
a more systematic or regular distribution of individuals would decrease
the nonlinearity.

Caughley (1977) suggested that even under less than ideal conditions a
presence–absence index may be useful for estimating density when a species
occurs on less than 20% of the plots (note that the relevant segment of the
graph in Fig. 3.1 is nearly a straight line). He acknowledged that animals
are usually neither randomly distributed nor completely or equally detectable, and therefore even the index–density relationship for species occurring on less than 20% of the plots is likely nonlinear. He pointed out,
however, that violations of distribution and detectability assumptions lead
to underestimation (i.e., negative bias) of relative density and that magnitude of this bias increases with increasing density; subsequent comparisons
between density indices either across areas or for the same area across
time would usually underestimate the true difference. Caughley felt this
approach was acceptably conservative because it would lead to fewer incorrect management decisions than would positively biased estimates. We do
not necessarily agree with this. One cannot conclude much with certainty
about different density indices without knowing their magnitudes of bias
or their degrees of constancy or proportionality across space, time, or both.

For instance, even if relative density is underestimated, the magnitude of bias from one estimate to the next could easily change over time; hence, true changes in densities could be masked or misinterpreted because of unknown shifts in bias among estimates (regardless of the conservatism of the estimates). In addition, these comments are given in the context of a single species; the problems mount when attempting to add more target species to a survey (see Chapter 4).

The other major problem with presence–absence surveys is that the investigator assumes the absence of the species when none are detected. The level of effort in searching for the animals must be constant across surveys if the results are to be compared. Thus, surveys where a quick "drive-through" of each plot is performed should not be compared to surveys with a thorough, intensive search of each plot. A strict definition of absence is needed prior to starting a survey, so that all the people performing the survey decide if a species is absent from a plot with the same rule, i.e., the same level of search effort.

3.2.1.2. Index of Relative Abundance or Density

Indices of relative abundance and density are commonly used to assess the status of fish and wildlife species. Examples include vocalization frequency (e.g., number of birds vocalizing or number of vocalizations recorded per unit of time), track or sign surveys, scent-station surveys, and harvest. However, one of the hazards of using this index is that no reliable relationship exists between the index and the parameter of interest, such as density. To demonstrate that a relationship exists between an index and density requires collecting data on both quantities simultaneously. These data are used to develop a regression to predict the index from the population density. This equation then is used to convert future observations of the index into estimates of population density. The general statistical problem is known as calibration or inverse regression (see, e.g., Draper and Smith, 1981, pp. 47–51). A "cheap" measure (the index) is calibrated against an "expensive" measure (population density). Such procedures are used in analytical chemistry to develop procedures that relate the concentration of a compound in a "known" sample to a cheap and quick procedure, e.g., litmus paper to test for acidity.

Use of an index for comparisons across time or space requires a critical assumption: the index is monotonically related to the variable it indexes. The simplest relationship would be a linear relation with a zero y intercept. Consider the functions in Fig. 3.2. Function A is a linear relationship between the index on the x axis and population density (which can be converted to population size) on the y axis. If the population density doubles,

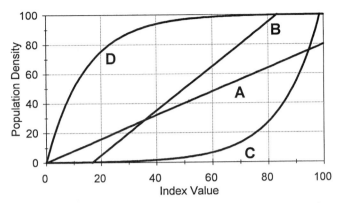

Figure 3.2   Four examples of potential relationships between an index on the *x* axis and population density on the *y* axis.

so does the index. Hence, the index provides a predictor of population size that contains as much information as if we were able to estimate the population directly. In contrast, function B does not have a *y* intercept of zero. Hence, as the value of the index goes to zero, some animals are still in the population, and the index provides a poor predictor of population size. Functions C and D are examples of indices where the index is a nonlinear relationship of the population size. In the case of function C, a change of 1 unit on the left side of the graph implies a much smaller change in the population density than 1 unit on the right side of the graph. The opposite is true for function D. These types of relationships of an index to a population might result from counts of signs involving territory marking. Function C implies that as the density increases and more animals are present, each individual marks its territory at a higher rate than it would at a lower density. Function D may be the result of many nonterritorial animals in the population, so that marking by territorial individuals is maxed out as an index because many in the population do not have a territory.

Statistical procedures are available to develop calibration curves for nonlinear functions such as C and D in Fig. 3.2. However, more data are required to develop the calibration curve because more parameters need to be estimated to describe the relation between the index and population density. For function A in Fig. 3.2, only one parameter (slope) is estimated, whereas function B will require two parameters (slope and intercept), and functions C and D will require at least two, and likely three or more to describe the more complex nonlinear relationships.

Without knowledge of the functional relationship between the index and population density, an index is of little value in making inferences to populations. Typically, investigators assume a linear relationship, as in function A of Fig. 3.2. However, all inferences made to the population are based on this untested assumption. Only function A provides a linear quantitative relationship between the index and the actual population. Thus, if the index doubles in magnitude, so has the population. This relationship does not hold for the other three indices displayed in Fig. 3.2.

A second problem can cause an index to have little value: high sampling variance. For an index to be useful, either the sampling variance of the index must be small or the measure must be easily obtained so that the sampling variance can be reduced with large sample sizes. Little effort should be needed to obtain a precise estimate of the index. Otherwise, the index is not cheap to measure, and the investigator would be better off to put effort into actually measuring the population. A really fine index would have much smaller sampling variation than direct estimation of the population. Unfortunately, such is seldom the case. To provide a useful index, the variable being measured must provide repeatable answers for a given population density.

Rotella and Ratti (1986) described an attempt to relate calling frequency (an index) of gray partridge to partridge density estimated via line transects. They evaluated convariates that might affect calling frequency: temporal, seasonal, and meteorological variables. They reported that morning counts of number of calling groups during spring or summer was a valid index of gray partridge density, although they only had 4 points in their regression of calling frequency and density. However, calling frequency during summer evenings was not directly related to population density, demonstrating the danger of using an untested index. Guthery (1989) criticized their results because there was not a significant correlation between number of calling groups and density of coveys [see Ratti and Rotella (1989) for a reply].

This example demonstrates the difficulty in actually validating an index against a direct estimate of density. Small sample sizes, and sampling variation in both the density and the index estimates, make conclusive results difficult. Moreover, we can never prove that an index is directly related to population density because statistical tests only reject null hypotheses. In addition, some specialized set of conditions in the future may invalidate the calibration function we have developed. Developing the relationship between the index and the population can become an end unto itself.

An example of a situation in which the investigator unknowingly may actually be using an index is "complete" counts (previous section) where the visibility of the animals is not 100%. As a result, the count is an index of density of the sampling unit. Another common example is the use of the minimum number of animals known alive (MNA) in a capture–recapture survey instead of a population estimate based on MNA. Skalski and Robson (1992, pp. 64–65) have suggested that if the assumption of constant capture probabilities can be met, the MNA index can be more efficient than using population estimates because the sampling variance of MNA is less than the sampling variance of a population estimate. However, violation of this assumption is often likely, so we generally do not agree with the recommendation to use MNA instead of an absolute population estimate.

Eberhardt and Simmons (1987) suggested calibration of an index through double sampling. Double sampling is the statistical methodology in which a cheap measure (the index) of the variable of interest (population density) is related to the variable of interest. Both ratio estimators (intercept of index and population density goes through the origin) and regression estimators (intercept not at zero) were considered by Eberhardt and Simmons (1987).

Because of the strong assumptions needed to use an index and the general lack of information to suggest that the assumptions are realistic, we strongly caution against their use except when there are no reasonable alternatives. In the absence of any information, an index may provide some useful knowledge. Also, funding constraints may limit the number of species whose numbers can be assessed using methods that produce unbiased abundance estimates. Thus, we have to accept that indices will be used.

3.2.2. ADJUSTING FOR INCOMPLETE DETECTABILITY

Methods whose estimators adjust for incomplete detectability are nearly always more costly and time consuming than those based on indices. Thus, such methods will probably only be applied to a single or relatively few species of particular interest for which unbiased estimates of abundance are required. However, these methods will only produce unbiased results when underlying assumptions are reasonably satisfied. This is an extremely important point. Methods discussed in this section are no better than indices if their underlying assumptions are not met; in fact, they would be worse because of the increased effort and cost expended to obtain them. Therefore, investigators always should use their knowledge of the biology of a species, as well as the survey conditions, to critically assess the validity of underlying assumptions of a proposed enumeration method. This section covers some of the more commonly encountered methods that have a valid "detectability correction" incorporated into their estimators.

3.2.2.1. Capture–Recapture and Related Estimators

Capture–recapture is a type of partial count in which a correction is developed in the sampling process to estimate the total population of animals on the sampling unit. Both closed and open capture–recapture models are available, plus a combination of the two applied together to provide a "robust" design. Closed models assume that the population is closed, i.e., that no animals are entering or leaving the population via births, deaths, immigration, or emigration. Open models provide estimates of births plus immigration and deaths plus emigration. In general, closed models provide more rigorous estimates of population size than open models. Hence, we will discuss their use relative to estimating population size on the sampling units making up a rigorous sampling procedure. However, open models are useful for monitoring age-specific survival, with a current example being northern spotted owls (Burnham et al., 1994).

Definitive summaries of capture–recapture estimation are provided by Seber (1982), with updates of the major work in Seber (1986, 1992). Other major reviews are presented in Otis et al. (1978), White et al. (1982), and Pollock et al. (1990).

Closed Capture–Recapture Estimators

The most widely known capture–recapture estimator is the Lincoln–Petersen or Petersen (1896) estimator. Assume a population of N animals. On occasion 1, n_1 are captured and marked. On occasion 2, n_2 are captured, of which m_2 are marked. Then, an estimate of the population size at time 1 is

$$\hat{N} = \frac{n_1 n_2}{m_2}. \tag{3.1}$$

The probability of being captured on occasion 1 is p_1, estimated by n_1/N, and the probability of capture on occasion 2 is p_2, estimated by $n_2/N = m_2/n_1$. Because the Lincoln–Petersen estimator is simple and straightforward, we use it as a starting point for discussing general issues related to closed capture–recapture estimators. However, in most situations, we strongly recommend using capture–recapture estimators based on ≥ 3 sampling occasions, as discussed later in this section.

There are three critical assumptions that must be met for the Lincoln–Petersen estimator to be valid, which we will briefly discuss in terms of mobile populations. Note that capture does not necessarily have to refer to physical handling or trapping; an animal can be detected via observation or other means.

The Population Is Closed, So That N Is Constant.   Closed population estimators require the population of interest to be both demographically and geographically closed (Fig. 3.3; Chapter 1). The point here is that N must be fixed during the sampling period to have any meaning. Demographic closure can be ensured, in part, by careful planning of the sampling period to minimize the likelihood of births and deaths. Controlling movements on and off the sampled area, however, offers a more challenging problem when studying mobile populations. This may be controlled in streams, in certain circumstances, by blocking off sampled areas with nets or other barriers. Unfortunately, this is rarely an option in terrestrial systems. A correction for movements across the boundary may be possible via radio-tagging animals and estimating the proportion moving on or off the sampled area (White, 1996). Minimizing the sampling period and maximizing the size of the sampled area compared to the average daily movements of the target animals may satisfy, at least approximately, the closure assumption.

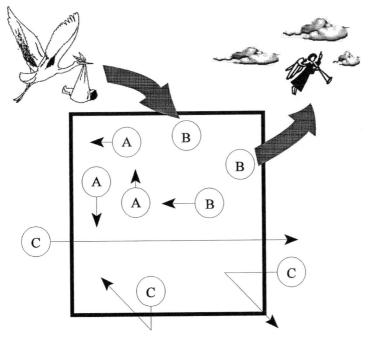

Figure 3.3   Three species (A, B, and C) exhibiting different degrees of closure within a plot or study area (area within square). Species A is both geographically and demographically closed. Species B is geographically closed but demographically open (i.e., 1 individual is born and another dies). Species C is geographically open.

All Animals Have the Same Probability of Being Caught in Each Sample. This assumption is extremely difficult to meet in just about any realistic situation. Violations are usually grouped into two categories, i.e., heterogeneity and behavior. These topics have been thoroughly reviewed elsewhere (e.g., Seber, 1982; White *et al.*, 1982), so we will discuss them only briefly here.

Capture heterogeneity refers to characteristics inherent to each individual that affect its probability of capture. Examples include age, sex, social dominance, trap placement (i.e., in relation to home range movements), and capture method. Juvenile animals in many species tend to move more than adults, such as movements off of natal grounds or movements as "floaters" in territorial species. Therefore, juveniles and adults may differ in their availability for capture depending on how often and far they move. Frequency and magnitude of movements may differ by sex as well. Also, socially dominant animals may restrict movements, and hence access to traps, of subdominant animals within a capture area. One could stratify animals by category, such as by sex and/or age, and analyze these strata separately, assuming equal catchability within each stratum. Unfortunately, many more data are required to have the minimum needed for precise estimates within each stratum. Moreover, classifying individuals as dominant or subdominant, or some other category potentially affecting equal catchability, may not be feasible.

Individuals are less susceptible to capture when traps are located near the edge of their home ranges than when traps are near the middle (Fig. 3.4). Also, individuals may be precluded from capture during a given sampling occasion if the trap in their daily movement area is already occupied, i.e., has already trapped another individual. Potential remedies for this situation are to use multiple traps per trap site, employ a trapping technique that

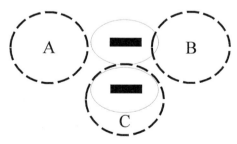

Figure 3.4 Differing capture probabilities among three individuals (A–C) in relation to two traps (dark rectangles) as shown by the degree of overlap between home ranges (dark dashed lines) and effective trapping areas (dotted lines). Individual A has essentially no chance of being captured, B has a slight chance, and C has a good chance.

captures multiple individuals at once (e.g., mist nets), and set out traps in high enough densities so that individuals are not "competing" for them. In any event, the number of traps (or observation routes for mark–resight, etc.) should adequately cover the area of interest. Too few traps will limit inferences because relatively few individuals of interest will have a reasonable chance of being sampled [see design recommendations in White *et al.* (1982) and Skalski and Robson (1992)].

The type of capture method can have a significant bearing on capture probabilities of animals, particularly if the same method is used over both sampling occasions. Many methods of capture are selective toward a particular portion of a population. A common example is the use of electroshocking for capturing fish. Electroshocking has been shown to be both size- and species-selective (Zalewski and Cowx, 1990), so that some individuals within a sampled area will essentially have little or no chance of being captured; magnitude of the effects of selectivity may vary by situation. The proportion of individuals missed by electrofishing in small streams may be minimal because of the small volume of water and relatively simple habitat structure, whereas those missed in large rivers may be substantial. Typically, the narrower the selectivity, the greater the potential for bias. A way around this is to use a different capture method for each sampling occasion. Even if selectivity exists for each capture method, as long as the selectivities of each method are independent, the resulting abundance estimates will be unbiased (Seber, 1982). Therefore, for two-sample capture–recapture studies, we recommend using different capture techniques for each sampling occasion.

Capture behavior refers to individual responses to trapping, including trap happiness and trap shyness. Trap happiness means that a previously trapped animal is even more likely to be trapped a second time. For instance, animals may return to baited traps for food. Trap shyness indicates the opposite situation, i.e., a previously captured animal is less likely to be caught again. This may arise if an individual is injured or mishandled during its capture.

All Previously Marked Animals Can Be Distinguished from Unmarked Animals. Either impermanent marks or indistinguishable marks will lead to biased population estimates. The result of either would be the same; the number of marked animals that were recaptured would be underestimated, which would lead to an overestimate of N. A possible remedy to loss of marks is to double-mark captured animals, i.e., apply two separate marks per individual (Seber, 1982). Indistinguishable marks may be avoided by using a proper marking method. Note that in two-sample mark–resight studies, one should mark individuals so that they can be distinguishable from unmarked individuals, but not so obviously that marked animals have a much higher probability of detection than unmarked ones.

Let us run through some simple examples to illustrate the effects of violations to the second and third assumptions just discussed. Suppose our area of interest contained $N = 1000$ animals. First, assume capture probabilities are equal within each sampling occasion (e.g., $p_1 = 0.4$ and $p_2 = 0.2$), the population is closed, marks are distinguishable, and no marks are lost. Next, suppose we initially capture $n_1 = 400$ animals (i.e., $N \times p_1$), mark them, and then capture a second sample of $n_2 = 200$ (i.e., $N \times p_2$) animals in which $m_2 = 80$ are marked (i.e., $n_1 \times p_2$). Note $(N - n_1) \times p_2$ is the number unmarked captured in sample 2. Substituting these values into Eq. (3.1) yields

$$\hat{N} = \frac{(400)(200)}{80} = 1000,$$

which is equal to the true abundance. Now suppose that the recapture probability of the marked animals was only 0.1, but we still assume it was the same for unmarked animals, i.e., 0.2. Thus, assume a violation of the assumption of equal capture probability within the second sample. The initial capture probability, and hence the initial number of animals captured and marked ($n_1 = 400$), would remain the same. However, now the 400 marked animals would only have a capture probability of 0.1, whereas that of the 600 unmarked animals would be 0.2. The number of animals captured in the second sample would only be $n_2 = 160$, or $400 \times 0.1 = 40$ marked plus $600 \times 0.2 = 120$ unmarked. Substituting these values into Eq. (3.1) gives us

$$\hat{N} = \frac{(400)(160)}{40} = 1600.$$

The adverse effect of heterogeneity in capture probabilities on the Lincoln–Petersen estimate is obvious. Loss of marks would have a similar effect. Even with equal capture probabilities within each sampling occasion, loss of marks would cause N to be overestimated. For instance, suppose 150 of 400 marked animals lost their marks between capture occasions in the initial example above. Thus, the number of animals in the second sample still would be 200, but the number marked in the second sample now would be $m_2 = 250 \times 0.2 = 50$, producing the following estimate,

$$\hat{N} = \frac{(400)(200)}{50} = 1600.$$

Again, satisfying the underlying assumptions of the Lincoln–Petersen method is critical for valid results.

Chapman (1951) showed that for $n_1 + n_2 \geq N$, the estimator in Eq. (3.1) is biased when sampling is performed without replacement, and that

$$\hat{N} = \frac{(n_1 + 1)(n_2 + 1)}{(m_2 + 1)} - 1 \tag{3.2}$$

is an unbiased estimator of N. He developed this estimator based on the hypergeometric distribution, so that samples at time 1 and 2 are assumed to be taken without replacement. Thus, Chapman's estimator is appropriate for any survey where sampling is performed without replacement, such as small mammal trapping or aerial surveys where animals are not counted twice. We will generally assume that sampling is performed without replacement, so that Eq. (3.2) is the correct estimator to use. Seber (1982) discussed situations where sampling with replacement is performed and the alternative estimators that should be used.

The $\text{Var}(\hat{N})$ for Chapman's estimator is estimated as

$$\hat{\text{Var}}(\hat{N}) = \frac{(n_1 + 1)(n_2 + 1)(n_1 - m_2)(n_2 - m_2)}{(m_2 + 1)^2(m_2 + 2)}, \tag{3.3}$$

and also is unbiased when $n_1 + n_2 \geq N$ for surveys where sampling is without replacement. This estimator and its associated variance usually provide confidence intervals larger than most surveys require (unless a high proportion of the population is captured on each occasion, as in Example 3.1).

Example 3.1. Lincoln–Petersen Population Estimator

A fishery biologist has marked 40 Woodhouse's toads at an isolated desert pond. Three days later, the biologists systematically resurveys the pond (not counting any of the same toads twice), and captures 50 toads, 35 of which are marked. From these data, he computes a population estimate.

The raw data provide the following inputs: $n_1 = 40$, $n_2 = 50$, and $m_2 = 35$. Thus,

$$\hat{N} = \frac{(40 + 1)(50 + 1)}{(35 + 1)} - 1 = 57.08 = 57$$

with the variance estimated as

$$\hat{\text{Var}}(\hat{N}) = \frac{(40 + 1)(50 + 1)(40 - 35)(50 - 35)}{(35 + 1)^2(35 + 2)} = 3.2705.$$

Taking the square root of the variance to obtain $\hat{\text{SE}}(\hat{N}) = \sqrt{\hat{\text{Var}}(\hat{N})} = 1.8084$, a 95% confidence interval can be computed as $\hat{N} \pm 1.96\,\hat{\text{SE}}(\hat{N})$, or 53.46 to 60.54.

Chapman's estimator has commonly been used in aerial mark–resight studies with two independent observers in a procedure called *double counting* (see review by Pollock and Kendall, 1987). Each observer independently maps exact locations of detected individuals or objects. In this case, n_1 is the number of detections by observer 1, n_2 is the number of detections by observer 2, and m_2 is the number of detections in common between observers 1 and 2. This procedure has the same three critical assumptions listed for the Lincoln–Petersen estimator. Again, a particularly difficult assumption to meet is equal detectability of animals or objects within each sampling occasion in the surveyed area. Individuals farther away from a flight line tend to be less detectable. One possible way to address this problem is to poststratify survey data into categories of more or less equal detection probabilities based on variables thought to affect detectability of animals, and compute a correction factor for each of these categories (Rivest *et al.*, 1995). Manly *et al.* (1996) took this a step further by describing an approach to modeling covariates potentially affecting detectability (e.g., distance from line, group size) to adjust for incomplete detectability. Their approach was to use two individuals operating independently within an aircraft. However, there is a selectivity problem with using observers in the same sighting platform (e.g., an aircraft) even if they are operating independently. That is, an individual or object that is not detectable to one observer will probably not be detectable to the other observer. As we discussed previously, this problem can probably be rectified by using two different counting techniques, such as a ground count and an aerial count. Unfortunately, from a logistic standpoint, an area requiring an aerial count is likely not conducive to performing a large-scale ground count.

One approach to improving the precision of a capture–recapture estimate is to incorporate more sampling occasions. Further, the set of assumptions required for two-sample estimators often is biologically unrealistic; therefore, numerous extensions have been developed for more than two sampling occasions. First, we would like to allow each animal to have its own capture probability, and not require all animals in the population to have an equal capture probability at either time 1 or 2. Second, we may not want to assume that marked animals have the same probability of capture on occasion 2 as the unmarked animals. Removal estimators have been developed to incorporate this assumption, i.e., estimates obtained from kill trapping of small mammals are possible, with the animals removed from the population. Otis *et al.* (1978) summarized the available literature and developed a set of eight models that incorporated various assumptions.

Program CAPTURE Estimators. Otis *et al.* (1978) assumed a null model (Model M_0), with assumptions more restrictive than the Lincoln–

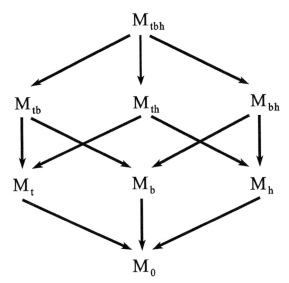

Figure 3.5 The hierarchy of models for closed capture–recapture estimation proposed by Otis *et al.* (1978) and White *et al.* (1982) used in program CAPTURE. This figure is modified from White *et al.* (1982).

Petersen model described above. Each animal is assumed to have a capture probability p that is equal for all animals on all occasions. This unreasonable assumption is relaxed by three models: Model M_t, where capture probabilities (p_t) vary by capture occasion; Model M_h, where each animal has its own capture probability (p_j) that is constant across occasions; and Model M_b, where each animal has an initial capture probability (p) and a recapture probability (c), thus allowing behavioral response to capture. Model M_t means *t*ime varying capture probabilities, Model M_h means *h*eterogeneous capture probabilities by individual, and Model M_b means *b*ehavioral response to capture. These simple models then were combined to form all possible pairs of two factors: Models M_{th}, M_{tb}, and M_{bh}, and then all three factors into one model, M_{tbh}. The relationships among capture–recapture models are shown in Fig. 3.5. Initially, estimators did not exist for Models M_{th}, M_{tb}, or M_{tbh}. Now, estimators have been developed for M_{th} and M_{tb}, but none are available for M_{tbh}. Program CAPTURE[1] (White *et al.*, 1982; Rexstad and Burnham, 1992) is available for computing these estimators. In addition, tests between models were developed (Otis *et al.*, 1978; White *et al.*, 1982) to assist with model selection. Program CAPTURE also imple-

[1] Available free via the Internet at www.cnr.colostate.edu/~gwhite/software.html.

ments a model selection procedure to assist users with identifying which of the eight models is most appropriate for their data.

The basic data of a capture–mark–recapture survey consist of the capture history matrix. This matrix consists of 0's (animal not captured or resighted) and 1's (animal captured or resighted). Rows of the matrix are animals, columns are capture occasions. As an example consider the following 5 × 10 matrix (5 capture occasions, 10 animals captured). By convention, the capture history matrix is generally designated the X_{ij} matrix:

$$X_{ij} = \begin{bmatrix} 1 & 0 & 1 & 0 & 1 \\ 1 & 1 & 1 & 0 & 0 \\ 0 & 1 & 1 & 1 & 0 \\ 0 & 1 & 1 & 0 & 0 \\ 0 & 1 & 0 & 0 & 1 \\ 0 & 0 & 1 & 0 & 1 \\ 0 & 0 & 1 & 1 & 0 \\ 0 & 0 & 0 & 1 & 1 \\ 0 & 0 & 0 & 1 & 0 \\ 0 & 0 & 0 & 0 & 1 \end{bmatrix}.$$

The animal represented in row 1 was captured on occasion 1, not captured on occasion 2, captured on occasion 3, not captured on occasion 4, and again captured on occasion 5. The last animal in the matrix at row 10 was not captured on the first 4 occasions, but finally was captured on occasion 5.

Note that some unknown number of animals are never captured, and are not included in the matrix. Their capture histories would consist of all 0's. The goal of population estimation is to determine the number of animals never captured based on the sample of captured animals. From the capture matrix, all the necessary statistics needed by the population estimators in program CAPTURE can be computed. To apply CAPTURE in the way the program was intended, the user first enters the capture history matrix in a format that is readable by the program. The details of how to do this were provided in White et al. (1982, Appendix A). Then, the model selection procedure is used to assess the sources of variation in the capture probabilities. The model selection procedure will recommend an appropriate model for the data. Typically, the user then will use this model to estimate population size.

For the eight models in CAPTURE, only seven have estimators, five of which are maximum likelihood estimators. Maximum likelihood is a statistical estimation method developed by the famous statistician and geneticist, Sir Ronald A. Fisher, in the early 1920's. Likelihood estimation has been the backbone of statistical estimation for more than 50 years. Estimators developed by this method are optimal, at least for large samples. More detail on the maximum likelihood estimation procedure is given in White *et al.* (1982).

To demonstrate how a maximum likelihood estimator is developed, we will use the binomial distribution, and estimate the probability of success, *p*. For the binomial distribution, the probability density function that describes the probability of observing *x* successes in *n* trials is

$$\binom{n}{x} p^x (1-p)^{(n-x)},$$

where $\binom{n}{x}$ is a binomial coefficient, defined as $n!/[x!(n-x)!]$, with the ! sign meaning factorial ($5! = 5 \times 4 \times 3 \times 2 \times 1$). This probability density function might be used to model the number of sightings (*x*) of *n* animals when each animal has a sighting probability *p*. If $n = 5$ and $p = 0.8$, then the probability of seeing all 5 animals on a survey is

$$\binom{5}{5} 0.8^5 (1-0.8)^{(5-5)} = 0.8^5 = 0.32768,$$

and the probability of seeing 4 of the 5 animals is

$$\binom{5}{4} 0.8^4 (1-0.8)^{(5-4)} = 5 \times 0.8^4 \times 0.2^1 = 0.4096.$$

The probability of seeing 3, 2, 1, and 0 of the 5 animals is 0.2048, 0.0512, 0.0064, and 0.00032, respectively. The sum of the six probabilities adds to 1; i.e., there are only six possible outcomes, so with probability 1, we will observe one of these outcomes (at least according to our model).

Now, we reverse the process, and let us assume that we want to know what the sighting probability (*p*) is for an observation. Suppose that we observed 4 animals, and we know that $n = 5$ animals are present. Then intuition says that we would estimate *p* as $\hat{p} = 4/5 = 0.8$. To develop a likelihood estimator to support our intuition, we first assume the above probability model. Then, the likelihood $\mathcal{L}(p|n, x)$ of our sample (read as the likelihood for parameter *p* given the observed data *n* and *x*) is computed as

$$\mathcal{L}(p|n, x) = \binom{n}{x} p^x (1-p)^{(n-x)}.$$

The maximum likelihood estimator of p is the value of p that maximizes this function. For the range of p from 0.7 to 0.9 in the following table,

p	Likelihood
0.7	0.36015
0.71	0.3684694
0.72	0.376234
0.73	0.3833763
0.74	0.3898255
0.75	0.3955078
0.76	0.4003461
0.77	0.40426
0.78	0.4071656
0.79	0.4089759
0.8	0.4096
0.81	0.4089438
0.82	0.4069096
0.83	0.4033957
0.84	0.3982971
0.85	0.3915047
0.86	0.3829057
0.87	0.3723834
0.88	0.3598172
0.89	0.3450823
0.9	0.32805

we find that the likelihood is maximized for $p = 0.8$. Typically, we denote the estimate of a parameter with a caret or hat, so $\hat{p} = 0.8$. A graph of the likelihood function is shown in Fig. 3.6. For this simple model, we can find the maximum likelihood estimator of p analytically by taking the derivative of the likelihood function and solving for the value of p that maximizes the likelihood. For most of the models in program CAPTURE, at least with more than two capture occasions, the analytical estimator is not feasible, so numerical search techniques, such as that demonstrated in the above example, are used.

For the models in program CAPTURE, estimates for more than one parameter are needed. The likelihood function is then multidimensional, but the same approach is used. To find the maximum likelihood estimates for two parameters, the values are found that maximize the likelihood. Another useful feature of the likelihood function is that the variance–covariance of the parameters can be computed from the second partial derivatives of the log of the likelihood function. This is the procedure used in program CAPTURE to compute the standard errors of the population estimates.

For capture–recapture estimates of population size, symmetric confi-

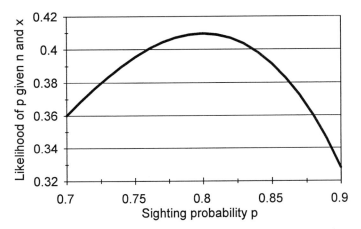

Figure 3.6 Likelihood of parameter p given the observed data $n = 5$ and $x = 4$. The maximum likelihood estimate of p is $\hat{p} = 0.80$, i.e., the value of p that maximizes the likelihood function is 0.80.

dence intervals based on either z or t distributions (see Chapter 1) do not perform well. For capture–recapture surveys, we know a reasonable lower bound on what the population can be, i.e., we know that we captured and handled some number of animals. Realistically, the lower confidence bound on N cannot be less than this number. Symmetric confidence intervals based on a z or t do not account for this lower bound. Thus, in program CAPTURE, two different procedures are recommended.

The first procedure that works for all the estimators in CAPTURE is to use a log transformation to construct the confidence interval (Rexstad and Burnham, 1992). We define MNA as the minimum number of animals known alive in the population, that is, the number of individuals that we actually captured and marked. Then the number of animals we never captured is estimated as $\hat{N} -$ MNA. If we assume that this value is lognormally distributed, then a confidence interval can be constructed as

$$\left[\text{MNA} + \frac{(\hat{N} - \text{MNA})}{C_l}, \text{MNA} + (\hat{N} - \text{MNA}) \times C_l \right],$$

where

$$C_l = \exp\left(1.96 \sqrt{\log\left[1 + \frac{\hat{\text{Var}}(\hat{N})}{(\hat{N} - \text{MNA})^2} \right]} \right).$$

Again, 1.96 is the z value for $\alpha = 0.05$ needed to construct a 95% confidence interval. An example of computing this confidence interval is given in Example 3.2a.

The second approach to computing asymmetric confidence intervals in program CAPTURE is based on the likelihood function. The steeper the peak of the likelihood function at its maximum, the better the estimate. This observation can be used to construct a profile likelihood confidence interval. At 1.92 units below the maximum of the log likelihood function for $\alpha = 0.05$, the values of the parameter are taken as the confidence interval. This procedure is illustrated in Example 3.2b, with a graph of the likelihood and the profile interval shown in Fig. 3.7. From the examples, we see that the log-based and profile confidence intervals do not match exactly. Simulations of these intervals with program CAPTURE have shown that in general, the two types of confidence intervals perform about the same, and both provide the desired coverage level. That is, both types of confidence intervals will provide 95% coverage.

Example 3.2. Log-Based and Profile Likelihood Confidence Intervals for the Lincoln–Petersen Population Estimator

a. Using the data from Example 3.1, we calculated that $\hat{N} = 57.08$, and $\hat{\text{Var}}(\hat{N}) = 3.2705$. Taking the square root of the variance to obtain $\hat{\text{SE}}(\hat{N}) = \sqrt{\hat{\text{Var}}(\hat{N})} = 1.8084$, a 95% ($\alpha = 0.05$) confidence interval can be computed as $\hat{N} \pm 1.96\,\hat{\text{SE}}(\hat{N})$, or 53.54 to 60.62. This interval is symmetric about $\hat{N}$.

To compute a log-based confidence interval, we note that MNA $= n_1 + n_2 - m_2 = 40 + 50 - 35 = 55$. Then,

$$C_l = \exp\left(1.96 \sqrt{\log\left[1 + \frac{3.2705}{57.08 - 55} \right]} \right) = 6.7205,$$

so that the 95% confidence interval is

$$\left[55 + \frac{(57.08 - 55)}{6.7205}, 55 + (57.08 - 55) \times 6.7205 \right],$$

or 55.3095 to 68.9786. We round the lower bound up to 56 and the upper bound down to 68, so that the interval endpoints are part of the interval. Thus, we conclude that the 95% confidence interval is 56–68, with these integers part of the interval. This interval demonstrates that the lower bound is much closer to the estimate because we think we captured 55 of the 57 animals in the population. However, we are less sure of just how many more animals might be in the population; hence, the upper bound is considerably above the estimate.

b. Using the data from Example 3.1, we computed $\hat{N} = 57.08$, and $\hat{\text{Var}}(\hat{N})$ is 3.2705. The likelihood function for the Lincoln–Petersen estimator when we sample without replacement is

$$\mathcal{L}(N|n_1, n_2, m_2) = \frac{\binom{n_1}{m_2}\binom{N - n_1}{n_2 - m_2}}{\binom{N}{n_2}}.$$

For $n_1 = 40$, $n_2 = 50$, and $m_2 = 35$, the follow values of the likelihood and log likelihood function are computed.

N	Likelihood	Log likelihood
55	0.1891501	−0.723193
56	0.3242573	−0.48911
57	0.3384791	−0.470468
58	0.2801207	−0.552655
59	0.2029688	−0.692571
60	0.1353125	−0.868662
61	0.085402	−1.068531
62	0.05195	−1.284417
63	0.030819	−1.511176
64	0.017978	−1.745259
65	0.010372	−1.984142
66	0.00594	−2.225985
67	0.00339	−2.469428
68	0.00193	−2.71345
69	0.0011	−2.957275
70	0.0006	−3.20031

We see that the maximum of the likelihood is at $\hat{N} = 57$, which corresponds to the estimate we obtained with the analytical estimator. The max of the log likelihood is always at the same value as the likelihood. The value of the log likelihood at the maximum is −0.4705. Subtracting 1.92 gives −2.3905, so we are looking in either direction from the maximum to find values of the log likelihood less than −2.3905. The likelihood is undefined for $N <$ MNA (obviously a reasonable biological assumption), so we have a lower bound at 55. The upper confidence interval is between 66 and 67, so we use 66 as the last value still in the interval. As with the log-based asymmetric interval, the upper bound is farther from $\hat{N}$ than the lower bound.

In the following paragraphs, we provide the details of the models currently incorporated into program CAPTURE. Considerably more detail is provided by Otis *et al.* (1978) and White *et al.* (1982).

Model M_0 Assumptions of Model M_0 are that all animals in the population have the same capture probability, p, on each occasion, and the marked animals are assumed to have the same recapture probability as the initial capture probability of animals not yet marked. This model is meant to provide a null model, and is not considered to be a useful model for estimating population sizes. The estimator for Model M_0 is a maximum likelihood estimator. This estimator was originally developed by Darroch (1958).

Model M_t Assumptions of Model M_t are that each animal has an identical capture probability on a capture occasion, but that the probability of

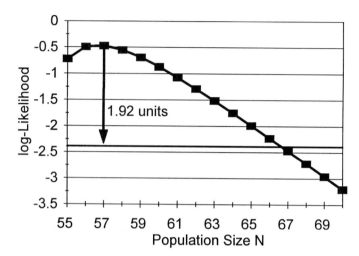

Figure 3.7 Demonstration of the profile likelihood confidence interval calculation. The horizontal line 1.92 units below the max defines the profile likelihood interval, with all values of N having log likelihood values above the line included in the confidence interval. For this likelihood, the 95% profile likelihood interval would be 55–66.

capture changes from occasion to occasion (hence, the time designator for this model). Marked animals are assumed to have the same capture probability as unmarked animals, so the model does not incorporate behavioral response to capture. The estimator for Model M_t is a maximum likelihood estimator. This estimator was originally developed by Darroch (1958). A special case for two occasions is the Lincoln–Petersen estimator.

Model M_b Assumptions of Model M_b are that each animal has an identical initial capture probability on all capture occasions, but that the probability of capture changes once the animal has been captured and marked. Marked animals are assumed to have a different capture probability than unmarked animals, so the model incorporates behavioral response to capture (and hence its name). The estimator for Model M_b is a maximum likelihood estimator. This estimator was originally developed by Zippen (1956, 1958), and is commonly applied to removal surveys to estimate population size.

Model M_h Assumptions of Model M_h are that each animal has its own capture probability, but that this probability does not change from occasion to occasion and does not change with capture (i.e., no behavioral response). The subscript h indicates heterogeneity, meaning the individual heterogeneity of capture probabilities allowed by this model. Such differences could arise from age, sex, or other factors. The estimator for Model M_h is not a maximum likelihood estimator, but was developed by Burnham and Overton (1978, 1979) with a jackknife procedure.

Model M_{tb} Assumptions of Model M_{tb} are that all animals have the same capture probability on an occasion, but this probability depends on whether the animal has been previously captured. A logistic model is used to model the capture probabilities across time, with an additive effect due to recapture. Thus, the probability of capture at time j is

$$\text{logit}(p_j) = \log\left(\frac{p_j}{1 - p_j}\right) = \mu + T_j + B,$$

where μ is the mean capture probability across all occasions for initial capture, T_j is the deviation from the mean for the jth occasion, and B is the effect of initial capture on the recapture probability. Thus, B is 0 for initial captures, but >0 for recaptures. This model provides a convenient structure for modeling both temporal and behavioral effects on capture probabilities.

Model M_{th} Assumptions of Model M_{th} are that each animal has its own capture probability, and that this probability can vary with time. Developed by Chao *et al.* (1992), this estimator is a jackknife estimator, and so does not have a profile likelihood confidence interval. The estimator provides the average capture probability on each occasion for all the animals in the population.

Model M_{bh} Assumptions of Model M_{bh} are that all animals have their own capture probability that is constant across capture occasions, but which will change with initial capture. This estimator is similar to Model M_b in that it uses only the first captures to estimate population size. Based on the individual heterogeneity assumed by this model, the average capture probability in the population should decline as the easily captured animals are "removed" from the population. Thus, the model starts with the assumption that the average capture probability is the same for all occasions. Then, the first capture occasion is allowed to have a different (assumed larger) average capture probability, with the remainder assumed the same. This process continues until only three occasions are assumed to have the same capture probability. The estimator requires at least four occasions to relax the assumption of individual heterogeneity. With only three occasions, this estimator is the same as Model M_b.

Model M_{tbh}   No estimator is available for this model. Typically, if the model selection procedure selects this model as the appropriate one, your best option is to use the next most complex model as ranked by the model selection procedure that incorporates two of the three factors, i.e., use one of the Models M_{tb}, M_{th}, or M_{bh}.

Simulation procedure   Program CAPTURE has a simulation capability to assist with the design of surveys (White *et al.*, 1978, 1982). This procedure is useful in designing surveys because the user can specify capture probabilities and a model thought to apply to the field situation, and then determine the precision of the estimates expected with the design. For instance, trade-offs between the number of capture occasions and the number of traps per occasion (more traps will presumably increase capture probabilities) can be examined.

More extensive examples of using the simulation procedure and design of surveys are given in White *et al.* (1982). As a general rule of thumb, White *et al.* (1982) recommended a minimum capture probability of 0.3 for a population of at least 100 animals for reliable estimates of abundance or density. More recently, Rosenberg *et al.* (1995) reported that reliable estimates of abundance could be obtained for low-density populations with low ($p \doteq 0.10$) heterogeneous capture probabilities and ≥ 12 capture occasions. Finally, Skalski and Robson (1992) have a chapter on the design of capture–recapture surveys, including a section on trap density related to animal movements.

Program MCAPTURE Estimators.   Because simultaneous capture–recapture surveys will likely be conducted on multiple sampling units, a logical extension of the capabilities of program CAPTURE is to estimate capture probabilities in common across sampling units (e.g., trapping grids) and to test this assumption. Program MCAPTURE (White, 1994) has been developed to provide this capability. All of the maximum likelihood estimators of program CAPTURE have been incorporated into MCAPTURE, with a general parameter specification capability to model capture probabilities within and among sampling units. Population estimates are estimated separately for each sampling unit. However, the use of common capture probabilities across sampling units will induce a sampling covariance between population estimates. Yet, the gain in precision of the population estimates from reducing the parameter space (i.e., fewer nuisance parameters in the form of capture probabilities specific to each sampling unit) offsets this sampling covariance. Further, model selection now incorporates capture–recapture data from all the sampling units, so a more efficient model selection scheme can be implemented. MCAPTURE has recently been incorporated into program MARK.[2]

Program NOREMARK Estimators. A modification of the standard capture–recapture framework is that marked animals are not "recaptured" on the second or later occasions, but identified by resightings. In capture–release–recapture surveys, animals generally are captured on a number of occasions, individually marked, and then released. Any unmarked animals caught on any capture occasion are marked and released back into the population, thereby increasing the size of the pool of marked animals. However, it is not necessary to mark the unmarked animals captured, i.e., resighting surveys can be used where the animals are not actually captured. The full-capture history of all marked individuals is recorded. This approach can be used with radio-marked animals to estimate population size (Bartmann *et al.*, 1987; Minta and Mangel, 1989; White and Garrott, 1990; Neal *et al.*, 1993; Bowden and Kufeld, 1995). White (1996a,b) has developed program NOREMARK[3] to compute population estimates under Models M_h and M_t, plus the case where the study area is not geographically closed. Simulation capabilities also are provided to design mark–resight experiments.

Skalski (1991) has developed a statistical framework for sign counts that permits abundance and density estimation for closed populations (e.g., with respect to ingress and egress). The estimation technique requires the capture, marking, and release of animals so the subsequent sign they produce is distinguishable from unmarked individuals. He has developed variance estimators using finite sampling theory for the cases of simple random sampling and stratified random sampling of field plots for animal sign in the landscape. The variance formulas are used to determine the effects of sampling effort on the subsequent precision of abundance estimates. He also provides sample size formulas to determine the joint levels of marking effort and areal sampling required for a prespecified level of sampling precision in sign-marking studies. We note that Skalski (1991) has taken the approach of a sampling effort based on plots. The Bowden estimator in program NOREMARK also would be appropriate for this type of survey, but without the use of plots.

Program EAGLES. Population size can be estimated even if the number of uniquely marked animals in the population is not known. Arnason *et al.* (1991) developed a method for estimating population size when the number of marked animals is unknown, but individually identifiable. Pro-

[2] Ibid.
[3] Ibid.

gram EAGLES[4] was developed to compute these estimates. Independent sightings of marked and unmarked animals are taken. The authors also provided criteria for calculating the number of sightings required to yield satisfactory estimates. Their estimator assumes that all animals in the population on a particular sampling occasion have the same sighting probability.

Removal Methods

The actual removal of individuals from a population of animals has long been used as a basis for population estimation especially with respect to harvest returns from sport hunting or commercial fishing. However, individuals do not necessarily have to be killed. Fish captured via electroshocking may be placed in holding tanks and then returned to their capture locations after sampling has been completed, and marked animals may be treated as "removed" from a population of unmarked animals (Seber, 1982).

Removal methods may be categorized by whether individuals are selectively or nonselectively removed (Lancia *et al.*, 1994). Nonselective removals further can be partitioned by the effort expended. Methods applying equal effort use specialized capture–recapture models, either M_b or M_{bh}, which were discussed previously. Note that the geographic closure assumption will be violated if removed individuals are replaced by individuals outside of the capture area (White *et al.*, 1982).

Nonselective removals based on unequal levels of effort use *catch per unit effort* (CPUE) models to compute abundance estimates. These models attempt to capitalize on the idea that effort expended to catch animals will increase over time because fewer individuals will be available (Lancia *et al.*, 1994). The cumulative number of animals removed is used in a linear regression model with the dependent variable corresponding to the CPUE observations. CPUE models have been used most often in fishery research, although terrestrial applications were presented in Novak *et al.* (1991), Laake (1992), and Lancia *et al.* (1996). More extensive discussions of CPUE methods are presented in Ricker (1958, 1975), Seber (1982), and Lancia *et al.* (1994). Bishir and Lancia (1996) discussed a generalized CPUE approach that could be applied to a broader set of circumstances (e.g., treating sighting and kill data in the same category). The CPUE technique assumes that there is a linear relationship between the cumulative number removed and the CPUE observations, the population is closed except for removed individuals, all removals are known, and each individual has an equal probability of being caught (Seber, 1982; Lancia *et al.*, 1994). Lancia *et al.* (1994) cited unpublished work by J. W. Bishir that concluded >70–80% of

---

[4] Available free via the Internet at www.cs.umanitoba.ca/~popan/.

the sampled population must be removed for unbiased and precise CPUE estimates to be likely. Assumptions underlying the CPUE method will probably be difficult to meet in most field situations, and hence they should be carefully evaluated and tested before this technique is applied.

Animals may be selectively removed according to age, sex, size, or other criterion, e.g., a bucks-only hunting season or allowable fish catches regulated by minimum body length. *Change in ratio* (CIR) methods use the estimated proportion of some class of selected individuals (e.g., antlered deer) pre- and postremoval, along with the known number of removals from each class, to generate abundance estimates. For valid estimates, this method requires a closed population except for removed individuals, an unbiased estimate of the proportion for each class of individuals, a known number of removals (i.e., if classes are based on harvest returns, all returns must be reported and there should be no crippling loss), the proportion removed from the first class must differ from the proportion in the population, and both classes of individuals are equally detectable (Seber, 1982; Lancia *et al.,* 1994). Similar to the CPUE method, >70–80% of one class must be removed from a population of 50 to 1000 individuals for unbiased and precise estimates to be obtained (Lancia *et al.* (1994). Moreover, assumptions underlying the CIR method may prove difficult to meet in most circumstances and should be critically evaluated for validity. We refer interested readers to Seber (1982) and Lancia *et al.* (1994) for a more thorough discussion of the CIR method.

Open Capture–Recapture Estimators

Open models allow the estimation of survival/emigration, births/immigration, and population size from capture–recapture data. Another term for open models that include population estimation is the Jolly–Seber model, after Jolly (1965) and Seber (1965) who simultaneously developed the theoretical framework. The penalty for this model's flexibility is that population estimates are not as reliable as from closed models for two reasons. First, by making fewer assumptions, the estimates are less precise. That is, more information must be gleaned from the data, rather than based on assumptions, so larger variances result. Second, the open capture–recapture methods of estimating population size are all based on Model M_t, and so do not incorporate individual heterogeneity or behavioral responses. However, open population models are still useful in monitoring populations (particularly survival), as has been demonstrated by extensive work with the northern spotted owl (Burnham *et al.,* 1994).

Pollock *et al.* (1990) provided an overview of estimation of population size from a series of nested models making assumptions about capture probabilities and survival rates constant or variable across time. Analyses are implemented in program JOLLY[5] (Pollock *et al.*, 1990) and program POPAN-4[6] (Arnason and Schwarz, 1995). Burnham *et al.* (1987) have developed a framework for testing survival rates and capture probabilities between groups of animals, and have supplied program RELEASE[7] for performing estimation and simulating experiments. Lebreton *et al.* (1992) developed further theory on the modeling of capture probabilities and survival rates with program SURGE.[8] The capabilities of RELEASE and SURGE have been incorporated into program MARK.

Population size is estimated at each occasion except the first by estimating the number of unmarked animals in the population as the number of unmarked animals captured divided by the capture probability for the occasion. The number of immigrants and/or births (or recruits) to the population also can be estimated as the difference of the population estimates over an interval, with the expected mortalities subtracted from the initial population estimate. Variance formulas are provided by Seber (1982) and by Pollock *et al.* (1990). A problem with the variance estimates (also for closed population models) is that they are positively correlated with the estimate of the parameter. Thus, small estimates have small variance, whereas large estimates have large variance, making underestimates look better than they really are. The different estimates also are correlated because they are estimated from the same data. Therefore, the estimator for the births in the ith interval can be seen to be negatively correlated with the estimate for the $(i + 1)$th interval. This is known as the sampling correlation of the estimates and is encountered quite generally where multiple parameters are estimated from single data sets.

The Jolly–Seber estimator for population size is sensitive to capture heterogeneity (i.e., differences in capture probability among individuals), causing a considerable underestimation of population size (Carothers, 1973; Gilbert, 1973), although Gilbert (1973) concluded this bias was negligible if capture probabilities were 0.5 or greater. Unfortunately, capture probabilities of individuals are seldom this high in many capture–recapture studies, particularly those involving small mammals (Menkens and Anderson, 1988). Further, the Jolly–Seber estimator of population size is sensitive to changes in animal behavior (trap shyness and trap happiness; Nichols *et al.*, 1984). Conversely, the estimator for survival rate is relatively insensitive to capture heterogeneity, and insensitive to permanent trap response. Pollock *et al.*

---

[5] Available free via the Internet at www.mbr.nbs.gov/software.html. Further descriptions of, and links to, additional software are located at mendel.mbb.sfu.ca/wildberg/cmr/cnr.html.

[6] See footnote 4.

[7] See footnote 1.

[8] See footnote 5 for information on how to obtain this program for a nominal fee.

(1990) discussed a range of extensions of the basic Jolly–Seber model, including models that take into account a behavioral response to trapping as well as models that allow different cohorts of the population to have different survival and capture probabilities. They also present information on the sampling intensity required to achieve specified levels of precision of the parameter estimates.

Robust Capture–Recapture Design

Although first used by Lefebvre *et al.* (1982), Pollock (1982) elaborated a robust capture–recapture design that incorporates features of closed model population estimation and open model survival rate estimation. Animals are marked in primary periods, each of which consists of a number of secondary periods (Kendall and Pollock, 1992). For example, a capture–recapture study is conducted each summer for 7 nights over a period of 5 years. Five population estimates can be constructed for each of the five 7-night capture occasions. Survival rates can be estimated for the four intervals with the open Cormack–Jolly–Seber model. The main advantages of this design are that it allows separate estimation of immigration and *in situ* recruitment numbers (Pollock *et al.*, 1993), reduces bias of population size estimators because models not assuming equal catchability can be used, and reduces the dependence of estimators of survival and population size. Kendall *et al.* (1997) developed program RDSURVIV[9] to compute estimates of demographic parameters under the robust design approach. If population size is the only parameter of interest, one may simply use program CAPTURE within secondary periods. Estimation of population size across primary periods remains sensitive to violations of assumptions underlying the Jolly–Seber model, particularly with respect to heterogeneity in capture probabilities and trap response. Program MARK computes both closed population estimates within secondary periods and survival rates across primary periods.

3.2.2.2. Distance Sampling

Distance sampling (Buckland *et al.*, 1993) is another type of partial count where a correction is developed to estimate the total population of animals within some defined area. Methods incorporating distance to an animal from a line or point have become useful in the last 2 decades to estimate population density, where distance to the animal is used to correct for

---

[9] See footnote 5.

visibility bias. A comprehensive review of the methodology is provided by
Buckland *et al.* (1993).

Consider a population of animals distributed according to some spatial
stochastic process in an area. The traditional approach has been to
sample with quadrats or strips. Distance sampling methods provide a new
approach. Consider a line placed randomly across the area to be sampled
(Fig. 3.8). The observer moves along the line, watching for animals. When
an animal is detected, the distance to the animal is recorded. These
distances are used to develop a sighting function (Fig. 3.9). Note that
"sighting" is not limited to visual detection; any method of detection
that allows reliable distance measurements will suffice. Based on the
distance sampling framework, one expects to detect more animals closer
to the line with a steady decrease in frequency of animals recorded as
distance from the line increases.

There are three critical assumptions underlying distance sampling (Buckland *et al.*, 1993). First, this method assumes that all animals on the line
are detected. Undetected animals occurring on the line cause density estimates to be biased low. In certain situations, specific methods (e.g., Buckland *et al.*, 1993, pp. 200–217; Alpizar-Jara, 1994; Manly *et al.*, 1996) may

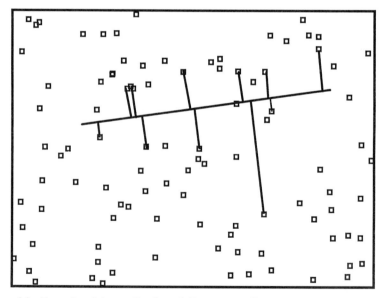

Figure 3.8 Example of the application of distance sampling to a population of animals.
Observed animals are connected to the line, whereas animals missed (not sighted) are not.
This figure is modified from van Hensbergen and White (1995).

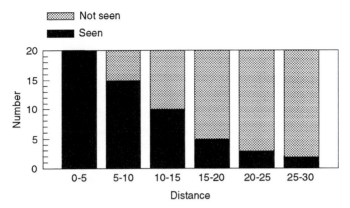

Figure 3.9 Distribution of sighting distances (solid bars) as a function of distance. If all the animals were seen in a fixed strip width, then a flat line should result. Because animals are missed (cross-hatched bars), the sighting function is assumed to be monotonically decreasing. This figure is modified from van Hensbergen and White (1995).

be used to adjust counts for less than perfect detectability on the line. In any event, there should be very little decrease in sighting probability near the line to achieve good estimates of density. The second key assumption is that detected individuals are recorded at their original locations. That is, undetected movements by animals in response to an observer are not allowed unless these movements are random, are slow compared to observer movements, or occur relatively infrequently (i.e., about 5% or less of the time; Buckland *et al.*, 1993). Systematic movements away from the line will cause density estimates to be biased low, whereas those toward the line will cause density estimates to be biased high. Substantial movements either toward or away from the line will likely be evident in the histogram of sighting distances (e.g., Buckland *et al.*, 1993, p. 33). Finally, the third critical assumption is that sighting distances are measured without error. Measurement errors that are relatively small and random still will produce reliable estimates of density at large sample sizes (Gates *et al.*, 1985); however, systematically overestimating or underestimating distances will produce biased results. A related problem occurs when estimated sighting distances are "rounded off" to the closest convenient value (e.g., 10, 15, 20), which results in distance observations recorded for only (or mostly) certain values. This is called *heaping,* and also leads to biased estimates of density. The effects of heaping may be reduced by *smearing* (Butterworth, 1982), an approach discussed in some detail by Buckland *et al.* (1993, pp. 319–322).

Distance sampling theory can handle a variety of situations. Distances can be grouped into intervals, or used individually, to estimate density. Animals can occur singly, or in clusters (i.e., flocks, schools, herds). Grouping detections into distance categories can be especially helpful if exact distances are difficult to obtain, i.e., exact distances need only be measured to the cutpoint of each category rather than to each detected individual. Further, animals often may move away from the line in response to the observer. This flight response likely decreases with increasing distance from the line. A possible approach to addressing this movement is to create distance categories so that the initial category is large enough to account for any movement (i.e., movements would only occur within the category), but small enough to ensure high detectability of individuals throughout this category. The feasibility of this approach will depend on the species and situation. Buckland *et al.* (1993) suggested that a minimum of five to seven distance categories were needed for the detection function to be adequately fitted. They also provided a list of recommended cutpoints for four to eight distance categories for both line and point transects (Buckland *et al.*, 1993, p. 328).

Within the sampling framework developed in Chapter 2, the line is the sampling unit. Distances measured across all the sampling units are used to estimate the sighting function, and then estimate density. Distance methods also allow the stratification of the sampling units. Thus, instead of living with a high sampling variance because of the high variation between sampling units, the investigator can stratify the area to be surveyed into multiple strata. Sighting functions can be estimated in common across the strata, or tested to see if they may differ across strata (Buckland, 1993, pp. 100–102).

The density estimate based on a line transect is defined as

$$\hat{D} = \frac{nf(0)}{2L},$$

where n is the number of sighting distances (i.e., number of objects seen), L is the length of line traversed, and $f(y)$ is defined in terms of the detection or sighting function $g(y)$ as

$$f(y) = \frac{g(y)}{\int_0^w g(y)\,dy},$$

where w is the strip width of the line transect survey; w can be a fixed width or an unlimited distance.

A specialization of line transect sampling is point transect sampling (also called the variable circular plot method; Reynolds *et al.*, 1980), i.e., the observer stands at one location and determines distances to animals from this location. The theory is similar to line transect sampling, but requires a slightly different sighting probability model because of the radial distances observed. The sampling unit is now the point. Program DISTANCE[10] (Laake *et al.*, 1993) handles point transect data with a model designed to account for the radial nature of the data. This program also handles stratification of the sampling units, i.e., lines or points, and can compute an overall estimate of density across strata.

The most difficult aspect of line (or point) transect estimation is fitting an appropriate detection function, $g(y)$, to the observed sighting distances. Burnham *et al.* (1980) recommended five different sighting functions: exponential, exponential polynomial, exponential power series, half-normal, and Fourier series. Their program TRANSECT fits these functions to observed distances and generates density estimates. More recently, Buckland *et al.* (1993) have recommended program DISTANCE, a replacement for TRANSECT, that includes the uniform, half-normal, negative exponential, and hazard functions as key functions, with modifiers to these key functions provided by cosine, simple polynomial, and Hermite polynomial functions. Thus, 12 different models can be fitted using maximum likelihood estimation methods. Akaike's Information Criterion (AIC; Akaike, 1973) is recommended to select the most parsimonious model to fit to the observed distances. The AIC is defined as

$$\text{AIC} = -2 \log \mathcal{L}(D | \text{observed distances}) + 2 \times \text{No. parameters estimated},$$

so the criterion is based on likelihood theory. According to this criterion, the model with the smallest AIC is the one used to make inferences from the data. AIC can be viewed as a trade-off between lack of fit (the log-likelihood portion reflecting bias) and the number of parameters fitted to the data (reflecting precision). As more parameters are fitted to the observed data, the lack of fit (i.e., bias) decreases. However, the penalty for too many parameters is loss of precision of the density estimate. Thus, AIC seeks to find the model with the optimum number of parameters to adequately fit the observed data by balancing model bias and precision (called the *principle of parsimony* by Box and Jenkins, 1970).

A problem with distance-based surveys arises where the animals occur in groups, or clusters. The method assumes that observations are made independently of one another. Where animals occur in clusters, this is unlikely to be the case because if one group member is seen the others are more likely to be seen. Also, larger clusters are probably more likely to be seen further from the line, so that sighting distance is not independent of cluster size. Buckland *et al.* (1993) reviewed some methods for dealing

---

[10] Available free via the Internet at nmml01.afsc.noaa.gov/distance/index.html. Note that "l01" in this address is the letter "ell" followed by the numbers zero and one.

with clustered populations. Program DISTANCE includes tests for the effect of cluster size on sighting probability.

The distance data need not be recorded as exact distances, but can be grouped. Thus, observers working from helicopters commonly record distances to mule deer into categories, such as 0–15 m, and increments of 10 m from 15 to 155 m (White *et al.,* 1989; Example 3.3). Observers working from fixed-wing aircraft often use markers on the struts of the aircraft to demarcate sighting distances into groups, e.g., surveys of pronghorn by Johnson *et al.* (1991).

Example 3.3. Estimate of Mule Deer Density with Helicopter Line Transects

White *et al.* (1989) presented the following distance data for mule deer in the Piceance Basin, Colorado (Fig. 3.10).

Distance interval (m)	Number of deer clusters
0–15	41
>15–25	18
>25–35	17
>35–45	20
>45–55	13
>55–65	9
>65–75	10
>75–85	14
>85–95	10
>95–105	6
>105–115	5
>115–125	6
>125–135	5
>135–145	4
>145	9

Program DISTANCE computed the density of deer clusters using three key functions: hazard, uniform, and negative exponential. The AIC for each of the models considered was:

Key/Adjustment	AIC
Hazard/no adjustment	872.40
Uniform/cosine	872.67
Neg. exp./no adjustment	869.21

No adjustments improved either the hazard or negative exponential keys. The minimum AIC is for the negative exponential model for which the density of clusters is estimated as 9.89 per km^2. With a mean cluster size of 2.54, this results in a deer density of 25.12 deer per km^2.

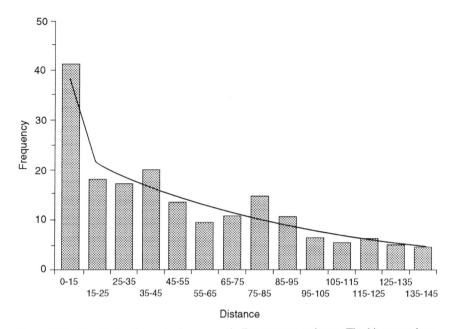

Figure 3.10 Results for the mule deer example line transect estimate. The histogram bars represent the frequency of deer clusters, and the solid line is the negative exponential sighting function fit by program DISTANCE. The goodness-of-fit test for these data gave a χ^2 value of 5.92 with 12 *df,* $P = 0.92$, so we conclude that the data are adequately fit by the negative exponential function.

For any strip transect survey, i.e., counting all the animals that occur in a strip of fixed width, line transects should be considered. The assumption in a strip survey is that all the animals are counted in the strip. To test this assumption, perpendicular distances can be taken, and their distribution tested. If indeed all the animals are counted, the distribution of the perpendicular distances should be uniform, with no decrease in the frequency of longer distances at or near the width of the strip. Burnham and Anderson (1984) concluded that for reasons of efficiency and validity, perpendicular distance data always should be taken with strip transect counts.

The book by Buckland *et al.* (1993) is a comprehensive reference on the application of line and point transects, and should be consulted prior to applying these techniques. They provide numerous examples, and include a chapter on the design of point and line transect surveys. They stress the need for a pilot survey in order to estimate either the overall transect length (L) or number of point transects (u) required to obtain some predetermined level of precision. The formula (Buckland *et al.,* 1993, p. 303) for estimating the overall transect length needed to satisfy a predetermined level of precision, $\hat{CV}(\hat{D})$, is

$$L = \left(\frac{b}{[\hat{CV}(\hat{D})]^2}\right)\left(\frac{L_0}{n_0}\right), \tag{3.4}$$

where b is called the dispersion parameter, $\hat{CV}(\hat{D})$ is the coefficient of variation for the estimated density, L_0 is the transect length on which the pilot data were collected and n_0 is the number of sightings made along L_0. The dispersion parameter is based on two sources of variation in distance sampling of individuals (assuming all individuals on a line or point can be detected): variation associated with numbers of sightings and variation in sighting distances. The overall transect should be partitioned into segments in order to obtain an empirical estimate for variation among sightings. If animals are randomly distributed, i.e., follow a Poisson distribution, then $b = 1$. The more clumped the underlying arrangement of animals, the greater the value of b. In addition, the narrower the shoulder of the detection function, the greater the value of b (Buckland *et al.*, 1993). Burnham *et al.* (1980) recommended an initial value of 3 if a previous estimate for b is not available. The formula (Buckland *et al.*, 1993, p. 307) used for calculating the number of point transects (u) needed to satisfy some level of precision is

$$u = \left(\frac{b}{[\hat{CV}(\hat{D})]^2}\right)\left(\frac{u_0}{n_0}\right), \tag{3.5}$$

where u_0 is the number of point transects surveyed during the pilot study.

3.2.2.3. Miscellaneous Methods

Line-Intercept Sampling

Becker (1991) presented a method of obtaining unbiased estimates of abundance that was based on probability of transects intercepting animal tracks. The first design applies when an animal's set of tracks can be identified and followed to its beginning and end. The second assumes that the number of animals crossing a transect line can be estimated based on movement data from a random sample of radio-marked animals. We will discuss the first design in more detail.

The probability of the kth track being intercepted depends on the horizontal distance (d_k) from one end of the track to the other compared to the width (W) of the x axis of a plot or study area boundary (Fig. 3.11). A repeated systematic random sample, based on a predetermined number of transects, is conducted on each plot or frame to obtain unbiased population estimates. Transects should be of equal length and spaced far enough apart so that tracks will not cross more than one transect.

The probability (p_k) that the kth track is included in 1 of the systematic samples of transects is computed using the formula (Becker, 1991, p. 732)

$$p_k = \frac{d_k}{(W/q)} \tag{3.6}$$

for tracks with horizontal distances less than or equal to W/q ($p_k = 1$ otherwise), where q is the number of transects surveyed in each systematic sample and each track is assumed to be associated with a single individual. The abundance estimate for each systematic sample is calculated using

$$\hat{N}_j = \sum_{k \in S_j} \frac{1}{p_k}, \tag{3.7}$$

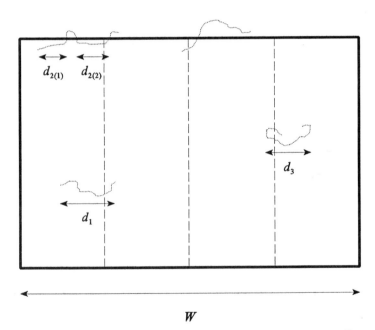

Figure 3.11 A single systematic sample of three transects within a study area. Four sets of tracks are intercepted by transects but horizontal distances of only three are included; not enough of the fourth set of tracks is contained within the boundaries to warrant its inclusion in the sample. Note that the horizontal distance (d_2) of the second set of tracks is a sum of two segments because the entire set of tracks is only partially contained within the boundaries.

where $\sum_{k \in S_j}$ means that the quantity is summed over all tracks contained in the jth systematic sample (S_j). The previous equation may be redefined in terms of groups if more than one individual is associated with a given set of tracks. The "1" in the numerator would be replaced with g_k, which is the size of the group associated with the kth track. In either case, the total abundance is estimated by (Becker, 1991, p. 732)

$$\hat{N} = \frac{\sum_{j=1}^{n_s} \hat{N}_j}{n_s} \tag{3.8}$$

with a variance estimator of

$$\hat{V}ar(\hat{N}) = \frac{\sum_{j=1}^{k} (\hat{N}_j - \hat{N})^2}{n_s(n_s - 1)}, \tag{3.9}$$

where n_s is the number of systematic samples. Use of these formulas is demonstrated in Example 3.4.

Example 3.4. Line-Intercept Sampling of Animal Tracks

Becker (1991) presented an example of line-intercept sampling for tracks applied to a wolverine population in southcentral Alaska. Four systematic samples were conducted on the study area that was 58 km wide (i.e., $W = 58$ km). Each sample was composed of $q = 3$ transects surveyed for tracks via airplane. Four sets of wolverine tracks, two of which were made by a group of two animals, were intercepted by transects during the course of the entire survey and yielded the following data (modified from Table 2 in Becker, 1991, p. 734).

Set No.	Group size	Horizontal distance (km) of tracks	Probability of interception
1	1	8.75	0.453
2	2	12.25	0.634
3	2	3.50	0.181
4	1	9.75	0.504

The probability that a given set of tracks was intercepted by a systematic sample of transects was computed using Eq. (3.6). For example, p_k for the first set of tracks was $p_1 = (8.75$ km$)/(58$ km$/3) = 0.453$. The first, second, and fourth set of tracks were encountered in the first systematic sample, the first and second set were intercepted in the second sample, and the third and fourth set were encountered in both the third and fourth systematic samples. We use these data in Eq. (3.7) (modified for groups, i.e., the numerator now reflects group size and may be greater than 1) to obtain abundance estimates for each systematic sample. The estimated abundance for the first systematic sample was

$$\hat{N}_1 = \frac{1}{0.453} + \frac{2}{0.634} + \frac{1}{0.504} = 7.35,$$

with $\hat{N}_2 = 5.36$, $\hat{N}_3 = 13.03$, and $\hat{N}_4 = 13.03$. Therefore, the overall abundance estimate [Eq. (3.8)] was $\hat{N} = (7.35 + 5.36 + 13.03)/4 = 9.69$ with a variance estimate [Eq. (3.9)] of

$$\hat{\text{Var}}(\hat{N}) = \frac{(7.35 - 9.69)^2 + \ldots + (13.03 - 9.69)^2}{4(4 - 1)} = 3.88.$$

Critical assumptions for this method are that all animals move during the course of the study, all animal tracks of the species of interest are readily recognizable, all animal tracks are continuous, animal movements are independent of the sampling process (i.e., animals do not move in response to the observer so track lengths are fixed), pre- and post-snowstorm tracks can be distinguishable, all animal tracks that intercept sampled transects are observed, the study area is rectangular in shape, and all the transects are oriented perpendicular to a specified reference axis (x axis) (Becker, 1991). Only that part of a set of tracks contained within the boundaries of the plot or study area is included in calculating horizontal distances. Tracks with more than half of their horizontal distance outside of the boundaries are not included in the survey (e.g., the set of tracks at the top and center in Fig. 3.11).

A number of the underlying assumptions of this method preclude it from general application, i.e., its proper use is likely limited to specific situations. For instance, larger animals tend to have larger daily movements and lower densities so surveys on foot are probably limited to studies of smaller animals because of logistic constraints. In addition, ground surveys would be difficult, and possibly dangerous, in areas of rugged terrain (i.e., areas susceptible to avalanches); use of snowmobiles (or other motorized vehicles) to survey lines and intercepted tracks will probably cause a flight response in the target species, which would violate a key assumption of fixed track lengths. Further, aerial surveys are limited to areas of open habitats; obstruction of tracks by overhead vegetation would preclude its use. Even in open habitats, tracks of the target species must be readily discernable from the air from tracks of other resident species. However, despite these and other related difficulties, Becker's method could be useful in situations where appropriate assumptions are satisfied.

Doubling Sampling

As previously mentioned, a *double sample* (also called a *two-phase sample*) is one in which an inexpensive, relative measure (i.e., index) is gathered in a large initial sample (first phase) followed by expensive, reliable measure in a small second sample (second phase; Cochran, 1977). Then, the index can be calibrated to yield reliable estimates using either ratio or regression estimators, or improved by some type of prediction equation (e.g., Rivest *et al.*, 1990). We will concentrate on the former approach.

In animal population studies, the calibration approach applies to those very limited circumstances where a cheap, relative count (e.g., aerial survey) is conducted over a large area and an expensive, complete count (e.g., complete ground count) is performed over a smaller subsection of this same area (Jolly, 1969; Eberhardt and Simmons, 1987). The abundance estimator (Jolly, 1969; Pollock and Kendall, 1987) in this case would be

$$\hat{N} = U \times \frac{\hat{\overline{N}}}{\hat{p}},$$

where $\hat{p} = \hat{\overline{N}}{}^*/\overline{N}{}^*$, $\hat{\overline{N}}$ is the average of incomplete counts for all surveyed plots, $\hat{\overline{N}}{}^*$ is average of incomplete counts for all subsampled plots, and $\overline{N}{}^*$ is the average of complete counts for all subsampled plots. Note there is both a bar and a caret on $\overline{N}$ and $\overline{N}{}^*$ to denote two levels of uncertainty. The bar is associated with uncertainty from less than all plots being counted, whereas the caret is associated with uncertainty from less than all animals being counted on selected plots. The ratio of the average of incomplete counts to the average of complete counts over the subsampled plots is the detectability correction ($\hat{p}$) used to adjust the overall abundance estimate. The previous estimator will be reliable only if complete counts are truly possible on the subsample of plots, each animal has the same chance of being detected, and plots are randomly chosen (Jolly, 1969). Jolly (1969) and Pollock and Kendall (1987) provided the variance estimator of $\hat{N}$. As a simple example, suppose we obtained an average incomplete count of 12 animals from a random sample of 10 plots out of a frame of 200 plots. Further assume that each animal had the same probability of detection. Then, we conducted complete counts on 2 plots randomly subsampled from the 10 previously surveyed. The average here was 20 animals. The earlier incomplete counts on these 2 plots yielded an average count of 15 animals. Thus, $\hat{p}$ would be 15/20 or 0.75 and $\hat{N} = 200 \times (12/0.75) = 3200$ animals.

As noted previously, complete counts of animal populations are rarely possible at any spatial scale. Even if complete counts are feasible, the constant detectability assumption for incomplete counts would be difficult to meet in most circumstances. Assuming both of these key assumptions can be satisfied, the choice between using a double sampling/calibration approach or a sample of censused plots will depend on the trade-off between cost and precision. A double sample is more cost efficient than complete counts on all plots, but adds a level of variation due to the incomplete counts. In any event, the calibration approach to double sampling will rarely be a reliable alternative in most animal population studies.

Sightability Models

Another way of estimating detectability (or, in this case, sightability) is to collect data on various factors potentially affecting it and model these factors to obtain an abundance estimate (Caughley *et al.*, 1976). For aerial surveys, such factors could include vegetational cover, weather, group size, snow conditions, and so forth (Samuel *et al.*, 1987). The sightability estimate then can be used to adjust abundance estimates (Steinhorst and Samuel, 1989). For example, Samuel *et al.* (1987) collected information on various factors potentially affecting visibility of radio-marked/color-marked and unmarked elk during aerial surveys. An initial flight with a fixed wing aircraft located the radio-collared animals in the study area. Then, a helicopter survey was performed in which various detection factors were recorded for each group of elk sighted that contained a radio-collared individual. Finally, radio-collared elk that were missed were located and the various detection factors recorded. The authors used a logistic regression approach for modeling detection factors of sighted and missed elk to develop an appropriate sightability model.

The main advantage of using sightability models is that once an appropriate model has been chosen, subsequent surveys need only record data for model variables, and therefore will not be as labor intensive or costly as the original set of surveys (Lancia *et al.*, 1994). Unfortunately, conditions under which the model was developed must be closely mirrored or the model will yield spurious estimates of sightability, which will lead to biased estimates of abundance. This problem may even be worse if the model is applied to a different area, i.e., an area where detection factors such as vegetational cover will likely differ from the original area.

LITERATURE CITED

Akaike, H. (1973). Information theory as an extension of the maximum likelihood principle. In *Second International Symposium on Information Theory* (B. N. Petrov and F. Csaki, Eds.), pp. 276–281. Akademiai, Budapest.

Alpizar-Jara, R. (1994). *A Combination Line Transect and Capture–Recapture Sampling Model for Multiple Observers in Aerial Surveys.* M.S. Thesis, North Carolina State Univ., Raleigh, NC.

Arnason, A. N., and Schwarz, C. J. (1995). POPAN-4: Enhancements to a system for the analysis of mark-recapture data from open populations. *J. Appl. Stat.* **22:** 785–800.

Arnason, A. N., Schwarz, C. J., and Gerrard, J. M. (1991). Estimating closed population size and number of marked animals from sighting data. *J. Wildl. Mgmt.* **55:** 716–730.

Barker, R. J., and Sauer, J. R. (1992). Modelling population change from time series data. In *Wildlife 2001: Populations* (D. R. McCullough and R. H. Barrett, Eds.), pp. 182–194. Elsevier Scientific, London.

Bartmann, R. M., Carpenter, L. H., Garrott, R. A., and Bowden, D. C. (1986). Accuracy of helicopter counts of mule deer in pinyon-juniper woodland. *Wildl. Soc. Bull.* **14:** 356–363.

Bartmann, G. C., White, G. C., Carpenter, L. H., and Garrott, R. A. (1987). Aerial mark-recapture estimates of confined mule deer in pinyon-juniper woodland. *J. Wildl. Mgmt.* **51:** 41–46.

Becker, E. F. (1991). A terrestrial furbearer estimator based on probability sampling. *J. Wildl. Mgmt.* **55:** 730–737.

Bishir, J. W., and Lancia, R. A. (1996). On catch-effort methods of estimating animal abundance. *Biometrics* **52:** 1457–1466.

Bowden, D. C., and Kufeld, R. C. (1995). Generalized mark-resight population size estimation applied to Colorado moose. *J. Wildl. Mgmt.* **59:** 840–851.

Box, G. E. P., and Jenkins, G. M. (1970). *Time Series Analysis: Forecasting and Control.* Holden-Day, London.

Buckland, S. T., Anderson, D. R., Burnham, K. P., and Laake, J. L. (1993). *Distance Sampling: Estimating Abundance of Biological Populations.* Chapman and Hall, London.

Burnham, K. P., and Anderson, D. R. (1984). The need for distance data in transect counts. *J. Wildl. Mgmt.* **48:** 1248–1254.

Burnham, K. P., Anderson, D. R., and Laake, J. L. (1980). Estimation of density from line transect sampling of biological populations. *Wildl. Monogr.* **72.**

Burnham, K. P., Anderson, D. R., and White, G. C. (1994). Estimation of vital rates of the northern spotted owl. In *Final Supplemental Environmental Impacts Statement on Management of Habitat for Late-Successional and Old-Growth Forest Related Species Within the Range of the Northern Spotted Owl,* Appendix J, Vol. II, Appendices. U.S.D.A. For. Serv. and U.S.D.I. Bur. Land Manage., Interagency SEIS Team, Portland.

Burnham, K. P., Anderson, D. R., White, G. C., Brownie, C., and Pollock, K. H. (1987). Design and analysis methods for fish survival experiments based on release-recapture. *Am. Fish. Soc. Monogr.* **5.**

Burnham, K. P., and Overton, W. S. (1978). Estimation of the size of a closed population when capture probabilities vary among animals. *Biometrika* **65:** 625–633.

Burnham, K. P., and Overton, W. S. (1979). Robust estimation of population size when capture probabilities vary among animals. *Ecology* **60:** 927–936.

Butterworth, D. S. (1982). On the functional form used for $g(y)$ for Minke whale sightings, and bias in its estimation due to measurement inaccuracies. *Rep. Int. Whaling Comm.* **32:** 883–888.

Carothers, A. D. (1973). The effects of unequal catchability on Jolly-Seber estimates. *Biometrics* **29:** 79–100.

Caughley, G. (1977). *Analysis of Vertebrate Populations.* Wiley, New York.

Caughley, G., Sinclair, R., and Scott-Kemmis, D. (1976) Experiments in aerial survey. *J. Wildl. Mgmt.* **40:** 290–300.

Chao, A., Lee, S. M., and Jeng, S. L. (1992). Estimating population size for capture-recapture data when capture probabilities vary by time and individual animal. *Biometrics* **48:** 201–216.

Chapman, D. G. (1951). Some properties of the hypergeometric distribution with applications to zoological censuses. *Univ. Calif. Publ. Stat.* **1:** 131–160.

Cochran, W. G. (1977). *Sampling Techniques,* third ed. Wiley, New York.

Darroch, J. N. (1958). The multiple recapture census: I. Estimation of a closed population. *Biometrika* **45:** 343–359.

Debinski, D. M., and Brussard, P. F. (1994). Using biodiversity data to assess species-habitat relationships in Glacier National Park, Montana. *Ecol. Appl.* **4:** 833–843.

Draper, N., and Smith, H. (1981). *Applied Regression Analysis,* second ed. Wiley, New York.

Eberhardt, L. L., and Simmons, M. A. (1987). Calibrating population indices by double sampling. *J. Wildl. Mgmt.* **51:** 665–675.

Gasaway, W. C., DuBois, S. D., Reed, J. D., and Harbo, S. J. (1986). *Estimating Moose Population Parameters from Aerial Surveys.* Biol. Papers Univ. of Alaska No. 22.

Gates, C. E., Evans, W., Gober, D. R., Guthery, F. S., and Grant, W. E. (1985). Line transect estimation of animal densities for large data sets. In *Game Harvest Management* (S. L. Beasom and S. F. Roberson, Eds.), pp. 37–50. Caesar Kleberg Wildl. Res. Inst., Texas A&I Univ., Kingsville, TX.

Gilbert, R. O. (1973). Approximations of the bias in the Jolly-Seber capture-recapture model. *Biometrics* **29:** 501–526.

Guthery, F. S. (1989). Test of a density index assumption for gray partridge: a comment. *J. Wildl. Mgmt.* **53:** 1132–1133.

Jackson, D. A., and Harvey, H. H. (1989). Biogeographic associations in fish assemblages: Local vs. regional processes. *Ecology* **70:** 1472–1484.

Johnson, B. K., Lindzey, F. G., and Guenzel, R. J. (1991). Use of aerial line transect surveys to estimate pronghorn populations in Wyoming. *Wildl. Soc. Bull.* **19:** 315–321.

Jolly, G. M. (1965). Explicit estimates from capture-recapture data with both death and immigration—stochastic model. *Biometrika* **52:** 225–247.

Jolly, G. M. (1969). The treatment of errors in aerial counts of wildlife populations. *East Afr. Agric. For. J. (Spec. Issue)* **34:** 50–55.

Kendall, W. L., Nichols, J. D., and Hines, J. E. (1997). Estimating temporary emigration using capture-recapture data with Pollock's Robust Design. *Ecology* **78:** 563–578.

Kendall, W. L., and Pollock, K. H. (1992). The robust design in capture-recapture studies. In *Wildlife 2001: Populations* (D. R. McCullough and R. H. Barrett, Eds.), pp. 31–43. Elsevier Applied Science, Essex, UK.

Kufeld, R. C., Olterman, J. H., and Bowden, D. C. (1980). A helicopter quadrat census for mule deer on Uncompahgre Plateau, Colorado. *J. Wildl. Mgmt.* **44:** 632–639.

Laake, J. L. (1992). Catch-per-unit-effort models: An application to an elk population in Colorado. In *Wildlife 2001: Populations* (D. R. McCullough and R. H. Barrett, Eds.), pp. 44–55. Elsevier Applied Science, Essex, UK.

Laake, J. L., Buckland, S. T., Anderson, D. R., and Burnham, K. P. (1993). *Distance User's Guide,* v2.0. Colorado Coop. Fish and Wildl. Res. Unit, Colorado State Univ., Fort Collins.

Lancia, R. A., Bishir, J. W., Conner, M. C., and Rosenberry, C. S. (1996). Use of catch-effort to estimate population size. *Wildl. Soc. Bull.* **24:** 731–737.

Lancia, R. A., Nichols, J. D., and Pollock, K. H. (1994). Estimating the number of animals in wildlife populations. In *Research and Management Techniques for Wildlife and Habitats* (T. A. Bookhout, Ed.), fifth ed., pp. 215–253. The Wildlife Society, Bethesda, MD.

Lebreton, J.-D., Burnham, K. P., Clobert, J., and Anderson, D. R. (1992). Modeling survival and testing biological hypotheses using marked animals: Case studies and recent advances. *Ecol. Monogr.* **62:** 67–118.

Lefebvre, L. W., Otis, D. L., and Holler, N. R. (1982). Comparison of open and closed models for cotton rat population estimates. *J. Wildl. Mgmt.* **46:** 156–163.

Manly, B. F. J., McDonald, L. L., and Garner, G. W. (1996). Maximum likelihood estimation for the double-count method with independent observers. *J. Agric. Biol. Environ. Stat.* **1:** 170–189.

Menkens, G. E., Jr., and Anderson, S. H. (1988). Estimation of small-mammal population size. *Ecology* **69:** 1952–1959.

Minta, S., and Mangel, M. (1989) A simple population estimate based on simulation for capture-recapture and capture-resight data. *Ecology* **70:** 1738–1751.

Neal, A. K., White, G. C., Gill, R. B., Reed, D. F., and Olterman, J. H. (1993). Evaluation of mark-resight model assumptions for estimating mountain sheep numbers. *J. Wildl. Mgmt.* **57:** 436–450.

Nichols, J. D. (1992). Capture-recapture models: Using marked animals to study population dynamics. *BioScience* **42:** 94–102.

Nichols, J. D., Hines, J. E., and Pollock, K. H. (1984). Effects of permanent trap response in capture probability on Jolly-Seber capture-recapture model estimates. *J. Wildl. Manage.* **48:** 289–294.

Novak, J. M., Schribner, K. T., Dupont, W. D., and Smith, M. H. (1991). Catch-effort estimation of white-tailed deer population size. *J. Wildl. Manage.* **55:** 31–38.

Otis, D. L., Burnham, K. P., White, G. C., and Anderson, D. R. (1978). Statistical inference from capture data on closed animal populations. *Wildl. Monogr.* **62:** 1–135.

Owen, J. G. (1990). An analysis of the spatial structure of mammalian distribution patterns in Texas. *Ecology* **71:** 1823–1832.

Petersen, C. G. J. (1896). The yearly immigration of young plaice into the Limfjord from the German Sea. *Rep. Danish Biol. Sta.* **6:** 1–48.

Pollock, K. H. (1982). A capture-recapture design robust to unequal probability of capture. *J. Wildl. Mgmt.* **46:** 752–757.

Pollock, K. H., and Kendall, W. L. (1987). Visibility bias in aerial surveys: A review of estimation procedures. *J. Wildl. Mgmt.* **51:** 502–510.

Pollock, K. H., Kendall, W. L., and Nichols, J. D. (1993). The "robust" capture-recapture design allows components of recruitment to be estimated. In *Marked Individuals in the Study of Bird Populations* (J.-D. Lebreton and P. M. North, Eds.), pp. 245–252. Birkhäuser Verlag, Basel, Switzerland.

Pollock, K. H., Nichols, J. D., Brownie, C., and Hines, J. E. (1990). Statistical inference for capture-recapture experiments. *Wildl. Monogr.* **107.**

Ratti, J. T., and Rotella, J. J. (1989). Test of a density index assumption for gray partridge: A reply. *J. Wildl. Mgmt.* **53:** 1133–1134.

Rexstad, E. A., and Burnham, K. P. (1992). *User's Guide for Interactive Program CAPTURE.* Colorado Coop. Fish and Wildl. Res. Unit, Colorado State Univ., Fort Collins.

Reynolds, R. T., Scott, J. M., and Nussbaum, R. A. (1989). A variable circular-plot method for estimating bird numbers. *Condor* **82:** 309–313.

Ricker, W. E. (1958). Handbook of computations for biological statistics of fish populations. *Bull. Fish. Bd. Can.* **119:** 1–300.

Ricker, W. E. (1975). Computation and interpretation of biological statistics of fish populations. *J. Fish. Res. Bd. Can.* **30:** 409–434.

Rivest, L.-P., Crepeau, H., and Crete, M. (1990). A two-phase sampling plan for the estimation of the size of a moose population. *Biometrics* **46:** 163–176.

Rivest, L.-P., Potvin, F., Crepeau, H., and Daigle, G. (1995). Statistical methods for aerial surveys using the double-count technique to correct visibility bias. *Biometrics* **51:** 461–470.

Rosenberg, D. K., Overton, W. S., and Anthony, R. G. (1995). Estimation of animal abundance when capture probabilities are low and heterogeneous. *J. Wildl. Mgmt.* **59:** 252–261.

Rotella, J. J., and Ratti, J. T. (1986). Test of a critical density index assumption: A case study with gray partridge. *J. Wildl. Mgmt.* **50:** 532–539.

Samuel, M. D., Garton, E. O., Schlegel, M. W., and Carson, R. G. (1987). Visibility bias during aerial surveys of elk in northcentral Idaho. *J. Wildl. Mgmt.* **51:** 622–630.

Scattergood, L. W. (1954). Estimating fish and wildlife populations: a survey of methods. In

Statistics and Mathematics in Biology (O. Kempthorne, T. A. Bancroft, J. W. Gowen, and J. L. Lush, Eds.), pp. 273–285. Iowa State College Press, Ames, IA.

Seber, G. A. F. (1965). A note on the multiple-recapture census. *Biometrika* **52:** 249–259.

Seber, G. A. F. (1982). *Estimation of Animal Abundance,* second ed. Griffin, London.

Seber, G. A. F. (1986). A review of estimating animal abundance. *Biometrics* **42:** 267–292.

Seber, G. A. F. (1992). A review of estimating animal abundance II. *Int. Stat. Rev.* **60:** 129–166.

Skalski, J. R. (1991). Using sign counts to quantify animal abundance. *J. Wildl. Mgmt.* **55:** 705–715.

Skalski, J. R., and Robson, D. S. (1992). *Techniques for Wildlife Investigations: Design and Analysis of Capture Data.* Academic Press, San Diego.

Steinhorst, R. K., and Samuel, M. D. (1989). Sightability adjustment methods for aerial surveys of wildlife populations. *Biometrics* **45:** 415–425.

van Hensbergen, H. J., and White, G. C. (1995). Review of methods for monitoring vertebrate population parameters. In *Integrating People and Wildlife for a Sustainable Future* (J. A. Bissonette and P. R. Krausman, Eds.), pp. 489–508, Proc. First Int. Wildl. Manage. Congress, The Wildlife Society, Bethesda, MD.

White, G. C. (1994). *User's Manual for Program MCAPTURE* [Project completion report to U.S.D.A. Forest Service].

White, G. C. (1996a). NOREMARK: Population estimation from mark-resighting surveys. *Wildl. Soc. Bull.* **24:** 50–52.

White, G. C. (1996b). *Program NOREMARK Software Reference Manual.* Dept. Fishery and Wildlife Biology, Colorado State Univ., Fort Collins, CO.

White, G. C., Anderson, D. R., Burnham, K. P., and Otis, D. L. (1982). *Capture-Recapture and Removal Methods for Sampling Closed Populations.* Los Alamos National Laboratory, Los Alamos, NM. [Rep. No. LA-8787-NERP]

White, G. C., Bartmann, R. M., Carpenter, L. H., and Garrott, R. A. (1989). Evaluation of aerial line transects for estimating mule deer densities. *J. Wildl. Mgmt.* **53:** 625–635.

White, G. C., Burnham, K. P., Otis, D. L., and Anderson, D. R. (1978). *User's Manual for Program CAPTURE.* Utah State Univ. Press, Logan, UT.

White, G. C., and Garrott, R. A. (1990). *Analysis of Wildlife Radio-Tracking Data.* Academic Press, San Diego, CA.

Zalewski, M., and Cowx, I. G. (1990). Factors affecting the efficiency of electric fishing. In *Fishing with Electricity* (I. G. Cowx and P. Lamarque, Eds.), pp. 89–111. Fishing News Books, Blackwell Scientific, Oxford, UK.

Zippen, C. (1956). An evaluation of the removal method of estimating animal populations. *Biometrics* **12:** 163–189.

Zippen, C. (1958). The removal method of population estimation. *J. Wildl. Mgmt.* **22:** 82–90.

Chapter 4

Community Surveys

4.1. **Number of Species Versus Their Density**

4.2. **Assessment of Spatial Distribution**
 4.2.1. Presence–Absence
 4.2.2. Other Methods for Assessing Distribution

4.3. **Assessment of Abundance or Density**

4.3.1. Presence–Absence
4.3.2. Index of Relative Abundance or Density
4.3.3. Abundance or Density Estimation

4.4. **Recommendations**
 Literature Cited

One approach to assessing status of wildlife species within an area is through collection of information at the community level. Before this can be attempted, however, "community" must be defined in quantifiable terms. Therefore, we operationally define a community as all individuals of every species of interest within one or more vertebrate groups (e.g., fish, herpetofauna, birds, and/or mammals) occurring in a specified area during a particular time period. The objective of a "community survey" then is to collect data on the distribution and perhaps abundance or density (or indices of these two) for all target species in an area. This is contrasted by a "single species survey" in which efforts are concentrated on only one species at a time. A "multispecies survey" refers to a survey of more than one species at a time and may or may not be expanded to include a community. In theory, one would like to collect population data concurrently for all species in either multispecies or community surveys.

In this chapter, we discuss the community survey approach to assessing both current status and long-term trends in distribution and numbers of species in an area. Compiling species lists will be considered part of collecting basic spatial distribution/occurrence data. We will assume that the area of interest is large enough so that it must be divided into sampling units (i.e., plots). A portion of these units are randomly chosen, using one of the plot selection methods described in Chapter 2, and then surveyed.

4.1. NUMBER OF SPECIES VERSUS THEIR DENSITY

An important characteristic of a community is the relationship between the density of a species' population, and the number of species in its community. As a general rule, many more species occur at low densities than at high densities. This general relationship between the frequency of species and their density is shown in Fig. 4.1. The ideas were first developed by Fisher *et al.* (1943) and Preston (1948). More recently, Pielou (1977) and Gotelli and Graves (1996) have discussed this topic.

The impact of the relationship in Fig. 4.1 on community surveys is that most of the species in a community are rare, and hence seldom, if ever, are observed. Only very intensive surveys will detect a rare species. Further, a series of surveys through time will likely suggest that many rare species are going extinct in the area, or colonizing the area, when in fact the density has not changed for these species. Rather, sampling variation causes the abundance of species to appear to be changing. As an example, suppose

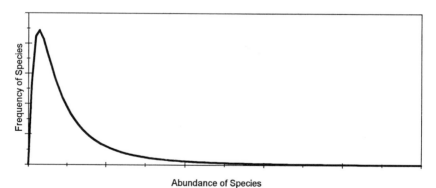

Figure 4.1 The frequency of occurrence of species as a function of their abundance. Most species in a community are rare, with only a very small number of species having high abundance. Species with low abundance are shown at the left side of the *x* axis, and because there are many rare species compared to abundant ones, there is a high frequency on the *y* axis. The most abundant species occur on the right side of the *x* axis, with very few of them as indicated by the low frequency.

the first three surveys of an area fail to detect the species. As we show in this chapter, such an incident is not unlikely. Then, a fourth survey detects the species. Because biologists are now more aware of the species, a fifth survey also detects the species. The suggestion from these five surveys is that the species has colonized the area, when in reality, its density may not have changed. Detecting changes in abundance of rare species is difficult because very intensive surveys must be conducted. Given that there are mostly rare species in a community, the possibility of detecting most species in a community is low.

4.2. ASSESSMENT OF SPATIAL DISTRIBUTION

A possible first step for evaluating the status of a collection of species could be to find out which species are present and how they are spatially arranged across the sampling frame. That is, which sampling units, if any, contain a given species? The simplest way to collect this information is to record the presence or absence of a species on surveyed plots. A more indirect approach is to collect distributional data as a by-product of obtaining either abundance/density indices or estimates. We will focus most of our discussion on the presence–absence approach because it is commonly applied at the community level.

4.2.1. PRESENCE–ABSENCE

Simply noting the presence of species on plots, while disregarding their abundance, will yield the proportion of plots that contain at least one member of each species over a surveyed area (called a *frequency index;* Dice, 1952). This information is easier and less costly to obtain than trying to assess how many individuals of each species are contained within each plot. However, even trying to assess whether a species is present or absent on a particular plot could be time-consuming. Failing to detect the presence of a species on a plot does not necessarily indicate absence.

The ability to document presence or absence of species in a target area depends on a number of factors such as spatial and temporal definitions of the target population, spatial arrangement of species, species abundances, plot size, and methods for selection and survey of plots. These various influences may be placed more or less into two categories based on spatial scale. The first category contains more frame-level factors affecting selection of plots occupied by the species of interest, whereas the second is composed of more plot-level factors affecting detectability of species within selected units. These categories, and their component factors, are interrelated and often overlap. Because validity of presence–absence information is directly influenced by both frame- and plot-level factors, we will consider these factors in some detail.

4.2.1.1. Frame-Level Factors Affecting Confirmation of Presence or Absence

Defining the Target Population

All else being equal, large areas will contain more species and individuals than small areas. Thus, size of a target population and resulting sampling frame will, in part, dictate the number of both species and individuals available for inclusion in a community survey. In addition, length of time a survey is conducted may have a large influence on number of species and individuals potentially included in a community survey. A survey over a long time period increases the likelihood of encountering additional species in the sampled area compared to a shorter survey. This is especially true for cyclic species, i.e., those that peak in numbers in more or less regular time intervals. Other complications include temporary migrants, i.e., lack of geographic enclosure that allows some species to move on and off the study area.

Spatial Pattern of Animals

How individuals are spatially arranged within a sampling frame directly influences their chances of being included in a selected plot. Species that are randomly distributed have an equal chance of being included in any plot, whereas those that are clumped or aggregated have a greater chance of being in some plots than others. In other words, the more clustered a species is, the fewer plots in which it will occur and the less likely it will be included in a given sample of plots. Therefore, highly clustered species are, on average, less likely to be encountered in a small sample from a sampling frame than less clustered ones.

There is usually at least some aggregation of animals in most animal populations (Cole, 1946). Why are most species spatially clustered? Many species are more likely to be in certain habitats than others. For instance, some species tend to occur in riparian zones but not in surrounding areas. Fish species may congregate more in pools than in riffles. Even habitat generalists may occur in high densities within some habitats (Van Horne, 1983). In addition, species that exhibit any type of grouping behavior will obviously be aggregated.

Commonness or Rareness of Species

How many members of each species are contained in a sampling frame will have significant bearing on their detectability. As we stated previously, there are usually many rare but few abundant species in the average natural community. This means that most of the species present in a sampling frame probably occupy relatively few plots, assuming plots are small relative to size of the frame. If species are rare and clustered they will occupy even fewer plots.

A given set of plots will very likely be dominated by a few common species with only an occasional rare or less common species present. In Fig. 4.2 we see that 95 individuals of species A occupy 34 plots, whereas 5 individuals of species B only occupy 3. Thus, the chance of selecting a plot that contained species A would be about 1 in 3, but that for species B would only be about 1 in 33. Assuming complete detectability of individuals, one would have to, on average, randomly select about 8 plots to be 95% confident of encountering species A, but about 98 plots to be 95% confident of detecting species B. Let us see how these values were derived. We know the probability of encountering species A is 1/3 based on its occurrence on 34 of 100 plots. Then, the probability that a plot does not

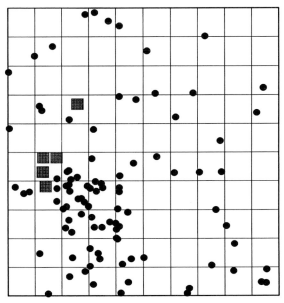

Figure 4.2 Clumped distributions of species A (95 individuals; circles) and species B (5 individuals; squares).

contain species A is 2/3. To be 95% confident that we will encounter species A in our sample of plots, we want the probability of never encountering species A to be 0.05, so

$$\left(\frac{2}{3}\right)^u = 0.05,$$

giving $u = 7.388$, or (rounding up) $u = 8$. The same calculation for species B yields $u = 97.356$.

Plot Design/Spatial Scale

On average, the larger the plot, the more species will be contained within it (Dice, 1952). Similarly, the more plots selected for a sample, the more likely a sample will contain more species. Further, certain plot shapes may be more efficient at including individuals than others. For instance, a sampling frame composed of long and narrow plots tends to contain fewer "empty" plots than one with square plots when the underlying distribution of species is clumped (see discussion in Chapter 2).

Plot size is dependent on the spatial scale at which a group of species is surveyed. Moreover, spatial scale is strongly related to a species' body size (i.e., body mass), and body size is correlated fairly strongly with home range size (McNab, 1963; Holling, 1992). Obviously, an optimal size for a plot will differ for mice and deer. There will be more home ranges of mice for a given plot of ground than there will be for deer, which translates into differences in numbers of individuals per unit area. All else being equal, the ability to detect a member of a small species (i.e., with many individuals) on a plot will be much greater than that for a large species (i.e., with a few individuals).

Plot Selection Method

The random selection method used for choosing which plots to survey could have a significant impact on the chance of encountering a given species. One could stratify an area by species' habitat affiliations (Fig. 4.3). However, this would require prior life history information, or at least historical records of occurrence, for the species in question. Species would have to be grouped by habitat preference with different sampling frames constructed by group. This would obviously remove the possibility of collecting distributional data in a single survey.

The advantage of using stratified random sampling to increase the chance of selecting a plot that contains a member of a target species is shown in

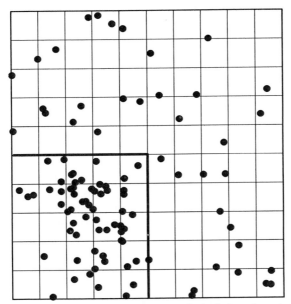

Figure 4.3 Two strata, separated by a thick line, hypothetically based on a habitat affiliation for a given species. The probability of selecting a plot containing an individual of a given species is much greater in the smaller stratum (preferred habitat) than for the larger stratum.

Fig. 4.3. We use the simplest case of a single species to demonstrate this advantage, but the same principle applies for groups of species as well. Note that there are 50 empty plots out of 100 in the sampling frame so that without stratification the chance of picking an occupied plot would be, on average, 1 in 2 or 50%. However, stratifying by habitat yields one stratum in which 20 of 25 (80%) of the plots contain at least one individual and another stratum that has only 30 of 75 (40%) plots that are occupied. Sampling effort then can be concentrated in the small stratum to increase the probability of selecting occupied units.

Stratification of the sampling frame by some attribute associated with spatial clustering of species may be effective when applied to a single species, but probably cannot be implemented for more than several related species. That is, the greater the number of target species, the greater the possible number of different cluster patterns. One species may be clustered in riparian habitat whereas another may be aggregated in grasslands. There is simply no way to have one optimum and worthwhile stratification design that can account for the many different patterns of distribution; the aggregation of a species is a species trait, and cannot be accommodated for all the different species with a single stratification design or plot size. Thus, one ends up with a suboptimal allocation of effort for all species. That is, the reasonable approach under these circumstances is to select a simple random sample of the entire sampling frame, which means there is no stratification and no optimum allocation of sampling effort.

4.2.1.2. Plot-Level Factors Affecting Confirmation of Presence or Absence

Up to this point, we have dealt mostly with factors affecting the probability of selecting occupied plots. We have largely ignored the detectability of species on these plots. However, there are few, if any, situations in which one can reasonably assume that all species are detected during a given survey. Moreover, consistently misrecording a "0" instead of a "1" for a species will yield biased and misleading results. Therefore, selecting plots that contain a host of species is not enough—one must be able to reliably confirm their presence as well.

Confirming presence of a species may be done by observing, hearing, or capturing an animal, or by noting a sign of its presence such as tracks, feces, or other physical cues. Confirming its absence is more problematic. One can never be sure that a species is truly absent unless it has a 100% chance of being detected when it is present. However, failure to detect a species over repeated surveys may provide some level of certainty regarding its absence. In this section we will discuss a general approach to computing the number of repeated surveys required to reach a specified level of certainty regarding presence or absence of a species, evaluate the usefulness of this approach, and discuss other factors related to species detection during plot surveys.

Number and Length of Plot Surveys

The greater the number of surveys conducted on a plot, the more likely a given species will be detected, and the more species that will be detected. This relationship also may be expressed in terms of time (Fisher *et al.*, 1943; Preston, 1948). That is, more species will likely be detected either when individual surveys are longer (e.g., operating a fish weir continuously for 7 days instead of 1) and/or there are more individual surveys conducted (e.g., operating a fish weir over three non-overlapping 7-day periods instead of a single 7-day period). Whether expressed in number or length of surveys, survey effort is an important factor influencing species detectability. For consistency, we will define this effort as number of surveys and assume that time for each survey is similar or the same.

The number of surveys necessary to confirm the presence or absence of a species on a given plot will depend on the number and detectability of the species. Species that are numerous and easy to detect will require fewer surveys to confirm their presence or absence than those that are not. If we assume that the detection probability of individuals of a given species is constant and known (or reliably estimated) and detections are independent of one another, then we can calculate the probability of failing to detect any individuals that are present via $(1 - p)^{N_i}$, where p is the detection probability and N_i is the number of individuals of a given species in the ith plot. For example, if a given plot had five individuals of a species and each individual had a 25% chance of detection, then the probability of failing to detect all individuals during a survey would be $(1 - 0.25)^5 = 0.237$. If we have some estimate of density of individuals for a given species, we could adjust the plot size to contain enough individuals so that there would be a high probability of detecting at least one individual. In the previous example, plots would have to be large enough to contain eight individuals for us to have a 90% probability of detecting at least one individual.

There are a number of difficulties in using the above approach. First of all, we usually do not have an estimate of density. Even if we did, densities will certainly vary among species. Second, we rarely have an estimate of the detection probability for a given species within a specific area and time period. Although detection probabilities for a few species may be available from previous work, most will probably have to be assessed beforehand for the vast majority of species. That is, selected plots would have to be intensively surveyed for presence of a particular species, and then results from occupied plots used to assess how many plot surveys were needed to detect this species. Such an effort would be very cost- and labor-intensive, which would defeat the advantage of conducting a presence–absence survey. Third, it is very unlikely that the detection probability will be constant across individuals within a given species. Moreover, detection probabilities will undoubtedly vary among species, which means that more surveys will be required for some species than others. Consequently, plots that normally would have required relatively few repeated counts for one species may need many more when other species of interest are added to the survey. Finally, detections will probably not be independent, especially for counts based on vocalizations of territorial species. Thus, computing the number and length of plot surveys for assessing presence–absence for a given species is far from straightforward; it becomes even more complex for multiple species.

Detection Method/Species of Interest

Attempting to gather presence–absence data at the community level requires a variety of detection methods. The effectiveness of these methods is dictated by characteristics of a given species. For instance, searching for animal sign, like tracks or feces, requires target animals large enough to produce sign that is readily visible. A trapping grid laid out for small mammals is obviously useless for trapping larger mammals or birds. Surveys based on species' vocalizations are best conducted during early daylight hours for songbirds (Skirven, 1981) and during late evening hours for owls (Morrell *et al.*, 1991). Finally, approaches for detecting terrestrial species will likely differ from those for aquatic species. There is no avoiding the need for a number, possibly a large number, of different methods for assessing a community.

Movements of Individuals

Animals do not recognize plot boundaries except in the rare case when these boundaries coincide with some impassable barrier. As discussed in a previous chapter, treating plot boundaries as barriers, or assuming that animal movements are small compared to plot size, is essential to satisfy the closure assumption. Even if presence is based on some type of physical sign, such as tracks, one individual could easily leave such sign in multiple plots particularly if its average daily movements are large compared to plot size (Fig. 4.4). Alternatively, species whose daily movements cover multiple plots will be much less likely to be detected on a single plot than species that have much more limited movements, and hence more individuals occurring on a given plot.

Some species may occupy different plots depending on time of day. For instance, some lake fish may seek deeper water during the day but may move up near the surface at night to feed (Bone *et al.*, 1995). In general, larger nocturnal species may be sedentary during the day, whereas they may roam fairly widely at night. Again, the multitude of behavioral traits and movement patterns in a community of species make it impossible for one or even a few survey designs to adequately assess status of each.

Nonsampling Error

Any time an incomplete count is conducted there is a chance for bias to occur. For presence–absence surveys, this means failing to record a species that is present or recording a species as present when it is in fact absent. Or it could simply be a case of misrecording a number on a data

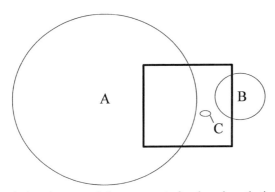

Figure 4.4 Boundaries of average daily movements for three hypothetical species (A, B, and C) in relation to size and placement of a selected plot (dark square). Area covered by species C is small compared to plot size, so there is little chance of double counting individuals across adjacent plots or missing their presence altogether (i.e., many individuals can "fit" within one plot). Conversely, fewer individuals of species B can fit into one plot, whereas one plot could fit inside the area of daily movements for species A, which makes both species susceptible to being either double-counted across adjacent plots or missed.

sheet. Whatever the reason, the result is incorrectly recording a 0 for a 1 or a 1 for a 0. We will focus on the first two causes because the third can be corrected through careful recording of data. A more complete discussion of bias is given in Chapter 1 and in chapters addressing specific animal groups.

Failure to detect the presence of a species could be due to varying efficiencies at which a method detects different species. For example, electrofishing is more effective on some fish species than others and on larger individuals than smaller ones (Zalewski and Cowx, 1990). Similarly, pitfall trapping for terrestrial species is size-biased (Gibbons *et al.*, 1977). Techniques based on visual and audio stimuli, such as for large, visible animals and singing birds, will be weighted toward more common and conspicuous species; hence, visual or audio signals from less common or rare species could possibly be "swamped" by more abundant ones.

Detection methods will usually function less effectively in different environmental circumstances. Inclement weather conditions could affect the behavior of target species in ways that make them less detectable. Weather also could hamper an observer's ability to detect species. Rainfall will lessen vocalization rates of birds and reduce visibility, and may affect activity patterns of a number of species (Verner, 1985); rainfall could trigger spring movements of temperate herpetofauna and thus enhance detectability (Vogt and Hine, 1982). Similarly, high wind could adversely affect detectability of songbirds by suppressing their singing rates while enhancing observability of soaring raptors by offering better flying conditions. This is another example of the fundamental difficulty of attempting to survey many species at once.

4.2.2. OTHER METHODS FOR ASSESSING DISTRIBUTION

The most basic information available for species occurrences is usually based on records of previous sightings, trapping returns, harvest returns, or other sources. That is, these data are haphazardly gathered so that they provide no basis for inferences about a specified population, but could be useful for providing some idea about which species may occur within a defined area. As mentioned previously, historical records could be used in stratifying a sampling frame by habitat if species were documented in certain habitats and not others. However, previous records of occurrence only provide rough baseline information; data need to be properly collected before any type of inferences may be made about a species, much less a community.

Distributional information also may be obtained during the collection of either indices or estimates of abundance/density on selected plots. Both differ from a presence–absence surveys in that their aim is to quantify numbers, and not simply presence, of individuals within each selected plot. In other words, spatial distributional data are obtained as a by-product of gathering more detailed information. Hence, abundance/density techniques require more time, effort, and money to acquire distributional information than one designed specifically to gather these data.

4.3. ASSESSMENT OF ABUNDANCE OR DENSITY

Attempting to assess either abundance or density of species at the community level presents even greater difficulties than those encountered in assessing distribution. We still must contend with problems affecting confirmation of presence or absence of species, but now also must address those affecting our ability to obtain useful counts of individuals for each species. One could use presence–absence data as an index to abundance or density, and thus avoid the need for plot counts, but this index must be linearly correlated with animal numbers to be useful. The more intensive relative index of abundance/density requires counts of animals to be done in a standardized fashion and assumes that changes in abundance or density are proportional across space and time. An even more labor-intensive and costly approach is to obtain unbiased estimates of abundance or density.

4.3.1. PRESENCE–ABSENCE

If presence–absence data are collected to assess distribution then one would like to be able to use the same data to assess density, too. This sounds straightforward enough but in actuality its validity depends on the relationship between proportion of occupied plots and average density of individuals within plots for a given species. If this relationship is linear or nearly linear then the index is a good indicator of density. If not, presence–absence data will yield misleading results. The relationship is, in fact, nonlinear (see Chapter 3).

Aside from the nonlinearity issue, another potential difficulty with using presence–absence data to assess abundance is that a more widely distributed species does not necessarily signify a more abundant one. As we discussed earlier, a species that is more clumped in distribution will occupy fewer plots than one that is more randomly distributed, even if both occur in the same numbers. Consequently, a lower estimated proportion of occupied plots ($\hat{p}$), and hence a lower density index, may or may not indicate lower numbers of animals. This applies to a community as well as to single species. In order to assess precision of the point estimate, $\hat{p}$, we need to calculate its estimated variance, $\hat{\text{Var}}(\hat{p})$. For simple random sampling, this may be done using the following formula derived from the binomial distribution (Cochran, 1977, p. 52):

$$\hat{\text{Var}}(\hat{p}) = \left(1 - \frac{u}{U}\right) \frac{\hat{p}(1 - \hat{p})}{u - 1}. \tag{4.1}$$

In Eq. (4.1), the $u - 1$ term in the denominator is often shown as just u. The difference is negligible for reasonable sample sizes. The formula in Eq. (4.1) is unbiased as $u \to \infty$. We may next wish to calculate a sample size required for a simple random sample to be within some prespecified percentage from the true proportion of a given species about $100(1 - \alpha)\%$ of the time, where e is relative error and α sets our confidence level. Note that a relative error of 10% of 100 would place our estimate between 90 and 110. This approach is useful for setting monitoring goals in which a certain level of precision is desired. The relevant formula (Cochran, 1977, pp. 75–76) is

$$u = \frac{\dfrac{t^2(1 - \hat{p})}{e^2 \hat{p}}}{1 + \dfrac{1}{U}\left(\dfrac{t^2(1 - \hat{p})}{e^2 \hat{p}} - 1\right)}, \tag{4.2}$$

where t is the critical value associated with a 2-tailed (i.e., $\alpha/2$) t distribution based on $u - 1$ degrees of freedom. Because the value of t is based on the number of selected plots (u), which is unknown beforehand, we must pick a value for t, insert this value into Eq. (4.1), and solve iteratively for u. The initial value used for t is the one that has infinite degrees of freedom at a specified $1 - \alpha/2$, which is equivalent to a z value. The final solution for u is obtained when successive iterations produce the same value. If the iterations converge on two values, then the larger of these values should be chosen. This process is demonstrated in both Example 4.1 and Example 4.2.

Because a clumped arrangement of animals occupies fewer plots, more plots are needed to satisfy precision requirements than for a random arrangement. For instance, Example 4.1 only requires 25 surveyed plots, whereas Example 4.2 requires more than twice that many. This idea may be extended to spatial arrangements of different species, i.e., more clustered species will require more plots to satisfy precision requirements. Thus, there is no way to achieve an optimal frame design and selection scheme when attempting to survey so many different species, each with their own spatial arrangement, as in a community approach.

Example 4.1. Estimating the Proportion of Occupied Plots, Its Associated Variance Estimate, and Number of Plots Needed To Be within 20% of the True Proportion about 90% of the Time for a Random Distribution of Animals

A biologist chose a simple random sample of $u = 10$ plots from a sampling frame consisting of 100 equal-sized plots (Fig. 4.5). He surveyed them for presence of species A, which produced the following results [Eqs. (1.2) and (4.1)]:

$$\hat{p} = \frac{0 + 0 + 1 + 0 + 1 + 1 + 1 + 1 + 1 + 1}{10} = 0.7$$

$$\text{V\^ar}(\hat{p}) = \left(1 - \frac{10}{100}\right)\frac{(0.7)(0.3)}{(10 - 1)} = 0.021.$$

He used this information to compute sample sizes [Eq. (4.2)] needed to meet the above stated requirements by using $e = 0.20$ and $\alpha = 0.1$, and solving iteratively for u by starting with $t_{(0.95,\,\infty)} = 1.645$:

$$u = \frac{\dfrac{(1.645)^2(0.3)}{(0.2)^2(0.7)}}{1 + \dfrac{1}{100}\left(\dfrac{[1.645]^2[0.3]}{[0.2]^2[0.7]} - 1\right)} = 22.65 \doteq 23.$$

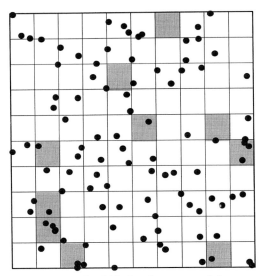

Figure 4.5 A simple random sample of 10 plots (gray squares) from a sampling frame with an underlying random distribution of individuals.

He uses this result to obtain a new t value, $t_{(0.95, 22)} = 1.717$, and inserts this new value into the same equation:

$$u = \frac{\dfrac{(1.717)^2(0.3)}{(0.2)^2(0.7)}}{1 + \dfrac{1}{100}\left(\dfrac{[1.717]^2[0.3]}{[0.2]^2[0.7]} - 1\right)} = 24.19 \doteq 25.$$

He repeats the same procedure with $t_{(0.95, 24)} = 1.711$, which results in u again equaling 25. Therefore, he must select 25 plots in order to satisfy his desired precision.

Example 4.2. Estimating Proportion of Occupied Plots, Its Associated Variance Estimate, and Number of Plots Needed To Be Within 20% of the True Proportion about 90% of the Time for a Clumped Distribution of Animals

A biologist chose a simple random sample of $u = 10$ plots from a sampling frame consisting of 100 equal-sized plots (Fig. 4.6). She surveyed them for presence of species A, which produced the following results [Eqs. (1.2) and (4.1)]:

$$\hat{p} = \frac{0 + 0 + 0 + 0 + 0 + 0 + 1 + 1 + 1 + 1}{10} = 0.4$$

$$\hat{V}ar(\hat{p}) = \left(\frac{100 - 10}{100}\right)\frac{(0.4)(0.6)}{(10 - 1)} = 0.024.$$

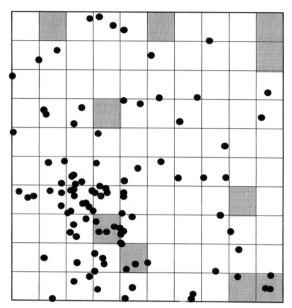

Figure 4.6 A simple random sample of 10 plots (gray squares) from a sampling frame containing a clumped distribution of individuals.

She used this information to compute sample sizes [Eq. (4.2)] needed to meet the above-stated requirements by using $e = 0.20$ and $\alpha = 0.1$, and solving iteratively for u by starting with $t_{(0.95, \infty)} = 1.645$:

$$u = \frac{\dfrac{(1.645)^2(0.6)}{(0.2)^2(0.4)}}{1 + \dfrac{1}{100}\left(\dfrac{[1.645]^2[0.6]}{[0.2]^2[0.4]} - 1\right)} = 50.62 \doteq 51.$$

She used this result to obtain a new t value, $t_{(0.95, 50)} = 1.676$, inserted this new value into the same equation, and computed a value of 51.55 or 52 plots. She repeated the same procedure with $t_{(0.95, 51)} = 1.675$, which resulted in u again equaling 52. Therefore, she must select 52 plots in order to satisfy her desired precision.

A modification of the presence–absence approach that may lead to unbiased density estimates is to reduce the size of plots so that each contains at most one individual from a given species (Dice, 1952). In this way, proportion of occupied plots could be converted to either number or density of individuals, assuming that animals are completely detectable. However, the cost of this approach is increased logistical effort to locate the numerous small plots. Because of differences in species' sizes and spatial scales, this approach is not plausible for a community survey.

4.3.2. INDEX OF RELATIVE ABUNDANCE OR DENSITY

Collecting information on selected plots to calculate indices of relative abundance or density requires more time and effort than a presence–absence approach, but less than directly gathering abundance or density information. Relative indices should be based on data collected in a standardized way to increase integrity of comparisons between two or more populations across space or the same population across time. The key assumption here is that either changes in or differences between relative indices must be proportional across space and/or time for valid inferences to be made.

In contrast to presence–absence techniques, relative index methods record all animals encountered during a survey of a plot, but do not account for potential differences in detectability either among individuals or species. Thus, these methods are subject to errors in detectability associated with counting multiple individuals in addition to those affecting confirmation of presence. Therefore, inferences based on relative indices are still problematic and also are more costly to obtain than presence–absence data.

Karr (1981) proposed a method, called the *index of biotic integrity* (IBI), for assessing and comparing the health and stability of fish communities, especially in degraded streams. The IBI incorporates species richness and composition information from relative index data with ecological data based on presence–absence of species in certain trophic categories to produce an overall ranking for the community of interest. A number of studies have evaluated the use of the IBI for midwestern and eastern streams (Fausch *et al.*, 1984; Angermeier and Karr, 1986; Leonard and Orth, 1986; Karr *et al.*, 1987), and reported that it was a useful index to stream degradation. The IBI differs from other indices in that it includes functional attributes of a community in its assessment criteria. Such an approach is seemingly well-matched for aquatic communities, but would be more difficult to apply to terrestrial systems. One possible exception is the use of guilds in breeding bird communities, although defining a "guild" is problematic (Verner, 1984; Szaro, 1986). Moreover, IBI suffers from the same problems as other indices regarding factors affecting detection of a given species within an area of interest.

4.3.3. Abundance or Density Estimation

The most costly and time-consuming way to assess population status is through collection of unbiased abundance/density estimates within selected plots. Methods that produce these estimates account for incomplete detectability of individuals. However, these techniques are essentially limited to use on a few of the more common species because of the necessity for a relatively large number of sampled individuals required to produce adequate estimates of abundance/density (see Chapter 3), and the high cost of collecting such information.

Another difficulty with conducting multispecies surveys, much less community surveys, is the lack of a precise search image. When two species require different methods of detection, such as on a line transect survey, combining methods will result in one or both being poorly sampled. Better results for each will usually result if an investigator concentrates on one species at a time. This problem becomes really difficult during aerial surveys, where the observer is expected to observe multiple species within a limited time period dictated by aircraft speed. The lack of one search image to concentrate on, the lack of time to study the habitat, and the resulting inefficiency of the search probably means that most of the species included in the survey are poorly sampled.

As mentioned before, the number of plots that must be selected to satisfy some predefined level of precision will vary among species depending on their spatial arrangements. To see this, we apply Eq. (1.6) and Eq. (1.8) from Chapter 1 to results of a simple random sample from the random distribution of individuals in Example 4.1, which yields an estimated variance of

$$\hat{\mathrm{Var}}(\hat{N}) = 100^2 \left(1 - \frac{10}{100}\right) \frac{2.044}{10} = 1840,$$

with

$$\hat{S}^2_{N_i} = \frac{(0 - 1.4)^2 + \cdots + (1 - 1.4)^2}{10 - 1} = 2.044,$$

whereas a simple random sample from the clumped distribution in Example 4.2 produces an estimated variance of

$$\hat{\mathrm{Var}}(\hat{N}) = 100^2 \left(1 - \frac{10}{100}\right) \frac{2.222}{10} = 2000,$$

with

$$\hat{S}^2_{N_i} = \frac{(0-1)^2 + \cdots + (3-1)^2}{10-1} = 2.222.$$

There will be even greater differences between species that have few individuals and are more restricted in distribution. From a monitoring standpoint, larger variances make it more difficult to detect a trend (see Chapter 5). Computing estimated sample sizes to satisfy the level of precision specified in these two examples yields $u = 43$ (random) and $u = 61$ (clumped). Thus, because there are so many possible different spatial arrangements and abundances/densities for species in a given community, attempting to incorporate these species into one or even a few sampling designs of similar sample sizes would be futile.

4.4. RECOMMENDATIONS

At first glance, a community survey would seem to be a logical approach for assessing status of all species of interest within some area during some time interval. A seemingly plausible first step may be to gather presence–absence data to assess species occurrences to serve as baseline information. Then, more specific information related to species numbers or densities could be collected later depending on funding availability. Unfortunately, although such an approach seems reasonable, in reality there are a multitude of problems associated with obtaining species occurrence data (see sections 4.2.1.1 and 4.2.1.2), much less abundance information.

Insufficient information on resident species, and the wide variety of them, make construction of usable sampling frames extremely difficult, if not impossible. Most species are rare, but a few are relatively abundant. Moreover, there are a number of different spatial scales that must be considered in designing frames. For the majority of species simply recording their presence on selected plots, if it can be confirmed at all, will undoubtedly require more than one, and possibly many, plot surveys. Sampling results, even in presence–absence surveys, will likely be dominated by a few common species. Hence, attempting to gather the simplest level of information (i.e., species distributions) is far from simple at the community level.

Despite their apparent appeal, community surveys are simply not a viable option for rigorous monitoring under current technology. Even if they were, we would not recommend collecting spatial distributional data separately from abundance or density information because distributional data alone are not adequate for detecting trends in animal numbers, which is the whole point of monitoring abundances of vertebrate species. Therefore, we generally advise a single-species or, in certain circumstances, a multispecies approach in which valid abundance estimates can be collected as part of an inferential monitoring program. Such approaches would necessarily have to concentrate on more common species under normal funding constraints or possibly on a rare species or species of special concern in certain situations. Other species would have to be "roughly" monitored using some type of index approach. The IBI method may be a suitable choice for stream communities, and probably other aquatic communities as well. Perhaps index-based monitoring approaches will be able to detect large trends in community numbers and composition, given enough data, but this is far from guaranteed because of the potentially confounding effect of biased estimates caused by incomplete detectability of individuals and species.

LITERATURE CITED

Angermeier, P. L., and Karr, J. R. (1986). Applying an index of biotic integrity based on stream-fish communities: Considerations in sampling and interpretation. *N. Am. J. Fish. Mgmt.* **6:** 418–429.

Bone, Q., Marshall, N. B., and Blaxter, J. H. S. (1995). *Biology of Fishes,* second ed. Blackie, London.

Cochran, W. G. (1977). *Sampling Techniques,* third ed. Wiley, New York.

Cole, L. C. (1946). A study of the cryptozoa of an Illinois woodland. *Ecol. Monogr.* **16:** 50–86.

Dice, L. R. (1952). *Natural Communities.* University of Michigan Press, Ann Arbor, MI.

Fausch, K. D., Karr, J. R., and Yant, P. R. (1984). Regional application of an index of biotic integrity based on stream fish communities. *Trans. Am. Fish. Soc.* **113:** 39–55.

Fisher, R. A., Corbet, A. S., and Williams, C. B. (1943). The relation between the number of species and the number of individuals in a random sample of an animal population. *J. Anim. Ecol.* **12:** 42–58.

Gibbons, J. W., Coker, J. W., and Murphy, T. M., Jr. (1977). Selected aspects of the life history of the rainbow snake (*Farancia erytrogramma*). *Herpetologica* **33:** 276–281.

Gotelli, N. J., and Graves, G. R. (1996). *Null Models in Ecology.* Smithsonian Institution Press, Washington, DC.

Holling, C. S. (1992). Cross-scale morphology, geometry, and dynamics of ecosystems. *Ecol. Monogr.* **62:** 447–502.

Karr, J. R. (1981). Assessment of biotic integrity using fish communities. *Fisheries* **6:** 21–27.

Karr, J. R., Yant, P. R., Fausch, K. D., and Schlosser, I. J. (1987). Spatial and temporal variability of index of biotic integrity in three midwestern streams. *Trans. Am. Fish. Soc.* **116:** 1–11.

Leonard, P. M., and Orth, D. J. (1986). Application and testing of an index of biotic integrity in small, coolwater streams. *Trans. Am. Fish. Soc.* **115:** 401–414.

McNab, B. K. (1963). Bioenergetics and the determination of home range size. *Am. Natur.* **97:** 133–140.

Morrell, T. E., Yahner, R. H., and Harkness, W. L. (1991). Factors affecting detection of great horned owls by using broadcast vocalizations. *Wildl. Soc. Bull.* **19:** 481–488.

Pielou, E. C. (1977). *Mathematical Ecology.* Wiley, New York.

Preston, F. W. (1948). The commonness, and rarity, of species. *Ecology* **29:** 254–283.

Skirven, A. A. (1981). Effect of time of day and time of season on the number of observations and density estimates of breeding birds. *Stud. Avian Biol.* **6:** 271–274.

Szaro, R. C. (1986). Guild management: an evaluation of avian guilds as a predictive tool. *Environ. Mgmt.* **10:** 681–688.

Van Horne, B. (1983). Density as a misleading indicator of habitat quality. *J. Wildl. Mgmt.* **47:** 893–901.

Verner, J. (1984). The guild concept applied to management of bird populations. *Environ. Mgmt.* **8:** 1–14.

Verner, J. (1985). Assessment of counting techniques. In *Current Ornithology* (R. F. Johnston, Ed.), Vol. 2., pp. 247–302. Plenum, New York.

Vogt, R. C., and Hine, R. L. (1982). Evaluation of techniques for assessment of amphibian and reptile populations in Wisconsin. In *Herpetological Communities* (N. J. Scott, Jr., Ed.), pp. 201–217. U.S.D.I. Fish Wildl. Serv. Washington, DC. [Wildl. Res. Rep. 13]

Zalewski, M., and Cowx, I. G. (1990). Factors affecting the efficiency of electric fishing. In *Fishing with Electricity* (I. G. Cowx and P. Lamarque, Eds.), pp. 89–111. Fishing News Books, Blackwell Scientific, Oxford, UK.

Chapter 5

Detection of a Trend in Population Estimates

5.1. **Types of Trends**
5.2. **Variance Components**
5.3. **Testing for Trend**
    5.3.1. Graphical Methods
    5.3.2. Regression Methods

5.3.3. Randomization Methods
5.3.4. Nonparametric Methods
5.4. **General Comments**
Literature Cited

The main objective of most biological monitoring surveys is to detect changes or a trend in the population size, survival, or recruitment of certain species. The purpose may be to link the observed decline in habitat with population size, such as reduction in the amount and patch size of sagebrush habitat with a decline in sage grouse population size. Conversely, monitoring may be performed to document that changes in grazing practices result in an increase in cutthroat trout populations.

In the first section of this chapter, we will discuss some types of trends that might occur in population size. To be effective in detecting a trend, we should have some ideas about what types of trends might occur in a time series of population sizes. In the next section, we will discuss the sources of variation that make the detection of a trend difficult because of the resulting stochasticity of the observations. Finally, we will examine some statistical approaches to detecting a trend. The most common approach is based on regression, which we will discuss. In addition, we will explore randomization of the data with regression and a nonparametric procedure based on ranks that are robust to violations of assumptions necessary for regression methods to be valid.

5.1 TYPES OF TRENDS

Figure 5.1 shows plots demonstrating different types of trends (adapted from Gilbert 1987, p. 205). The top graph is simply a random sequence of points (Fig. 5.1a), where no trend is present. Figure 5.1b is also a time series with no trend, but with more random noise. We always expect some random noise in our measurements, so the idea that fluctuations along the sequence of measurements are due to random (unassignable) sources is introduced immediately. Next (Fig. 5.1c) is a linear downward trend with randomness. Contrast this to an exponential downward trend with randomness (Fig. 5.1d). In Fig. 5.1c, the decline is a constant amount between years, with random noise associated with each measurement. A model that describes this trend might be (Gerrodette, 1987)

$$N_i = N_1[1 - (i - 1)r], \tag{5.1}$$

where N_i is the number of animals in the population at time i, and r is the rate constant, such that an rN_1 reduction takes place each time interval. In Fig. 5.1d, the decline is proportional to the number currently present (Gerrodette, 1987),

$$N_i = N_1(1 - r)^{i-1}, \tag{5.2}$$

so that the decline at each time interval is rN_i and hence depends on the population size at time i.

To clarify the difference between Eqs. (5.1) and (5.2), consider the following example of the model in Eq. (5.1): $N_1 = 100$ and $r = 0.1$. Then the sequence of population sizes from Eq. (5.1) for N_2 to N_5 are 90, 80, 70, and 60. Each year, $rN_1 = 10$ fewer animals exist in the population. A plot of N_i vs i is a straight line. For the same values of N_1 and r, the population sizes from Eq. (5.2) for N_2 to N_5 are 90, 81, 72.9, and 65.6. Each year, 10% fewer animals exist in the population. A plot of $\ln(N_i)$ versus i is a straight line.

Another possibility that has to be considered is a cycle (Fig. 5.1e), and then a cycle with a downward trend (Fig. 5.1f). Cycles of population size are well documented (Keith, 1963). With a population that is cyclic, we might conclude that a trend exists during the upward or downward portions of the curve, but that no trend is present during the peak or trough periods.

Keep in mind some of the complicating factors that may cause a time series of population estimates to appear to have a trend when they do not. Changes in methodology or personnel can cause changes in the estimates of population size through time, but these suggested changes are just due to changes in monitoring procedures. For example, early estimates in the time series may have been made with population estimation methods lack-

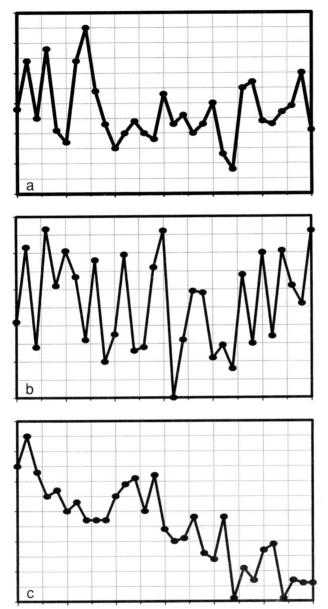

Figure 5.1 (a) Random observations with no trend. (b) Random observations with greater fluctuation than (a). (c) Randomness with a linear trend. (d) Randomness with an exponential (multiplicative) trend. (e) Cyclic trend with random variation. (f) Cycle with downward trend and randomness.

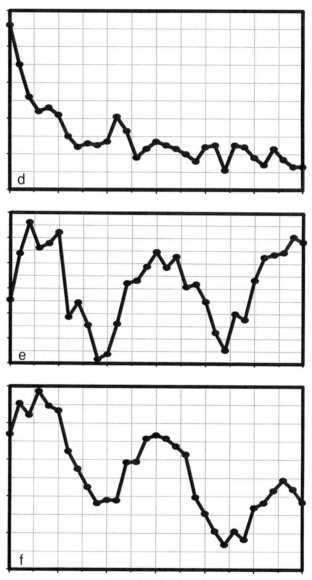

Figure 5.1 (*Continued*)

ing the rigor of the methods used later in the series. As a result, the early estimates are biased high, so that a decline seems obvious.

5.2. VARIANCE COMPONENTS

In this section, we discuss sources of variation that must be considered to make inferences from data when trying to detect trends. Three sources of variation must be considered: sampling variation, temporal variation in the population dynamics process, and spatial variation in the dynamics of the population across space. The latter two sources often are referred to as process variation, i.e., variation in the population dynamics process associated with environmental variation (such as rainfall, temperature, community succession, fires, or elevation). Methods to separate process variation from sampling variation will be presented.

Detection of a trend in a population's size requires at least two abundance estimates. For example, if the population size of Mexican spotted owls in Mesa Verde National Park is determined as 50 pairs in 1990, and as only 10 pairs in 1995, we would be concerned that a significant negative trend in the population exists during this time period, and that action must be taken to alleviate the trend. However, if the 1995 estimate was 40 pairs, we might still be concerned, but would be less confident that immediate action is required. Two sources of variation must be assessed before we are confident of our inference from these estimates.

The first source of variation is the uncertainty we have in our population estimates. We want to be sure that the two estimates are different, i.e., the difference between the two estimates is greater than would be expected from chance alone because of the sampling errors associated with each estimate. Typically, we present our uncertainty in our estimate as its variance, and use this variance to generate a confidence interval for our estimate. Suppose that the 1990 estimate of $\hat{N}_{90} = 50$ pairs has a sampling variance of $\text{Var}(\hat{N}_{90}) = 25$. Then, under the assumption of the estimate being normally distributed with a large sample size (i.e., large degrees of freedom), we would compute a 95% confidence interval as $50 \pm 1.96\sqrt{25}$, or 40.2–59.8. If the 1995 estimate was $\hat{N}_{95} = 40$ with a sampling variance of $\text{Var}(\hat{N}_{95}) = 20$, then the 95% confidence interval for this estimate is $40 \pm 1.96\sqrt{20}$, or 31.2–48.8. Based on the overlap of the two confidence intervals (Fig. 5.2), we would conclude that by chance alone, these two estimates are probably not different. We also could compute a simple test as

$$z = \frac{\hat{N}_{90} - \hat{N}_{95}}{\sqrt{\text{Var}(\hat{N}_{90}) + \text{Var}(\hat{N}_{95})}}, \tag{5.3}$$

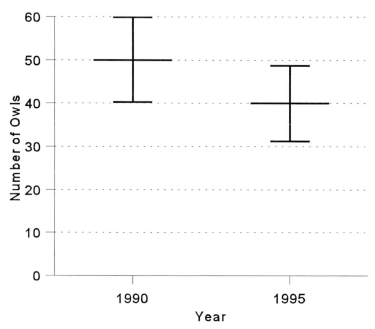

Figure 5.2   The 95% confidence intervals plotted with the 1990 and 1995 population estimates.

which for this example results in $z = 1.491$, with a probability of observing a z statistic this large or larger of $P = 0.136$. Although we might be alarmed, the chances are that 13.6 times out of 100 we would observe this large of a change just by random chance.

A variation of the previous test is commonly conducted for several reasons: (1) we often are interested in the ratio of two population estimates (rather than the difference) because a ratio represents the rate of change of the population, (2) the variance of $\hat{N}$ is usually linked to its estimate by $\hat{Var}(\hat{N}) = \hat{N}C$ (e.g., Skalski and Robson, 1992, pp. 28–29), and (3) $\ln(\hat{N})$ is more likely to be normally distributed than $\hat{N}$. Fortuitously, a log transformation provides some correction to all three of the above reasons and results in a more efficient statistical procedure. Because

$$Var[\ln(\hat{N})] = \frac{Var(\hat{N})}{\hat{N}^2},\qquad(5.4)$$

we construct the z test as

$$z = \frac{\ln(\hat{N}_{90}) - \ln(\hat{N}_{95})}{\sqrt{\hat{Var}[\ln(\hat{N}_{90})] + \hat{Var}[\ln(\hat{N}_{95})]}}\qquad(5.5)$$

to provide a more efficient (i.e., more powerful) test.

Suppose we had made a much more intensive effort in sampling the owl population, so that the sampling variances were one-half of the values observed (which would generally take about 4 times the effort). Thus, $V\hat{a}r(\hat{N}_{90}) = 12.5$ and $V\hat{a}r(\hat{N}_{95}) = 10$, giving a z statistic of 2.108 with probability value of $P = 0.035$. Now, we would conclude that the owl population was lower in 1995 than in 1990, and that this difference is unlikely due to variation in our samples, i.e., that an actual reduction in population size has taken place.

This leads us to the second variance component associated with determining whether a trend in the population is important. We would expect the size of the owl population (and any other population, for that matter) to fluctuate through time. How can we determine if this reduction is important? The answer lies in determining what the variation in the owl population has been for some period of time in the past, and then if the observed reduction is outside the range expected from this past fluctuation. Consider the example in Fig. 5.3, where the true population size (no sampling variation) is plotted. The population fluctuates around a mean of 50, but values more extreme than the range 40 to 60 are common. Note that a decline from 76 to 29 pairs occurred from 1984 to 1985, and that declines from over 50 pairs to under 40 pairs are fairly common occurrences. Thus, based on our previous example, a decline from 50 to 40 is not at all unreasonable given the past population dynamics of this hypothetical population.

Figure 5.3 Actual number of pairs of owls that exist each year. In reality, we never know these values, and can only estimate them.

To determine the level of change in population size that should receive our attention and suggest management action, we need to know something about the temporal variation in the population. The only way to estimate this variance component is to observe the population across a number of years. The exact number of years will depend on the magnitude of the temporal variation. Thus, if the population does not change much from year to year, a few observations will show this consistency. On the other hand, if the population fluctuates a lot, as in Fig. 5.3, many years of observations are needed to estimate the temporal variance. For the example in Fig. 5.3, we could compute the temporal variance as the variance of the 15 years. We find a variance of 265.7, or a standard deviation of 16.3 (Example 5.1). With a SD of 16.3, we would expect roughly 95% of the population values to be in the range of ± 2 SD of the mean population size. This inference is based on the population being stable, i.e., not having an upward or downward trend, and being roughly normally distributed. For a normal distribution, 95% of the values lie in the interval ± 2 SD of the mean. Therefore, a change of 2 SD, or 32.6, is not a particularly big change given the temporal variation observed over the 15-year period. Such a change should occur with probability greater than 1/20, or 0.05.

A complicating problem with estimating the temporal variance of a population's size is that we are seldom allowed to observe the true value of the population size. Rather, we are required to sample the population, and hence only obtain an estimate of the population size each year, with its associated sampling variance. Thus, we would need to include the 95% confidence bars on the annual estimates. As a result of this uncertainty from our sampling procedure, we would conclude that many of the year-to-year changes were not really changes because the estimates were not different. This complication leads to a further problem. If we compute the variance with the usual formula when estimates of population size replace the actual population size shown in Fig. 5.3, we obtain a variance estimate larger than the true temporal variance because our sampling uncertainty is included in the variance. For low levels of sampling effort each year, we would have a high sampling variance associated with each estimate, and as a result, we would have a high variance across years. The noise associated with our low sampling intensity would suggest that the population is fluctuating widely, when in fact the population could be constant (i.e., temporal variance is zero), and the estimated changes in the population are just due to sampling variance.

This mixture of sampling and temporal variation becomes particularly important in population viability analysis (PVA). The objective of a PVA is to estimate the probability of extinction for a population, given current size, and some idea of the variation in the population dynamics (i.e., temporal variation). If our estimate of temporal variation includes sampling variation, and the level of effort to obtain the estimates is relatively low, the high sampling variation causes our naive estimate of temporal variation to be much too large. When we apply our PVA analysis with this inflated estimate of temporal variance, we conclude that the population is much more likely to go extinct than it really is, and hence the importance of separating sampling variation from process variation.

Typically, we estimate variance components with analysis of variance (ANOVA) procedures. For the example considered here, we would have to have at least two estimates of population size for a series of years to obtain valid estimates of sampling and temporal variation. Further, typical ANOVA techniques assume that the sampling variation is constant, and so do not account for differences in levels of effort, or the fact that sampling variance is usually a function of population size. For our example, we have an estimate of sampling variance for each of our estimates, obtained from the population estimation methods considered in this manual. That is, capture–recapture, mark–resight, line transects, removal methods, and quadrat counts all produce estimates of sampling variation. Thus, we do not want to estimate sampling variation by obtaining replicate estimates, but want to use the available estimate. Therefore, we present a method of moments estimator developed in Burnham *et al.* (1987, Part 5). Skalski and Robson (1992, Chapter 2) also present a similar procedure, but do not develop the weighted estimator presented here.

Example 5.1. Population Size, Estimates, Standard Error of the Estimates, and Confidence Intervals for Owl Pairs in Fig. 5.3.

Year	Population	Estimate	Standard error	Lower 95% CI	Upper 95% CI
1980	44	40.04	5.926	28.42	51.66
1981	48	50.51	11.004	28.94	72.08
1982	61	61.36	15.278	31.42	91.31
1983	48	47.6	11.062	25.92	69.28
1984	76	95.51	18.988	58.3	132.72
1985	29	33.81	8.803	16.56	51.06
1986	60	34.39	5.804	23.01	45.76
1987	59	38.52	11.168	16.63	60.41
1988	76	84.57	21.312	42.8	126.34
1989	42	30.04	6.918	16.48	43.6
1990	29	20.29	7.529	5.54	35.05
1991	68	68.42	17.969	33.2	103.64
1992	42	45.51	13.225	19.6	71.44
1993	27	27.01	6.137	14.98	39.04

| 1994 | 72 | 71.12 | 14.511 | 42.67 | 99.56 |
| 1995 | 54 | 51.45 | 8.054 | 35.66 | 67.24 |

The variance of the $n = 16$ populations is 265.628, whereas the variance of the 16 estimates is 450.376. Sampling variation causes the estimates to have a larger variance than the actual population. The difference of these two variances is an estimate of the sampling variation, i.e., $450.376 - 265.628 = 184.748$. The square root of 184.748 is 13.592, and is the approximate mean of the 16 reported standard errors.

To obtain an unbiased estimate of the temporal variance, we must re-move the sampling variation from the estimate of the total variance. Define σ^2_{total} as the total variance, estimated for $n = 16$ estimates of owl pairs ($\hat{N}_i, i = 1980, \ldots, 1995$) as

$$\hat{\sigma}^2_{total} = \frac{\sum\limits_{i=1980}^{1995} (\hat{N}_i - \overline{N})^2}{(n - 1)} = \frac{\sum\limits_{i=1980}^{1995} \hat{N}_i^2 - \frac{\left(\sum\limits_{i=1980}^{1985} \hat{N}_i\right)^2}{n}}{(n - 1)}, \tag{5.6}$$

where the symbol indicates the estimate of the parameter. Thus, $\hat{N}_i$ are the estimates of the actual populations, N_i, and $\hat{\sigma}^2_{total}$ is an estimate of the total variance σ^2_{total}. For each estimate, $\hat{N}_i$, we also have an associated sampling variance, $\hat{\sigma}_i^2$. Then, a simple estimator of the temporal variance, σ^2_{time}, is given by

$$\hat{\sigma}^2_{time} = \hat{\sigma}^2_{total} - \frac{\sum\limits_{i=1980}^{1995} \hat{\sigma}_i^2}{n}, \tag{5.7}$$

when we can assume that all of the sampling variances, $\hat{\sigma}_i^2$, are equal. The above equation corresponds to Eq. (2.6) of Skalski and Robson (1992). When the $\hat{\sigma}_i^2$ cannot all be assumed to be equal, a more complex calcula-tion is required (Burnham et al., 1987, Section 4.3) because each estimate must be weighted by its sampling variance. We take as the weight of each estimate the reciprocal of the sum of temporal variance plus the sampling variance, $1/(\hat{\sigma}^2_{time} + \hat{\sigma}_i^2)$. That is, $\text{Var}(\hat{N}_i) = \hat{\sigma}^2_{time} + \hat{\sigma}_i^2$, so $w_i = 1/\text{Var}(\hat{N}_i) = 1/(\hat{\sigma}^2_{time} + \hat{\sigma}_i^2)$. Then, the weighted total variance is com-puted as

$$\hat{\sigma}^2_{total} = \frac{\sum\limits_{i=1980}^{1995} w_i(\hat{N}_i - \overline{N})^2}{(n - 1)\sum\limits_{i=1980}^{1995} w_i}, \tag{5.8}$$

with the mean of the estimates now computed as a weighted mean,

$$\overline{N} = \frac{\sum\limits_{i=1980}^{1995} w_i \hat{N}_i}{\sum\limits_{i=1980}^{1995} w_i}. \tag{5.9}$$

We now know that the theoretical variance $\overline{N}$ is

$$\mathrm{Var}(\overline{N}) = \mathrm{Var}\left(\frac{\sum\limits_{i=1980}^{1995} w_i \hat{N}_i}{\sum\limits_{i=1980}^{1995} w_i} \right) = \frac{1}{\sum\limits_{i=1980}^{1995} w_i} \tag{5.10}$$

and the empirical variance estimator is Eq. (5.8). Setting these two equations equal,

$$\frac{1}{\sum\limits_{i=1980}^{1995} w_i} = \frac{\sum\limits_{i=1980}^{1995} w_i (\hat{N}_i - \overline{N})^2}{(n-1) \sum\limits_{i=1980}^{1995} w_i} \tag{5.11}$$

or

$$1 = \frac{\sum\limits_{i=1980}^{1995} w_i (\hat{N}_i - \overline{N})^2}{(n-1)}. \tag{5.12}$$

Because we cannot solve for $\hat{\sigma}^2_{\text{time}}$ directly, we have to use an iterative numerical approach to estimate $\hat{\sigma}^2_{\text{time}}$. This procedure involves substituting values of $\hat{\sigma}^2_{\text{time}}$ into Eq. (5.12) via the w_i until the two sides are equal. When both sides are the same, we have our estimate of $\hat{\sigma}^2_{\text{time}}$. Using this estimate of $\hat{\sigma}^2_{\text{time}}$, we can now decide what level of change in $\hat{N}_i$ to $\hat{N}_{i+1}$ is important and deserves attention. If the change from a series of estimates is greater than $2\sqrt{\hat{\sigma}^2_{\text{time}}}$, we may want to take action.

Typically, we do not have the luxury of enough background data to estimate $\hat{\sigma}^2_{\text{time}}$, so we end up trying to evaluate whether a series of estimated population sizes is in fact signaling a decline in the population when both sampling and process variance are present. Note that just because we see a decline of the estimates for 3–4 consecutive years, we cannot be sure that the population is actually in a serious decline without knowledge of the mean population size and the temporal variation prior to the decline. Usually, however, we do not have good knowledge of the population size prior to some observed decline, and make a decision to act based on biological perceptions. Keep in mind the kinds of trends displayed in Fig. 5.1. Is the suggested trend part of a cycle, or are we observing a real change in population size? In this discussion, we have only considered temporal variation. A similar procedure can be used to separate spatial variation from sampling variation.

5.3. TESTING FOR A TREND

Regression procedures are commonly used to test a series of population estimates for a decline. Graphic methods can be the most compelling approach. We explore the use of regression estimators based on linear models, and then on exponential or multiplicative models. Finally, we examine nonparametric methods.

5.3.1. GRAPHICAL METHODS

Graphical methods are useful aids to portray the results of formal statistical tests of trends. In general, the formal test procedures can be viewed as methods that assign a probability level to the validity of the trends observed in graphs. Hence, we encourage the use of graphics to display time series. Spreadsheet programs provide easy to use graphics, and skill with their use should be acquired. A time series graph is shown in Fig. 5.4 for data from Table 5.1.

Besides the obvious graph of population size or other variables against time (i.e., years, months), cumulative sum (CUSUM) charts also can be

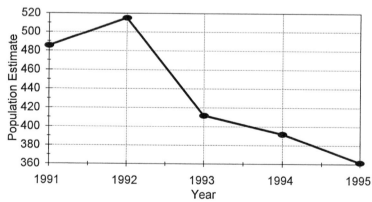

Figure 5.4   Time series of Lincoln–Petersen estimates of a meadow vole population.

Table 5.1

**Time Series of Five Annual
Population Estimates of Meadow
Voles on a 2.5-ha Grid**

Year	$\hat{N}$	$\hat{Var}(\hat{N})$
1991	485.35	2266.55
1992	515.13	4744.02
1993	411.65	2471.72
1994	391.86	2856.44
1995	361.50	2892.75
Mean	433.10	3349.40

effective for displaying trends in the time series. Changes in the mean can be detected by keeping a cumulative total of deviations from a reference value or of residuals from a realistic stochastic model of the process (Gilbert, 1987). References providing details for this approach are Page (1961, 1963), Ewan (1963), Gibra (1975), Wetherill (1977), Berthouex *et al.* (1978), and Vardeman and David (1984). For illustration, consider the data in Table 5.1, plotted as a time series in Fig. 5.4 and as a CUSUM plot in Fig. 5.5. The mean of 433.10 is subtracted from each observation, and then the cumulative sum of these differences is plotted against time.

5.3.2. REGRESSION METHODS

Regression fits a model to observed data. Typically, a straight line is fit, and is called a linear regression. Gerrodette (1987) examined the linear

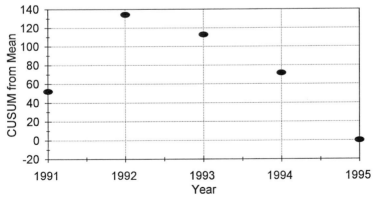

Figure 5.5 Cumulative sum (CUSUM) of deviations from the mean of 433.10 plotted with time for data in Table 5.1.

and exponential models in some detail. An exponential model is a model that assumes that the decrease (or increase) in the next observation 1 time unit ahead is proportional to the magnitude of the current observation. In contrast, a linear model decreases (or increases) by a constant amount between observations one time unit apart. Exponential models can be made into linear models by taking logarithms of the observations. Linear and exponential models are shown in Fig. 5.6.

Regression models are fit by minimizing the sum of the squared distances from the line to the observed data. Thus, the linear regression model is $N_i = \beta_0 - \beta_1 t + \varepsilon_i$, where t is time (independent variable), N_i is population size (dependent variable), β_0 is the intercept (a parameter to be estimated), and β_1 is the slope of the line (another parameter to be estimated). The ε_i are the distances between the prediction (i.e., the line) and the observed population size. The sum of the ε_i^2 is minimized to produce the estimates $\hat{\beta}_0$ and $\hat{\beta}_1$. For this model, the estimate of β_1 is computed as

$$\hat{\beta}_1 = \frac{\sum\limits_{i=1}^{n} (t_i - \bar{t})(N_i - \bar{N})}{\sum\limits_{i=1}^{n} (t_i - \bar{t})^2},$$

where the bar over a symbol indicates the mean or average of the quantity. To decide if the slope is significantly less than zero, i.e., if the slope is truly negative or if the value observed is likely to have occurred by chance alone,

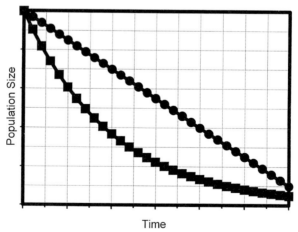

Time

Figure 5.6 Linear (dots) and exponential (squares) models of trends.

Example 5.2. Regression of the Lincoln–Petersen Estimates of Population Size Versus Time

For the untransformed data from Table 5.1, the slope of the regression line is -37.10 with SE $= 9.97$. This approach corresponds to the linear model, i.e., a constant decrease in each year. These values give a t statistic for the test of the null hypothesis of the slope equal to or greater than zero of -3.719 with 3 degrees of freedom (5 observations minus 2 parameters estimated in the model). We are interested in the alternative hypothesis of the slope less than zero, so perform a one-sided test. The probability of observing a t value this small is 0.0169, so we reject the null hypothesis in favor of the alternative that the slope is less than zero, and hence the population has declined during the interval 1991 to 1995.

The estimate of the slope of -37.10 for this model suggests that 37.1 individuals are being lost from the population each year. We can also use the estimate of the slope and its SE to compute a confidence interval on the annual loss. The confidence interval will be the estimate $\pm$ the t statistic with 3 df and $\alpha = 0.05$ times the SE, giving $-37.1 \pm 3.182 \times 9.97$, or -68.8 to -5.36.

we perform a statistical test (Example 5.2). To compute the probability level of this test with the usual procedure, several important assumptions are required. First, the ε_i are all assumed to be normally distributed with mean zero and variance σ^2. That is, each of the ε_i are assumed to have the same variance and mean. Second, they must be statistically independent of each other, i.e., the magnitude of ε_1 cannot somehow be related to the magnitude of ε_2 and so on for all values of i. If these assumptions can be met, a t statistic can be computed to test the null hypothesis that $\beta_1 = 0$, versus the alternative hypothesis that $\beta_1 < 0$. We have specified a one-sided alternative because we are typically interested in only a single direction of the trend. When a statistical test is performed, four scenarios are possible (Table 5.2). First, the null hypothesis can be true, and we can accept the null hypothesis. Second, with the null hypothesis true, we can reject it, making a Type I error. Statistical tests are constructed such that this error

Table 5.2

Four Possible Outcomes from a Statistical Test, and the Probability Level for the Two Decisions That Result in Errors

Decision	Reality	
	Null hypothesis true	Alternative hypothesis true
Accept null hypothesis	No error—correct result	Type II error—Probability β
Reject null hypothesis	Type I error—Probability α	No error—correct result

should only occur with a known probability, α. Thus, we have probability α of making a Type I error and concluding that a trend exists when in fact it does not. The third possible scenario is that the alternative hypothesis is true, and we reject the null and conclude the alternative is true. The fourth scenario is that the alternative hypothesis is true, but we fail to reject the null, and hence make a Type II error of failing to reject a false null hypothesis. This scenario can occur with probability β (readers should not confuse this symbol with the one for slope, β_1). Typically, we set α, but have little knowledge of β. The power of a statistical test is its ability (probability) to reject a false null hypothesis, and is equal to $1 - \beta$. A test with high power has a large chance of rejecting the null hypothesis when the hypothesis is false.

What if the assumption of normally distributed ε_i is not met? Then, we are generally more likely to fail to reject the null hypothesis when it is false than we should, i.e., the probability of a Type II error, β, increases. The assumption of normality is likely to be incorrect for several reasons. Unequal effort in collecting the population estimates can mean that the sampling variation is unequal across the N_i estimates. Also, because the variance of estimates of N_i is almost always linked to the magnitude of the population size, the variance may be unequal even when about the same level of effort was used to produce each of the estimates. This last scenario is common, and is one of the main reasons that we will generally recommend that the exponential model with a logarithmic transformation be applied (Example 5.3). First, this model makes more biological sense to us. Populations do not decline by a constant amount each year, but would be more likely to decline by a constant rate. The actual amount of decline will depend on how many individuals are present. Take the following simple example. Assume that the population has no recruitment, and that the annual survival rate is 0.9. Then, each year, 90% of the previous year's population is still present. Thus, population dynamics dictate that an exponential model is logical. Second, the logarithmic transformation often corrects the problem of the ε_i having the same variances. By taking the log(N_i), the variances are stabilized, and we meet the assumptions of the statistical test.

Example 5.3. Regression of the Logarithm of Lincoln–Petersen Estimates of Population Size versus Time

Based on the natural logarithms of the data in Table 5.1, the slope of the regression line is -0.0863 with SE $= 0.0212$. This approach corresponds to the exponential model, i.e., a constant rate of decrease each year. These values give a t statistic for the test of the null hypothesis of the slope equal to or greater than zero of -4.069 with 3 degrees of freedom.

We are interested in the alternative hypothesis of the slope less than zero, so we perform a one-sided test. The probability of observing a t value this small is 0.0134, so we reject the null hypothesis in favor of the alternative that the slope is less than zero, and hence the population has declined during the interval 1991 to 1995.

The estimate of the slope of -0.0863 for this model suggests that $\exp(-0.0863) \times 100$ or 91.73% of the individuals in the population at time i remain at time $i + 1$. We also can use the estimate of the slope and its SE to compute a confidence interval on the annual loss rate. The confidence interval will be the estimate $\pm$ the t statistic with 3 df and $\alpha = 0.05$ times the SE, giving $-0.0863 \pm 3.182 \times 0.0212$, or -0.1538 to -0.0188. This is a -15.38 to -1.88% decline each year.

When we compute the statistical test, we generally set the value of α at 0.05 or 0.10. However, we do not always consider the value of β. The message of Gerrodette (1987), and of follow-up articles by Link and Hatfield (1990) and Gerrodette (1991), is that the power of the statistical test should be considered. When the null hypothesis is rejected, we know that for a valid test, the probability of this occurring in error is α. However, when we fail to reject the null hypothesis, we have no idea of our chances of error without knowledge of β. We should compute the approximate power of the test prior to collecting the data, so that we can evaluate whether we have any chance of detecting an important decline with the amount of effort we are able to provide the project. "Operationally, power analysis is important during the planning of experiments to avoid wasted time and effort on a program that is unlikely to yield useful information" (Gerrodette, 1987, p. 1364).

When we fail to reject the null hypothesis, another option to evaluate the power of the statistical test is to compute a confidence interval for the slope, β_1, exactly as has been done in Examples 5.2 and 5.3. Because the test failed to reject the null, this confidence interval will include 0. If the interval is narrow, i.e., the length of the interval is small, then we conclude that a test with good power was conducted. But how do you decide what is a narrow interval? This is a biological question. The answer depends on what range of values for the slope estimates is biologically meaningful. The location of the mean slope relative to 0 also is important to interpreting biological significance.

An important rate of decline for a species with a large reproductive potential will differ from one with a low reproductive potential. In Fig. 5.7, the average rates of decline for populations that produced the confidence intervals on the left and in the middle are both low, but the width of the interval on the left is much smaller, indicating more power. Further, the interval on the left barely contains 0, so even though we failed to reject the null hypothesis, the true slope is probably negative, which leads us to

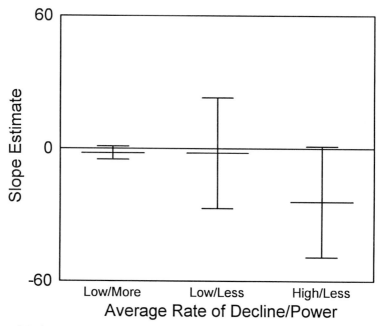

Figure 5.7 Slope estimates and corresponding confidence intervals for three populations characterized by the magnitude of their average rates of decline (low or high) and the width of their intervals (more power or less power).

suspect that the population is in decline. Conversely, the confidence interval in the center of the Fig. 5.7 is much wider and contains many values of $\hat{\beta}_1$ on both sides of 0. Thus, we are not very sure about the direction of the slope, much less its magnitude. Finally, note that the average rate of decline contained in the confidence interval on the right is much greater (i.e., has a much more negative value) than either of other two intervals, but has the same width as the one in the middle, indicating low power. Yet, because the right confidence interval barely contains 0, we suspect the population is in decline. The problem is that we cannot be sure of the magnitude of the rate of decline. Conversely, we suspect a low rate of decline for the confidence interval on the left because of its narrow range of values located near 0.

Although a population with a negative slope or growth rate will eventually go extinct, its persistence will obviously depend on the magnitude of the decline. The population that produced the confidence interval on the left side of Fig. 5.7 likely is declining at a low rate, depending on the species being studied. The one on the right may be declining anywhere from a low to high rate. Because of the width of the interval, we cannot narrow it down further. The middle and right confidence intervals indicate poorly designed studies, although the study represented by the left interval was not particularly well designed because 0 was included in its interval. The interval on the right contains more information regarding a population decline than the middle one because of its much lower (i.e., more negative) slope estimate and the location of the upper end of its confidence interval relative to 0.

5.3.3. RANDOMIZATION METHODS

Even though a logarithmic transformation may correct some of the problems with the regression models discussed previously, the inferences made from such an approach still may be questionable. The assumption of normally distributed random errors for the regression residuals is needed to obtain valid inferences from the t statistics computed to test the null hypothesis that the slope is equal to 0. To avoid making this assumption, and hence strengthen the validity of the inferences, computer-intensive methods have been derived. These methods have been the recent topic of a great deal of statistical research. Good (1994), Edgington (1995), and Manly (1997) provided overviews of these approaches. In the following section, we will discuss how to perform a permutation or randomization test to test for trends.

The general idea is to randomize the data to determine how likely the observed slope is to occur by chance alone. Suppose that the estimates of population size were relabeled randomly with years. That is, suppose all the "year" labels were put in a hat. Then, a year is drawn from the hat to label the first population estimate, and so on until all the estimates have new labels. Then a new estimate of the slope is computed from this relabeled data. If a real decline is occurring in the population, we would expect our original slope, computed from the actual data, to be less than the new slope computed from the relabeled data. If we randomly relabel the data many times, we can compute a probability level for our observed slope. For example, if 3 estimates of the population are available, there are 3! (read 3 factorial, or $3 \times 2 \times 1 = 6$) possible arrangements of the data. They are 123, 132, 213, 231, 312, and 321. If our original data happen to be the case where $\hat{N}_1 > \hat{N}_2 > \hat{N}_3$, then the other 5 possibilities (132, 213, 231, 312, and 321) all have to have larger (i.e., less steep) slopes, and hence our observed slope (123) has a probability of 1/6 given the observed data. This example is a simple permutation test, where the data are permuted into all possible arrangements. For $n = 4$ data points, 4! = 24 arrangements are possible, with 5! = 120, 6! = 720, and 7! = 5040. For 7 or more data points, we can sample the set of possible permutations to compute the probability of observing a slope as small or smaller than our observed slope. We sample the set of possible permutations because the number becomes so large that computing all possible regressions is not possible, and little if any information is lost by sampling the set of permutations (Example 5.4).

Example 5.4. Permutation Test of the Regression of the Lincoln–Petersen Estimates of Population Size Versus Time

As shown in Example 5.2, the slope of the regression line is −37.10. When the 5! = 120 possible regression are computed, we find that 2 of the permutations give estimates of the slope less than the observed value of −37.0980. The smallest 5 estimates are −40.0755, −38.0965, −37.0980, −37.0398, and −35.1190. Thus, we conclude that 3/120 = 0.025 is the probability of observing a slope as steep or steeper than −37.10.

The same procedure can be used with the log-transformed analysis, where the slope was estimated as −0.0863. Again, only 2 of the random permutations has steeper slopes than the observed value of −0.0863, with the smallest 5 estimates consisting of −0.0922, −0.0873, −0.0863, −0.0842, and −0.0813. Again, we conclude that 3/120 = 0.025 is the probability of observing a slope as steep or steeper than −37.10. If 7 or more points had been included in the time series, we would have randomly sampled 1000 of the possible permutations to compute the probability.

Typically, the permutation or randomization procedure will not be more efficient (i.e., greater power) than the parametric procedure unless the assumptions of the parametric regression procedure are badly violated, i.e., extreme heterogeneity of variance or nonnormality of residuals. However, the benefit of the permutation procedure is that the inference does not require these assumptions to be valid. The critical assumption needed is that the estimates are a representative sample of the population, which is also required for the parametric procedure. Thus, the permutation procedure provides confirmation of the parametric procedure. For sample sizes of 7 or more population estimates, the randomization procedure will have about the same power as the parametric procedure when the assumptions of the parametric procedure are met. If the assumptions are not met, the randomization procedure may actually have better power.

For small sample sizes, i.e., number of population estimates <6, the permutation procedure will not be as powerful as the parametric procedure because the sampling distribution of the parameter is small. Thus, for only 3 population estimates, the smallest probability level that can be generated by the permutation procedure is 1/6 = 0.16667. The parametric procedure is not restricted by this limitation. Hence, if the assumptions of the parametric procedure can be met, you should use it for small samples.

5.3.4. NONPARAMETRIC METHODS

Because regression methods make some fairly difficult assumptions to meet, and because the randomization and permutation procedures require some computer expertise to compute, we also present a nonparametric procedure for testing trends based on ranks. Gilbert (1987) presents the Mann–Kendall test (Mann, 1945; Kendall and Gibbons, 1990) as a nonparametric alternative to regression. This rank procedure has an additional advantage in that exact estimates of population size are not necessary. We only have to be able to rank the observations. Thus, we may not know the exact estimates of population size for 5 years, but as long as these years can be ranked, we can still compute a statistical test of trend.

Consider the scenario where a population estimate, N_i, is obtained for each year, $i = 1, ..., n$. List the data by the order they were collected, i.e., $\hat{N}_1, \hat{N}_2, ..., \hat{N}_n$. The test does not assume that estimates are made in all years, i.e., some years can be missing and do not appear in the analysis. Then, compute the signs ($+$ or $-$) of all possible differences of the N_i values. There are $n(n - 1)/2$ differences $\hat{N}_j - \hat{N}_k$, where $j > k$. These differences are $\hat{N}_2 - \hat{N}_1, \hat{N}_3 - \hat{N}_1, ..., \hat{N}_n - \hat{N}_1, \hat{N}_3 - \hat{N}_2, \hat{N}_4 - \hat{N}_2, ..., \hat{N}_n - \hat{N}_{n-2}, \hat{N}_n - \hat{N}_{n-1}$. A useful method of constructing these differences is shown in Table 5.3. For each difference, determine the sign, i.e., whether

Table 5.3

Format To Compute the Differences in Data Values Needed for Computing the Mann–Kendall S Statistic To Test for Trend in Population Estimates [Adapted from Gilbert (1987)]

| Data values listed in order collected over time | | | | | | | No. of $+$ signs | No. of $-$ signs |
$\hat{N}_1$	$\hat{N}_2$	$\hat{N}_3$	$\hat{N}_4$	...	$\hat{N}_{n-1}$	$\hat{N}_n$		
	$\hat{N}_2 - \hat{N}_1$	$\hat{N}_3 - \hat{N}_1$	$\hat{N}_4 - \hat{N}_1$	...	$\hat{N}_{n-1} - \hat{N}_1$	$\hat{N}_n - \hat{N}_1$		
		$\hat{N}_3 - \hat{N}_2$	$\hat{N}_4 - \hat{N}_2$	...	$\hat{N}_{n-1} - \hat{N}_2$	$\hat{N}_n - \hat{N}_2$		
			$\hat{N}_4 - \hat{N}_3$	...	$\hat{N}_{n-1} - \hat{N}_3$	$\hat{N}_n - \hat{N}_3$		
				$\ddots$	$\vdots$	$\vdots$		
					$\hat{N}_{n-1} - \hat{N}_{n-2}$	$\hat{N}_n - \hat{N}_{n-2}$		
						$\hat{N}_n - \hat{N}_{n-1}$		
						$S =$	Sum of $+$ signs	Sum of $-$ signs

the more recent estimate is greater than the older estimate. Thus, for observations j and k

$$\text{sign}(\hat{N}_j - \hat{N}_k) = 1 \text{ if } \hat{N}_j > \hat{N}_k$$
$$= 0 \text{ if } \hat{N}_j = \hat{N}_k$$
$$= -1 \text{ if } \hat{N}_j < \hat{N}_k,$$

where $j \neq k$. Then, compute the Mann–Kendall statistic,

$$S = \sum_{k=1}^{n-1} \sum_{j=k+1}^{n} \text{sign}(\hat{N}_j - \hat{N}_k),$$

which is the number of positive differences minus the number of negative differences. If S is a large positive number, measurements taken later in time tend to be larger than those taken earlier. Likewise, if S is a large negative number, measurements taken later in time tend to be smaller, so a decline is expected. Take as the null hypothesis that no trend is present, versus the alternative hypothesis of a downward trend. To determine whether to reject the null hypothesis, look up the value of S in Table 5.4, given the number of observations (n). The null hypothesis is rejected in favor of the alternative hypothesis if S is negative and if the probability value for the absolute value of S is less than the a priori specified α significance level of the test. Similarly, to test for an upward trend, you reject the null hypothesis if S is positive and if the probability value in the table corresponding to the computed S is less than the a priori specified α value. If a two-tailed test is desired, i.e., if we want to detect either an upward or downward trend, the value in the table corresponding to the computed value of S is doubled, and the null hypothesis is rejected if that doubled value is less than the a priori α level.

The nonparametric test based on ranks will not be as efficient as the regression procedures because the actual differences in the population estimates are not used, only their signs. However, this procedure provides a useful tool for situations where a quick test is desired without complicated calculations, and where only the relative ranking of the population estimates are available (Example 5.5).

Another procedure that often is discussed in the literature is to replace observations by their ranks and then perform a regression on the ranks (Good, 1994). One advantage of replacing the population estimates with their ranks is that the effect of outliers is reduced. That is, suppose that the sequence of observations 100, 999, 90, 80, and 70 is observed. The value 999 will totally invalidate a parametric regression. The randomization test also will be somewhat shaky because of this wild observation. However,

Table 5.4
Probabilities for the Mann–Kendall Nonparametric Test for Trend

S	Value of n 4	5	8	9	S	Value of n 6	7	10
0	0.625	0.592	0.548	0.540	1	0.500	0.500	0.500
2	0.375	0.408	0.452	0.460	3	0.360	0.386	0.431
4	0.167	0.242	0.360	0.381	5	0.235	0.281	0.364
6	0.042	0.117	0.274	0.306	7	0.136	0.191	0.300
8		0.042	0.199	0.238	9	0.068	0.119	0.242
10		0.0083	0.138	0.179	11	0.028	0.068	0.190
12			0.089	0.130	13	0.0083	0.035	0.146
14			0.054	0.090	15	0.0014	0.015	0.108
16			0.031	0.060	17		0.0054	0.078
18			0.016	0.038	19		0.0014	0.054
20			0.0071	0.022	21		<0.001	0.036
22			0.0028	0.012	23			0.023
24			<0.001	0.0063	25			0.014
26			<0.001	0.0029	27			0.0083
28			<0.001	0.0012	29			0.0046
30				<0.001	31			0.0023
32				<0.001	33			0.0011
34				<0.001	35			<0.001
36				<0.001	37			<0.001
					39			<0.001
					41			<0.001
					43			<0.001
					45			<0.001

the nonparametric test based on ranks will give exactly the same value as calculated for Example 5.5.

Example 5.5. Nonparametric Test of the Regression of the Lincoln–Petersen Estimates of Population Size versus Time

To compute the test statistic S, we must compute all possible signs of the differences of the 5 estimates in Table 5.1. The following table is a convenient format to perform this calculation.

Time Date	1 485	2 515	3 412	4 392	5 362	No. of + signs	No. of − signs
		515–485	412–485	392–485	362–485	1	3
			412–515	392–515	362–515	0	3
				392–412	362–412	0	2
					362–392	0	1

$$S = 1 - 9 = -8$$

Looking up the value 8 in Table 5.4, we find that the probability of observing a value of S as extreme as 8 is 0.042. We also note that if the first and second values have been reversed in the data, S would have been 10, with probability of 0.0083 of a value this extreme.

5.4. GENERAL COMMENTS

For all the tests for a trend in a series of observations of population sizes, the null hypothesis is that there is no trend. Failure to reject the null hypothesis means we will probably decide to do nothing to assist or manage the population because our data did not provide support for any action. We incorrectly assume that because we failed to reject the null hypothesis of the test, no decline is occurring in the population. However, validity of this decision depends on the power of the test we have performed, and power of the test depends on precision of the data used in the test. Poor precision from an inadequate survey means no power to reject the null hypothesis, and hence we take a default decision of no action. Surveys to detect a trend in a population must have adequate effort to obtain precise enough population estimates to have a reasonable chance (say 80%) to detect a biologically important trend in the population. Surveys with less effort waste time and money because they have no chance of providing useful information.

LITERATURE CITED

Berthouex, P. M., Hunter, W. G., and Pallesen, L. (1978). Monitoring sewage treatment plants: Some quality control aspects. *J. Qual. Tech.* **10:** 139–149.

Burnham, K. P., Anderson, D. R., White, G. C., Brownie, C., and Pollock, K. H. (1987). Design and analysis methods for fish survival experiments based on release-recapture. *Am. Fish. Soc. Monogr.* **5.**

Edgington, E. S. (1995). *Randomization Tests,* third ed. Marcel Dekker, New York.

Ewan, W. D. (1963). When and how to use Cu-Sum charts. *Technometrics* **5:** 1–22.

Gerrodette, T. (1987). A power analysis for detecting trends. *Ecology* **68:** 1364–1372.

Gerrodette, T. (1991). Models for power of detecting trends—A reply to Link and Hatfield. *Ecology* **72:** 1889–1892.

Gibra, I. N. (1975). Recent developments in control chart techniques. *J. Qual. Tech.* **7:** 183–192.

Gilbert, R. O. (1987). *Statistical Methods for Environmental Pollution Monitoring.* Van Nostrand Reinhold, New York.

Good, P. (1994) *Permutation Tests: A Practical Guide to Resampling Methods of Testing Hypotheses.* Springer-Verlag, New York.

Keith, L. B. (1963). *Wildlife's Ten-Year Cycle.* Univ. Wisconsin Press, Madison, WI.

Kendall, M. G., and Gibbons, J. D. (1990). *Rank Correlation Methods,* fifth ed. Edward Arnold, London.

Link, W. A., and Hatfield, J. S. (1990). Power calculations and model selection for trend: A comment. *Ecology* **71:** 1217–1220.

Manly, B. F. J. (1997). *Randomization, Bootstrap and Monte Carlo Methods in Biology,* second ed. Chapman and Hall, London.

Mann, H. B. (1945). Non-parametric tests against trend. *Econometrica* **13:** 245–259.

Page, E. S. (1961). Cumulative sum charts. *Technometrics* **3:** 1–9.

Page, E. S. (1963). Controlling the standard deviation by cusums and warning lines. *Technometrics* **5:** 307–315.

Skalski, J. R., and Robson, D. S. (1992). *Techniques for Wildlife Investigations: Design and Analysis of Capture Data.* Academic Press, San Diego.

Vardeman, S., and David, H. T. (1984). Statistics for quality and productivity. A new graduate-level statistics course. *Am. Stat.* **38:** 235–243.

Wetherill, G. B. (1977). *Sampling Inspection and Quality Control,* second ed. Chapman and Hall, New York.

Chapter 6

Guidelines for Planning Surveys

6.1. Step 1—Objectives
6.2. Step 2—Target Population and Sampling Frame
6.3. Step 3—Plot Design and Enumeration Method
6.4. Step 4—Variance among Plots
6.5. Step 5—Plot Selection, Plot Reselection, and Survey Frequency
  6.5.1. Plot Selection
  6.5.2. Plot Reselection
  6.5.3. Frequency of Surveys

6.6. Step 6—Computing Sample Sizes
6.7. Step 7—Power of a Test To Detect a Trend
6.8. Step 8—Iterate Previous Steps
6.9. Example
Literature Cited

Performing an inadequate series of surveys only results in wasted time and money because no solid conclusions about the trend in a population can be made. Thus, in this chapter we provide guidelines for how to determine the level of effort required to conduct a survey. The goal is to allocate resources (meaning time and money) to each of the components of the survey: (1) number of population estimates (surveys) across time; (2) number of plots for each survey; and (3) amount of effort to allocate to each sampling unit in enumerating the number of animals on each plot.

Allocation of effort depends on the amount of variability that we encounter for each of the three components. Because the precision of surveys is affected by enumeration variation, and because the power of the test for trends is affected by the precision of the population estimates used to

171

construct the test, the process is somewhat iterative. In the following sections, we will walk through the steps for designing a series of surveys for a monitoring program, and then provide a concrete example to flesh out the process. We strongly recommend that biologists consult with a statistician with a background in sampling biological populations, or a quantitative biologist with a strong statistical background, before planning a survey or monitoring program. We also emphasize that knowledge of the ecology of the target species is fundamental to good survey design.

6.1. STEP 1—OBJECTIVES

The first step in undertaking any type of monitoring program is to clearly define its goals. How many species will be monitored? Will it be a baseline study whose purpose is to gather background information on a little-known species or will it attempt to monitor a species for important decreases in abundance over time? Will the program be short-term or long-term? Are abundance estimates required or is presence–absence information adequate? Is there one species of interest or is the goal to monitor more than one species concurrently? These and related questions may sound self-evident, but too often time, money, and effort have been wasted for lack of clearly defined objectives. In a multispecies scenario, funding will likely limit the number of species that can be intensively monitored, and therefore some criteria for prioritizing species will have to be used.

6.2. STEP 2—TARGET POPULATION AND SAMPLING FRAME

After the objectives have been clearly defined, we then must delineate the population of interest, i.e., set up the target population. This process entails decisions on how large an area should be surveyed for the species of interest. Thus, the main emphasis is geographical. However, other considerations may result, such as eliminating elevational zones. For instance, if we are concerned about white-tailed prairie dogs in northwestern Colorado, we would not want to include areas at greater elevations than where prairie dogs would exist.

Once the target population has been identified, a sampling frame needs to be created that will contain a listing or mapping of plots (or a defined boundary that contains the area of interest) that covers as much of the target population as possible. Ideally, the sampling frame should contain all of the target population so that we can make direct inferences to it. An important point here is that we can only make statistically based statements about the animals within the sampling frame or sampled population. For instance, results from a sampling frame containing a small area within Colorado cannot be properly applied to other areas of the state.

We basically have two options if our sampling frame differs from the target population: (1) assume that the difference is not important and treat the two as equivalent; or (2) redefine the target population so that it coincides with the sampling frame. The first option would be acceptable if the area of nonoverlap between the frame and the target population was small compared to the total area covered by the target population, or if it is known with reasonable certainty that the area of nonoverlap does not significantly differ in spatial distribution and abundance for the species of interest. Assuming that nonoverlapping areas more or less reflect the same composition as areas within the frame becomes more risky with increasing size of this nonoverlapping component. When in doubt, the best approach is to redefine the target population.

6.3. STEP 3—PLOT DESIGN AND ENUMERATION METHOD

The best size and shape for plots composing the sampling frame depend upon such factors as counting method used within plots, potential edge effect, underlying distribution of animals, spatial scale of the species of interest, and costs. We refer the reader to the discussion of plot design in Chapter 2 for further details. Because all of these factors are related to enumeration method, we focus on this aspect in this section.

The enumeration method we use within each selected plot is as important as how we select the plots themselves. Both procedures are part of the process of obtaining (hopefully) unbiased and precise estimates of abundance for the target population. The size of the sampling unit is dictated in part by the method used to count the animals on the unit. Thus, if a capture–recapture estimator is to be used to estimate the population size on the unit, a large enough plot must be selected so that a valid capture–recapture estimate can be constructed, i.e., enough animals must be captured and marked to obtain a valid estimate. If animals can be completely counted (i.e., censused) on the unit, the size of the unit will be dictated by logistics. Overly large plots would be inefficient, whereas undersized plots would require too much travel time among plots.

A second consideration regarding the counting method is the variance associated with it. If a complete count can be performed, there is no variance associated with this level of the survey. However, such is seldom the case, so we must consider the magnitude of enumeration variance for a given counting method. The enumeration variance will increase the variance among the sampling plots, so we must have some idea of the expected enumeration variance before we can decide on the number of sampling units to include in the sample.

As an example, we may decide to use a capture–recapture estimation procedure to estimate the abundance of meadow voles on each 1-ha plot. We expect to find about 15 individuals per plot (i.e., a density of 15 voles/ha), and with the level of trapping effort and the number of occasions we expect to trap, the expected $\text{Var}(\hat{N})$ is about 40. Note that we did not put a caret on the Var portion of this term because we are assuming that the variance of $\hat{N}$ is 40 and are not concerned about its estimate at this time.

Some simple relationships have been developed to relate the $\text{Var}(\hat{N})$ to N. These relationships are usually based on the coefficient of variation (CV) of N instead of $\text{Var}(\hat{N})$ because the CV will remain relatively constant over a range of N. For example, Seber (1982, p. 60) suggests that for Lincoln–Petersen estimates, the relation

$$CV(\hat{N}) = \frac{1}{\sqrt{\dfrac{n_1 n_2}{N}}} \tag{6.1}$$

approximately holds and provides a rough estimate of $CV(N)$ given N and the expected number of animals captured on each occasion.

6.4. STEP 4—VARIANCE AMONG PLOTS

The variance among the sampling plots is a function of both the inherent variation in numbers among the plots [$\text{Var}(N)$] due to the underlying spatial distribution of animals and the enumeration variation [$\text{Var}(\hat{N})$]. That is, if the N animals on all plots can be counted completely, then the variance of $\hat{N}$ is zero ($\text{Var}(\hat{N}) = 0$). Then the only variation among the plots sampled is due to the true population variation, $\text{Var}(N)$. Assuming that $\text{Var}(\hat{N}) \neq 0$, then the total variation among the plots sampled is $\text{Var}(N) + \text{Var}(\hat{N})$. We have discussed how we can get a rough guess of $\text{Var}(\hat{N})$ in the previous section. How can we obtain an estimate of $\text{Var}(N)$, or the plot to plot variation with no enumeration variance?

There are basically four ways to obtain estimates of among-unit variance: (1) conduct a pilot study, (2) use estimates from previous studies conducted on or near the study area, (3) use estimates obtained from similar studies in other areas, or (4) assume some underlying model such as the negative binomial distribution or use Taylor's power law (Taylor, 1961, 1965). Conducting a pilot study has the advantages of field-testing the proposed methodology, obtaining a better idea of survey costs, and, possibly, obtaining better estimates of variance. The advantage of the other three is in their much reduced costs compared to conducting a pilot study. Which approach is best must be decided on a case-by-case basis, although we generally recommend conducting a pilot study because it allows proposed counting techniques to be evaluated before a large quantity of time and effort have been invested. We now will discuss the fourth option in more depth.

If animals are randomly distributed across the landscape, then from theory we know that the Var(N) should be equal to N. This is because the number of animals per plot should be distributed as a Poisson variable, and the mean and variance of a Poisson variable are the same. A random distribution does not mean an even or uniform distribution, but that each animal has an equal probability of occupying any point in space and that the presence of one individual does not influence the distribution of another (Southward, 1966, pp. 24–25). We know that a completely random distribution of animals is unlikely because of landscape heterogeneity and social interactions, i.e., the variance will depend on spatial variability in animal abundance. In general, natural plant or animal populations are rarely distributed at random, and organisms are usually found to be clustered together more than a Poisson distribution would predict. This means that there will be more empty plots and more plots with many individuals than expected with a Poisson distribution, so that the between-plot variance will be greater than the mean of the plots. Typically, we want to increase the variance considerably more than expected with the Poisson distribution. When the Poisson distribution is not applicable, a number of alternative distributions have been put forward to account for the spatial patterns observed [cf. Greig-Smith (1964), Southwood (1966), Pielou (1969), and King (1969) for references].

In particular, the negative binomial distribution has been proposed, and is discussed by Southward (1966, pp. 25, 34), Seber (1982, p. 25), and Skalski and Robson (1992, pp. 38–41). The negative binomial is widely used because of its flexibility. For the negative binomial,

$$\text{Var}(N) = E(N) + \frac{E(N)^2}{R}, \tag{6.2}$$

where $E(N)$ is the expected or mean of N for the plot size being considered and R is a constant to inflate the among-unit variance term. This model includes the Poisson distribution as a special case, because when $R \rightarrow \infty$, this relationship reduces to the Poisson distribution. As an example, Skalski and Robson (1992, p. 41) reported that $R = 31.75$ for data from small mammal populations on 1-ha grids in grass or shrub-steppe habitats of Washington and Colorado. Species used to construct this value were Ord's kangaroo rat, least chipmunk, montane vole, northern grasshopper mouse, silky pocket mouse, Great Basin pocket mouse, deer mouse, harvest mouse, and 13-lined ground squirrel.

Another approach to estimate the Var(N) from N explored by Skalski and Robson (1992, p. 41) and also Southward (1966, p. 9) is Taylor's power law (Taylor, 1961, 1965), where

$$\text{Var}(N) = \alpha E(N)^{\beta}. \tag{6.3}$$

Using the same data as described above, Skalski and Robson (1992) reported that Taylor's power law suggested the relationship

$$\text{Var}(N) = 11.4E(N)^{0.63} \tag{6.4}$$

for small-mammal abundance on 1-ha study plots.

For both the negative binomial and Taylor models, the Var(N) is greater than $E(N)$ as predicted by just the Poisson distribution. For the likely situation where you have no idea what the variation among plots might be, using one of these relationships may provide a reasonable guess for the purpose of designing a study.

6.5 STEP 5—PLOT SELECTION, PLOT RESELECTION, AND SURVEY FREQUENCY

6.5.1. PLOT SELECTION

How we choose our sample of plots will largely dictate the strength of our inferences about the species of interest. A nonrandom selection of plots will likely result in biased estimates of abundance with measures of precision of unknown reliability. Conversely, choosing plots using an imprecise random selection procedure, on average, will yield unbiased estimates of abundance, but inflated estimates of precision. For example, a species of interest could occur mostly in a specific habitat that occupies a small area within a frame. A simple random sample of plots in this situation, on average, is more likely to contain a wider range of plot counts than a carefully implemented stratified random sample, which would select more plots in strata with higher variability (i.e., usually higher species densities). A wider range of counts means greater variability and hence less precision. Thus, our choice in sampling schemes could have an important bearing on our ability to detect important decreases in numbers of animals over time.

6.5.2. Plot Reselection

The discussion in the preceding paragraphs was oriented more toward obtaining abundance estimates within a given time interval. What about selecting samples over time? Should we select a new sample of plots each time (i.e., a *complete replacement sample*)? Should we survey the same initial sample of plots at each sampling occasion (i.e., a *complete remeasurement sample*)? Or should we use a combination of these two (i.e., *sampling with partial replacement*; Schreuder *et al.,* 1993)? The answer depends on the survey situation.

Let us first consider the focus of our inferential monitoring program. We are interested in estimating the true average change, or trend, in animal numbers in a defined area over some stated time period. However, we are rarely able to census the sampling frame, so we cannot obtain the true abundance values within each subinterval of time (e.g., year). Therefore, counts are conducted on a random sample of plots to obtain an abundance estimate for a given year. In sampling across time, we are obtaining one estimate of the true trend from all possible estimates of trend, which are generated from all possible combinations of random samples among sampling occasions. The expected value of the sampling distribution of these estimates will be equal to the true trend if the design is unbiased.

On average, the most precise trend estimates are those based on a random sample of plots during the first year, with subsequent counts conducted on the same plots in subsequent years. The use of the same plots reduces the magnitude of among-unit variation because new plots are not chosen each year. In other words, only variability among one set of plots contributes to the among-unit variance component rather than variability from multiple sets of new plots sampled during each year of a study. Further, setting up "permanent" monitoring plots will likely reduce survey costs including travel to and from plots. Ideally, such a scheme will produce unbiased estimates of trend as long as chosen and unchosen plots are not treated differently by managers, i.e., inferences to the sampled population still are valid based on chosen plots.

A number of authors (e.g., Goodall, 1952; Greig-Smith, 1964; Usher, 1991) have suggested that surveys on an identical set of plots over time suffer from autocorrelation, i.e., an abundance estimate obtained in 1 year would be dependent on the estimate from the previous year(s). They further reasoned that resulting counts would lack independence and therefore would lead to biased trend estimates. Thus, a new set of plots should be chosen for each sampling period (or some slight modification of this). This viewpoint is incorrect. A high correlation among counts is exactly what we want when we are attempting to detect a trend or change in abundance estimates (Cochran, 1977, p. 345). Dependency among estimates becomes an issue only when choosing an appropriate trend analysis. Therefore, permanent plots should be used for collecting trend data (Patterson, 1950; Cochran, 1977) except when these plots are selectively treated differently than unchosen plots or there is a major change in animal distribution or numbers (or both) that is not captured by them (Schreuder *et al.,* 1993).

In reality, a single, initial sample of plots may not be treated the same as unchosen plots over time. For instance, managers may avoid applying some management scheme or treatment (e.g., logging) to permanent plots, but continue to do so on other plots. That is, some of the sampled plots may have been managed differently if they had not been part of the original sample. Or, perhaps less likely, characteristics of sampled plots may change due either directly or indirectly to the enumeration method. Building roads, erecting survey structures, or introducing other habitat modifications may affect the abundance or even presence of a species on a given plot if the modifications are severe enough. Effects of these modifications do not have to be immediate; changes may not be readily discernible over a short period of time. We will focus on the former scenario because it seems much more common.

What do we do if the unchosen plots are selectively treated differently than chosen plots? In this case, sampling with partial replacement (Patterson, 1950; Ware and Cunia, 1962; Scott 1984; Scott and Kohl, 1994) would be the appropriate method. When possible, we suggest estimating the area of selective treatment, stratifying the frame based on this area, and using the proportion of the selectivity treated area as a basis for reselecting the number of plots from the original sample. That is, suppose our initial sample contained 10 plots from a sampling frame of 100 plots. Now suppose that 20 unchosen plots (i.e., 20% of the frame) were selectively treated differently from the 10 permanent plots. We would stratify our frame into selectively treated (i.e., 20 plots) and nonselectively treated (i.e., 80 plots) areas. Then, we would randomly choose 2 plots (20%) from the 10 permanent plots to use as replacement units. Next, we would randomly choose 2 plots from the selectively treated stratum. This would give us 8 permanent plots in the nonselectively treated stratum and 2 in the selectively treated stratum. We would use any abundance estimates collected prior to the treatment as the baseline trend data for both strata (K. Burnham, personal communication). Note that at least 2 plots should be chosen for each stratum for variance estimates to be computed. If selective treatment of unchosen plots is suspected, but not quantified, then we also recommend sampling with partial replacement to obtain unbiased trend estimates.

Finally, good area coverage is as critical for precise estimates across time as it is during a single time period. For instance, if the area of interest consists of features that may have differential effects on the population of interest over time, these features should be the basis for stratification. For example, a clearcut area would change much more rapidly, with a concomitant change in species composition and numbers, than a mature forest over a 25-year period. Hence, to obtain better estimates of trend for a given sample, we would initially stratify on these habitats and allocate plots accordingly. Even in a seemingly homogeneous habitat, stratification into, say, 4 strata will probably provide good sample coverage in the event part of the area undergoes catastrophic change or simply changes in animal distributions or numbers. Note that even a random sample estimator with poor coverage will be unbiased, but the given sample trend estimate may be far from the true trend.

6.5.3. FREQUENCY OF SURVEYS

How often surveys should be conducted within a sample of plots over time depends on the variability in numbers, monitoring objectives, and funding availability. For an unrealistic case in which population numbers follow a precise linear trend over time, one could simply take two samples, one at the beginning of a study and one at the end, in order to detect a trend. However, animal populations are much more variable than this. Hence, more sampling occasions would be required for populations that undergo large changes in numbers (i.e., large variation). Sampling intervals also depend on the goals of the monitoring program. More samples within a given time interval will enable one to detect a smaller rate of change [see Eq. (6.8)]. Finally, funding constraints may limit the number of sampling occasions that can be conducted. Conducting too many surveys is a waste of time and money. Thus, survey frequency is very much situation-specific.

6.6. STEP 6—COMPUTING SAMPLE SIZES

After we have obtained estimates for both among-unit and enumeration variances, our next step is to use this information in computing the optimum number of plots that should be sampled given associated survey costs. The formulas for calculating number of samples are specific to the plot selection method used. After this step has been completed, we then must determine how precise our abundance estimates will be at some specified level of confidence. That is, we may wish to obtain abundance estimates $(\hat{N})$ that are within $\pm 10\%$ of the true abundance (N) about 95% $(\alpha = 0.05)$ of the time. After we have obtained this estimate, we can compute its estimated coefficient of variation $\hat{C}V(\hat{N})]$ and use this in a formula discussed in the next step.

6.7. STEP 7—POWER OF A TEST TO DETECT A TREND

From previous steps, we have a rough guess of our sampling variation. This is the critical information we need to determine what will be the power of the test to detect a trend in our surveys. As discussed in Chapter 5, Gerrodette (1987) proposed both a linear and multiplicative model for detecting trends. In the following analysis taken from Gerrodette (1987), we assume, as he does, that the regression line is fitted to equally spaced population estimates $\{y_i\} = \{\hat{N}_1, \hat{N}_2, \ldots, \hat{N}_n\}$ for a linear model,

$$\hat{N}_i = \hat{N}_1[1 + r(i - 1)] + \varepsilon_i, \tag{6.5}$$

or $\{y_i\} = \{\ln \hat{N}_1, \ln \hat{N}_2, \ldots, \ln \hat{N}_n\}$ for a multiplicative (often called exponential) model:

$$\ln(\hat{N}_i) = N_1(1 + r)^{i-1} + \varepsilon_i. \tag{6.6}$$

He also assumed three models that relate the sampling variation to the population estimate $(\hat{N}_i)$ for the $i = 1, \ldots, n$ surveys: $CV(\hat{N}_i)$ proportional to $1/\sqrt{\hat{N}_i}$, $CV(\hat{N}_i)$ constant, and $CV(\hat{N}_i)$ proportional to $\sqrt{\hat{N}_i}$. To compute the power (probability that the null hypothesis will be rejected when the null hypothesis is false), we need to specify the Type I error rate (α) or the probability of rejecting the null hypothesis when it is true, the Type II error rate (β) or the probability of failing to reject the null hypothesis when it is false, r or the actual trend in the n population estimates, and CV_1 or the coefficient of variation of the first observation. Based on these assumptions, the relationship between the parameters to determine the power of the test is shown in Table 6.1. In these equations, $z_{\alpha/2}$ are the z statistics for $\alpha/2$ and z_β are the z statistics for β. For $\alpha = \beta = 0.05$, $z_{\alpha/2} = 1.960$ and $z_\beta = 1.645$. For $\beta = 0.2$, or a power of 80%, $z_\beta = 0.842$, and for $\beta = 0.1$, or a power of 90%, $z_\beta = 1.282$. If only a one-sided test is desired, i.e., we are only interested in testing whether the null hypothesis that the slope is 0 or positive versus the alternative hypothesis that the slope is negative (trend is down in the population), then the term $z_{\alpha/2}$ can be replaced by z_α.

Table 6.1

Formulas from Gerrodette (1987) Relating Power to Detect a Trend in Regression of Time (Equally Spaced Observations) versus Population Estimates

Model	CV_i	Equation for power
Linear	$\dfrac{1}{\sqrt{\hat{N}_i}}$	$r^2 n(n-1)(n+1) \geq 12CV_1^2(z_{\alpha/2}+z_\beta)^2\left[1+\dfrac{r}{2}(n-1)\right]$
	constant	$r^2 n(n-1)(n+1) \geq 12CV^2(z_{\alpha/2}+z_\beta)^2\left(1+r(n-1)\left[1+\dfrac{r}{6}(2n-1)\right]\right)$
	$\sqrt{\hat{N}_i}$	$r^2 n(n-1)(n+1) \geq 12CV^2(z_{\alpha/2}+z_\beta)^2$
		$\times\left(1+\dfrac{3r}{2}(n-1)\left[1+\dfrac{r}{3}(2n-1)+\dfrac{r^2}{6}n(n-1)\right]\right)$
Mult.	$\dfrac{1}{\sqrt{\hat{N}_i}}$	$[\ln(1+r)]^2 n(n-1)(n+1) \geq 12(z_{\alpha/2}+z_\beta)^2\left(\dfrac{1}{n}\sum_{i=1}^{n}\ln\left[\dfrac{CV_1^2}{(1+r)^{i-1}}+1\right]\right)$
	constant	$[\ln(1+r)]^2 n(n-1)(n+1) \geq 12(z_{\alpha/2}+z_\beta)^2[\ln(CV^2+1)]$
	$\sqrt{\hat{N}_i}$	$[\ln(1+r)]^2 n(n-1)(n+1) \geq 12(z_{\alpha/2}+z_\beta)^2\left(\dfrac{1}{n}\sum_{i=1}^{n}\ln[CV_1^2(1+r)^{i-1}+1]\right)$

The equations in Table 6.1 are rather formidable, and usually require numerical iteration to solve them. Thus, Gerrodette (1987) provides some further simplifications. For the exponential or multiplicative model with constant CV, the equation from Table 6.1 can be simplified to

$$r^2 n^3 \geq 12CV^2(z_{\alpha/2}+z_\beta)^2, \tag{6.7}$$

which can easily be solved for any of the five parameters. For small to moderate values of r, n, and CV, this equation serves as a useful approximation for any of the equations. For the common case where $\alpha = \beta = 0.05$,

$$(z_{\alpha/2}+z_\beta)^2 = (1.960 + 1.645)^2 = 13.0,$$

and so an even simpler form useful for relating r, n, and CV is

$$r^2 n^3 \geq 156(CV)^2 \tag{6.8}$$

Link and Hatfield (1990) criticized Gerrodette (1987), for not using t statistics in place of z statistics. Their comments are particularly valid for small n, say <10. However, the power analysis becomes even more difficult to understand if you have to optimize the equations in Table 6.1 with t statistics instead of z statistics. Fortunately, Gerrodette[1] has provided program

---

[1] Available in a "zipped" or condensed file via the Internet at: ftp://ftp.im.nbs.gov/pub/software/CSE/wsb21515/trends.zip.

TRENDS to compute the numerical solutions to the equations in Table 6.1. The power calculation/sample size problem can be summarized in five parameters (Gerrodette, 1987, 1991): n, the number of sampling occasions; r, the rate of change in abundance that occurs between each sampling occasion; CV1, the coefficient of variation of the first estimate of abundance in the series; α, the significance level (probability of Type I error); and the statistical power ($1 - \beta$, where β is the probability of Type II error). The value of any parameter can be estimated if the other four are specified. The relations among these parameters are affected by a number of factors: (1) whether change is linear or exponential (multiplicative), (2) whether change is positive or negative, (3) whether the statistical test is one- or two-sided, (4) how the precision of the estimates depends on abundance, and (5) whether the standard normal (z) or Student's (t) distribution is used in the calculations. Four of the five parameters and the five relationships among the parameters provide the input to program TRENDS.

Program TRENDS can be run either interactively or from an input file named TRENDS.INP. As you provide values interactively, an input file is created so that runs can be repeated, or single values can be changed easily. Before computations are started, the program displays the input values and gives you the opportunity to change any of them. Output is displayed on the screen and also saved in a file named TRENDS.OUT.

The following example, taken from the User's Manual for program TRENDS, shows user input and program output in detail. User input, shown in bold lettering, is case insensitive.

```
TRENDS                          [Header appears]

One of the following parameters can be computed:

   (1)   number of samples   (n)
   (2)   rate of change (+/- r)
   (3)   initial coeff. of variation (CV1)
   (4)   significance level (alpha)
   (5)   power (1-beta)

Enter value for index of parameter to be computed: 5

                           [Power is selected, so program
                           now prompts for other values]

Enter value for number of samples (n)            : 5
```

Enter value for rate of change (+/-r)   : **.1**

Enter value for initial coeff. of variation (CV1): **.13**

Enter value for significance level (alpha)   : **.05**

Enter value for 1- or 2-tailed test   : **2**

Enter value for model (1=linear, 2=exponential) : **2**

   1=CV proportional to 1/sqrt(A)
   2=CV constant with A
   3=CV proportional to sqrt(A)

Enter value for pattern of CV with abundance *A*   : **1**

   1=use z distribution (variance assumed known)
   2=use t distribution (variance est. from residuals)

Enter value for distribution index (1=z, 2=t)   : **1**

                            [Input is complete; program
                            displays input values and
                            prompts for change]

You have specified the following input values:

(1) 5   number of samples (n)
(2) .100   rate of change (+/- r)
(3) .130   initial coeff. of variation
                              (CV1)
(4) .050   significance level (alpha)
(5) (to be computed)   power (1-beta)
(6) 2   1- or 2-tailed test
(7) 2   model (1=linear, 2=exponential)
(8) 1   pattern of CV with abundance A
(9) 1   distribution index (1=z, 2=t)

To proceed with these values, press ENTER;
to change one of them, enter the line number: ⟨**ENTER**⟩

```
*** PROGRAM TRENDS OUTPUT ***


Model of change: EXPONENTIAL
Pattern of variance: CV PROPORTIONAL TO 1/SQRT(A)
Calculation based on Z distribution
Alpha =        .050 (2-tailed)
Rate  =        .100
CV    =        .130
Sample size =    5
```

Given these parameters, power is estimated to be   .72

Do you wish to try another calculation? (Y/N) **y**

                        [We will try another
                         combination]

You have specified the following input values:

```
(1)  5                  number of samples (n)
(2)  .100               rate of change (+/- r)
(3)  .130               initial coeff. of variation
                         (CV1)
(4)  .050               significance level (alpha)
(5)  (to be computed)   power (1-beta)
(6)  2                  1- or 2-tailed test
(7)  2                  model (1=linear, 2=exponential)
(8)  1                  pattern of CV with abundance A
(9)  1                  distribution index (1=z, 2=t)
```

To proceed with these values, press ENTER;
to change one of them, enter the line number: **2**

                        [We ask to modify the rate of
                         change]

Enter value for rate of change (+/- r)   : **-.1**

                        [The rate is changed to a 10%
                         decline]

You have specified the following input values:
(1)   5   number of samples (n)
(2)   -.100   rate of change (+/- r)
(3)   .130   initial coeff. of variation
                                  (CV1)
(4)   .050   significance level (alpha)
(5)   (to be computed)   power (1-beta)
(6)   -2   1- or 2-tailed test
(7)   2   model (1=linear,
                                    2=exponential)
(8)   1   pattern of CV with abundance A
(9)   1   distribution index (1=z, 2=t)

To proceed with these values, press ENTER;
to change one of them, enter the line number: ⟨**ENTER**⟩

*** PROGRAM TRENDS OUTPUT ***

Model of change: EXPONENTIAL
Pattern of variance: CV PROPORTIONAL TO 1/SQRT(A)
Calculation based on Z distribution
Alpha =   .050 (2-tailed)
Rate =   -.100
CV   =   .130
Sample size =   5

Given these parameters, power is estimated to be .64

                        [Note the asymmetry in power
                         between increasing and
                         decreasing trends]

Do you wish to try another calculation? (Y/N) **n**

Results for this run are stored in TRENDS.OUT.

                        [Exit TRENDS, return to
                         system control]

   The estimate of power from program TRENDS must be considered
a maximum estimate of power, i.e., the real power of a planned set of

surveys will be less than the value predicted. This is because program TRENDS does not account for the temporal variation in population change, i.e., the true population is expected to follow the trend line exactly, and not exhibit the temporal variation portrayed in Fig. 5.1. The variability of the true population about the trend line is assumed to be 0 in program TRENDS. This omission becomes obvious when you consider that the only source of variation provided to the program is the coefficient of variation for the first estimate. No other variance components are provided.

To reduce the effect of not incorporating the variation of the true population size about the trend line on the estimate of power, the coefficient of variation of the first population estimate should be inflated to account for the variation of true population size about the trend line.

A competing program called MONITOR is available from James Gibbs.[2] Program MONITOR was designed to explore interactions among the many components of a monitoring program and to evaluate how each component influences the monitoring program's power to detect trends. Program MONITOR assumes a constant variance for the estimates of population size, and so is not as flexible as program TRENDS concerning the relationship between population size and its variance. Program MONITOR computes the power using Monte Carlo simulation models, so, depending on the random number used to initiate the run, slightly different estimates of power are achieved for identical problems.

The power to detect a trend is an important topic of research. Two recent examples that demonstrate the ideas developed here are provided for raptors (Hatfield *et al.*, 1996) and for cougars (Beier and Cunningham, 1996).

6.8. STEP 8—ITERATE PREVIOUS STEPS

Once you have computed the power of your planned series of surveys, you will undoubtedly want to make some changes. Most likely, you will find that your power is far too low, but even if your power exceeds 95%, you would want to change your allocation of resources in order to not waste extra effort in oversampling the population. Thus, the last step is to iterate the previous steps to fine-tune your surveys. You may want to put more effort into sampling the plots, and sample fewer plots, or vice versa depending on the values of $Var(N)$ and $Var(\hat{N})$.

---

[2] Available from the Internet at ftp://ftp.im.nbs.gov/pub/software/monitor.

Example 187

6.9. EXAMPLE

We are interested in assessing the population status of boreal toads. We have identified the area to be sampled as alpine wetlands in the Mt. Zirkel Wilderness Area. These wetlands have been mapped, and numbered, so that we can draw a sample of wetlands to survey from this sampling frame. A total of 300 areas are identified, ranging in size from small ponds and marshes <0.05 ha to larger marshes and lakes that exceed 2 ha.

Because of the logistical difficulty of visiting wetlands, requiring horse-back and foot travel, we decide to use Lincoln–Petersen estimates to enumerate the number of toads on our sampling units. We plan to spend 6 days at a time per crew in the wilderness area to spend alternate nights marking toads, and then surveying the pond again 3 nights later to determine the proportion of toads that are marked. Based on a preliminary survey, estimates of toads range from 0 to over 30 per wetland. We expect to capture about 50% of the toads each night, making n_1 and n_2 both $0.5 \times 30 = 15$. Thus, using Eq. (6.1), we expect $CV(\hat{N})$ to range from 0 to

$$\frac{1}{\sqrt{\dfrac{15 \times 15}{30}}} = 0.365$$

Because our estimate of the population will be the sum of all our sampling units, we anticipate a CV for the total of 0.30.

We want to determine how large of a downward trend (r) per year with the multiplicative model [Eq. (6.6)] would have to occur for the next $n = 4$ years, assuming that we survey each year. With the input of n and CV and using a two-sided test with $\alpha = 0.05$ and $\beta = 0.05$ (i.e., power $= 0.95$), we can use Eq. (6.8) to solve for r. We find that we could only detect a trend of ± 0.468.

Program TRENDS give a less optimistic answer when the t statistic is used, and the simplifications used to derive Eq. (6.8) are removed. The following is the input for TRENDS.

```
You have specified the following input values:

(1)  4                  number of samples (n)
(2)  (to be computed)   rate of change (+/- r)
(3)  .300               initial coeff. of variation
                        (CV1)
(4)  .050               significance level (alpha)
(5)  .950               power (1-beta)
(6)  -2                 1- or 2-tailed test
```

```
(7)   2                        model (1=linear, 2=exponential)
(8)   2                        pattern of CV with abundance A
(9)   2                        distribution index (1=z, 2=t)
```

The estimated rate of decline detectable from TRENDS is

```
*** PROGRAM TRENDS OUTPUT ***

Model of change: EXPONENTIAL
Pattern of variance: CV CONSTANT WITH A
Calculation based on T distribution
Alpha =         .050 (2-tailed)
Power =         .950
CV    =         .300
Sample size =    4

Given these parameters, minimum detectable rate r is
estimated to be -.64
```

The conclusion is the same in both cases: with so little effort put into estimating the toad population on each occasion, the amount of decline that must occur before we are confident we can detect it is much greater than our biological conscience should stand. Thus, we decide to increase our effort to capture toads on each occasion by increasing the number of people involved in the surveys, and estimate that we can get 85% of them each night. This effort results in a $CV(\hat{N})$ of 0.21, so we guess that we can get a CV for the total population of about 0.16. The estimated detectable rate, r from program TRENDS is now -0.42, still leaving us feeling helpless. Thus, we decide to do 6 years of surveys, instead of 4. The new estimate is -0.17, much better, but still marginal. However, 8 years gives us a much better estimate: -0.10.

Two lessons should be learned from this example. First, the large enumeration variation inherent in the Lincoln–Petersen estimate limits your chances of ever detecting a small change in the population. Even though 50% of the population was captured on each occasion, the expected $CV(\hat{N})$ was still large. Second, to detect small rates of change in a population, a larger number of years has a big impact. Basically, the more occasions, the more leverage to detect a shallow slope of the trend.

LITERATURE CITED

Beier, P., and Cunningham, S. C. (1996). Power of track surveys to detect changes in cougar populations. *Wildl. Soc. Bull.* **24:** 540–546.

Cochran, W. G. (1997). Sampling Techniques, Third ed. Wiley, New York.

Gerrodette, T. (1987). A power analysis for detecting trends. *Ecology* **68:** 1364–1372.

Gerrodette, T. (1991). Models for power of detecting trends—A reply to Link and Hatfield. *Ecology* **72:** 1889–1892.

Goodall, D. W. (1952). Some considerations in the use of point quadrats for the analysis of vegetation. *Aust. J. Sci. Res.* **5:** 1–41.

Greig-Smith, P. (1964). *Quantitative Plant Ecology,* second ed. Butterworths, London.

Hatfield, J. S., Gould, W. R., IV, Hoover, B. A., Fuller, M. R., and Lindquist, E. L. (1996). Detecting trends in raptor counts: Power and Type I error rates of various statistical tests. *Wildl. Soc. Bull.* **24:** 505–515.

King, L. J. (1969). *Statistical Analysis in Geography.* Prentice–Hall, Englewood Cliffs, NJ.

Link, W. A., and Hatfield, J. S. (1990). Power calculations and model selection for trend: A comment. *Ecology* **71:** 1217–1220.

Patterson, H. D. (1950). Sampling on successive occasions with partial replacement of units. *J. R. Stat. Soc. B* **12:** 241–255.

Pielou, E. C. (1969). *An Introduction to Mathematical Ecology.* Wiley-Interscience, New York.

Schreuder, H. T., Gregoire, T. G., and Wood, G. B. (1993). *Sampling Methods for Multiresource Forest Inventory.* Wiley, New York.

Scott, C. T. (1984). A new look at sampling with partial replacement. *For. Sci.* **30:** 157–166.

Scott, C. T., and Kohl, M. (1994). Sampling with partial replacement and stratification. *For. Sci.* **40:** 30–46.

Seber, G. A. F. (1982). *Estimation of Animal Abundance,* second ed. Griffin, London.

Skalski, J. R., and Robson, D. S. (1992). *Techniques for Wildlife Investigations—Design and Analysis of Capture Data.* Academic Press, San Diego.

Southwood, T. R. E. (1966). *Ecological Methods.* Methuen, London.

Taylor, L. R. (1961). Aggregation, variance and the mean. *Nature* **189:** 732–735.

Taylor, L. R. (1965). A natural law for the spatial disposition of insects. *Proc. XII Int. Congr. Entomol.* **12:** 396–397.

Usher, M. B. (1991). Scientific requirements of a monitoring programme. In *Monitoring for Conservation and Ecology* (F. B. Goldsmith, Ed.) pp. 15–32. Chapman and Hall, London.

Ware, K. D., and Cunia, T. (1962). Continuous forest inventory with partial replacement of samples. *For. Sci. Monogr.* **3:** 1–40.

Chapter 7

Fish

7.1. Sampling Design
    7.1.1. Overall Goals
    7.1.2. Suggestions
7.2. Fish Collection Methods
    7.2.1. Ichthyocides
    7.2.2. Underwater Observation
    7.2.3. Active Gear
    7.2.4. Passive Gear
    7.2.5. Electrofishing
    7.2.6. Hydroacoustics
    7.2.7. Comparisons among Gears
7.3. Estimating Populations
    7.3.1. Complete Counts

7.3.2. Presence–Absence
7.3.3. Indices of Relative
    Abundance
7.3.4. Capture–Recapture and
    Removal Estimators
7.3.5. Distance Sampling
7.4. Example
7.4.1. Background
7.4.2. Pilot Study
7.4.3. Sample Size Calculations
7.5. Dichotomous Key to
    Enumeration Methods
    Literature Cited

This chapter provides an overview of potential methods to survey fish. We begin by summarizing some appropriate sampling designs, move to an overview of fish collection methods, and finish with a review of available methods to generate abundance estimates. We have provided a dichotomous key to enumeration methods at the end of the chapter to serve as a very general guide to biologists for choosing a method that will suit their needs. Biologists should critically evaluate assumptions underlying any enumeration method considered before implementing it in a full-scale monitoring program.

7.1. SAMPLING DESIGN

7.1.1. OVERALL GOALS

7.1.1.1. Defining the Target Population and Sampling Frame

The goal of any sampling design is to select samples such that appropriate inference can be made to some larger target population, assuming that the sampled and target populations are the same (see Chapter 1). For example, one possible target population could be defined as all fish of a particular species in the state of Colorado. The species may inhabit different bodies of water (Colorado River cutthroat trout in different streams, for example) throughout the state. Because it is impossible to estimate numbers of fish in every stream, a sample of streams must be selected. This list of streams is called the sampling frame. Then we are interested in the fish within each selected stream. Unfortunately, it is probably impossible to sample the entire length of even one stream (Bohlin, 1982), so we must subsample within streams. In statistical terms, we face a multistage sampling problem (Hankin, 1984; Hankin and Reeves, 1988; see Chapter 2). The first stage occurs when we select which streams (or lakes) to survey and the second when we select which locations to sample within a particular stream. A third stage may occur if we cannot completely count fish from a selected stream unit but must rely on incomplete counts such as from a capture–recapture or removal method.

Regarding first-stage sampling (among bodies of water), the simplest conceptual approach is to develop a list or sampling frame of all waters in the state with the species of interest, and to use simple random sampling (SRS) to select which ones to study (Chapter 2 provides details on other sampling schemes). However, for practical reasons, the list of potential waters to study may be narrowed. For example, the list may be narrowed to include only those waters with road access. The statistical consequence is that inferences are now to "all waters with road access' rather than "all waters in Colorado." Another option would be to define the population of interest as fish in "all waters currently known to contain strong (or weak) fish populations." The point is that, statistically, you can only make statements about the fish in the sampling frame from which you drew the random sample (i.e., the sampled population). The real importance of these decisions is administrative in nature. For example, assume that the target population was defined based on road access. If, after conducting a high-quality study, it was concluded that the species was declining, the investigating agency could only assert that declines are occurring in streams with road access. From a statistical standpoint it would be incorrect to assert that the species is declining in Colorado. Would state-wide regulatory action be warranted? Could such actions be successfully defended in court? These are not statistical questions, but the statistical design of the studies could well determine how they are answered. These issues must be kept in mind when defining the first-stage target population to which inferences are to be made. It is imperative that this population is defined before the study starts, and that a valid sampling scheme is used to select which streams or lakes to study.

There is one circumstance in which the above recommendations do not apply, and that is when managers are concerned about a species' status in specific lakes or streams. Thus, studies are designed to determine if the species is declining in streams x, y, and z, but no conclusions are drawn about the species in any other streams. In essence, all first-stage units (i.e., streams) are being measured. Thus, the first-stage units now would be the stream segments within every stream of interest. If studies show that the species is declining in the selected waters, management actions directed at those waters can be fully supported by the data. The question is this: is the study being used to determine the need for state-wide management action, or simply for local actions targeted at specific populations? If the latter is the case, selection of study waters can be made on nonstatistical grounds (e.g., perceived risk to the resource), but inferences can be made only to the measured waters.

Once a sample of streams or lakes is selected for study from the sampling frame, second-stage (within water body) sampling begins. Again, the units within this frame should be explicitly defined. An alternative is to randomly place sampling units within the defined boundaries of the lake or stream (i.e., "undefined" frame; see Chapter 1), although this would probably be better used for lakes or other deep water bodies because of the sharp differences among habitat units in streams. For lakes, defining a sampling frame may not be difficult because the areal extent is defined by the shoreline. However, defining the boundaries for streams and rivers can be less certain. The investigator must delineate each stream by specifying the exact reach of stream from which samples will be drawn.

7.1.1.2. Minimizing Unexplained Variation with Stratification

We have to this point assumed that all lakes or streams, and all reaches within streams or locations within lakes, are the same in terms of their fish populations. Clearly this is not true because some waters have high fish densities, whereas others do not. If obvious differences among and within waters are ignored when the sampling scheme is developed, these differences will increase the uncertainty (variance) of the final population estimate (Bohlin, 1982; Bohlin *et al.*, 1989). Large variances will make it difficult to detect population trends through time. Stratification is a process whereby sampling units are divided into groups (strata) with similar characteristics, and then samples are drawn separately from each stratum. The general principles of stratification were outlined in Chapter 2. Here we provide some specific guidelines from sampling fish populations.

Before going further, it pays to be clear on the distinction between multistage sampling and stratification. There is room for confusion because both involve dividing a sampling frame into smaller subsets. To see the difference, consider the following example. You want to estimate the average weight of tree leaves in a small forest. With multistage sampling, you would randomly select some number of trees for measurement (first stage), and then from each tree randomly select some number of leaves to be weighed (second stage). Now consider that there are pine and oak trees in the woods. Because the weight of pine needles and oak leaves probably differs on average, it would make sense to first stratify trees by species, randomly sample trees of each species, and then randomly sample branches from each chosen tree (assuming a complete measurement of leaves within each selected branch). Average weight of leaves for each species is calculated separately. Note that this design incorporates both stratification (by tree species) and multistage sampling (trees are randomly selected within each stratum and branches are randomly selected from chosen trees). The advantage is that variation in leaf weight due to tree species is explicitly accounted for by the stratification, rather than being "lumped" together with all other sources of variation (leaf-to-leaf differences within trees, tree-to-tree variation within species, etc.). Reducing unexplained variation by stratification will ultimately increase the certainty of our estimated average leaf weight.

With regard to fish surveys, stratification can be done at each stage in the multistage sampling design. For example, a list of all waters in Colorado containing cutthroat trout is prepared. It would make sense to stratify this list into, say, lakes and streams because this species occurs in both. The stream classification could be further stratified, for example, into those affected by mining effluent versus those that are not. Note two things: first, each of these stratifications was made because it seemed reasonable to suspect that fish abundance (the parameter we are interested in estimating) would differ consistently among strata. Second, all of these stratifications apply at the first stage of a multistage design. Within each stratum, a sample of streams (or lakes) will be selected for measurement.

Stratification also can be helpful at the second stage. For example, a single stream may be stratified by location (upper, middle, and lower reaches) and habitat type (pools, riffles, and runs). Again, the goal is to minimize variation within each stratum, but to maximize it between strata. Once the stream is stratified, second-stage sampling determines which locations will be sampled within each location/habitat combination.

The formulas for combining data from different strata to develop a total population estimate (and its estimated variance) are provided in Appendix C. Guidelines for determining the number of samples to take within each stratum also are provided.

7.1.2. SUGGESTIONS

This section provides some specific suggestions for appropriate sampling designs. Discussions are grouped according to whether surveys are conducted in small, wadeable, streams or larger rivers and lakes.

7.1.2.1. Small Streams

The first step in designing an adequate sampling design for small streams is to define the entire number of streams to which inferences are to be made. As noted above, this could include all streams in the state that contain the target species, or a subset based on access or other considerations. Again, you will only be able to make statements about the sampling frame of streams, so be explicit about what that frame is.

Once the streams of interest have been defined, you will probably want to stratify them into subgroups. The goal is to define strata so that abundances of the target species within strata are as similar as possible. This could be done on the basis of total length of stream (assuming densities are about equal among streams, longer ones will hold more fish). If densities are likely to be predictably different among streams (based on elevation, habitat conditions, or other factors), streams should be grouped accordingly (Fig. 7.1). We strongly recommend that pilot work be done to help in identifying appropriate strata. A good stratification scheme can drastically reduce unexplained variation and lead to much greater statistical power for detecting population trends.

After defining strata, a sample of streams from each stratum is selected for study. Various sampling protocols and methods for determining sample sizes in each stratum are provided in Chapter 2 and Appendix C. The simplest method is to divide samples among strata based on the number (or total length) of streams in each stratum. For example, if one stratum contains 15% of all streams in the total identified population of streams, then 15% of all streams selected for measurement should be from that stratum. This method is the best that can be done with limited information.

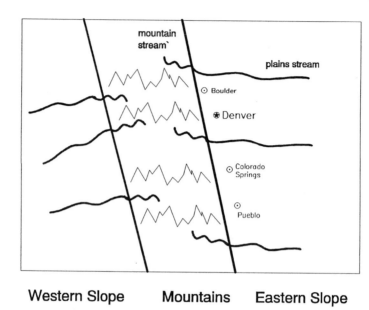

Western Slope   Mountains   Eastern Slope

Figure 7.1 A stratification example. Assume the species of interest lives in streams and rivers throughout Colorado. If it is known that fish densities are consistently different in three regions of the state, three strata could be defined as Eastern Slope, Mountains, and Western Slope, and a sample of streams selected from each.

However, if pilot work has provided estimates of within- and among-strata variation in fish abundance, then more sophisticated methods of allocating samples among strata are available (see Chapter 2).

Once the streams to be measured are identified, a sampling protocol for estimating fish abundance within streams must be decided upon. The first step is to define the exact reach of stream for which the estimate is to be made. All fish within this reach will be the defined target population we wish to estimate. In most cases, the stream reach of interest will be too long to sample in its entirety (Bohlin, 1982), and you will want to stratify before selecting areas for detailed measurements. Typically, in small streams stratification is based on habitat type (pool, riffle, run; see Bisson *et al.,* 1982, for definitions) because different types typically support different fish densities. Note that this leads to sampling units (individual pools, riffles, and runs) of differing size (Hankin, 1984; Hankin and Reeves, 1988).

Once within-stream strata are defined, the second stage of multistage sampling begins wherein a sample of habitats of each type is selected for detailed measurement. For clarity, sampling units will be called habitat units. Our goal is to select habitat units for measurement in such a way as to allow extrapolation to the unmeasured units within each stratum.

Hankin and Reeves (1988) suggested that systematic sampling is best for small streams. The main advantage is that other sampling schemes (random or proportional) require a complete list of sampling units before going to the field. Generating the list can require a significant amount of work before actual sampling begins. This requirement is avoided with systematic sampling in which every k_sth habitat unit (for example, every k_sth pool) is measured. Using the pool stratum as an example, the field crew starts at one end of the target stream reach, generates a random number i between 1 and k_s, and measures the ith pool and then every k_sth pool thereafter. The same procedures are used for other habitat strata. The value of k_s for each stratum is determined based on pilot work and professional judgment, as outlined in Chapter 2.

Hankin and Reeves (1988) also provided methods for estimating the size (surface area) of individual habitat units, and thus the total surface area within and among strata. The advantage is that these "auxiliary" data may allow fish abundance to be estimated with smaller variances than if habitat-unit size is ignored [see Hankin (1984) for details about making estimates using habitat-unit size as an auxiliary variable]. For this to be true, there must be a consistent relationship between the surface area of a habitat unit (within strata) and the number of fish in it. That is, density (fish/m^2) must be constant. Although density often is not constant even within strata, the surface area data can be collected relatively quickly using methods outlined in Hankin and Reeves (1988), and at the same time fish data are gathered. If these data indicate a relationship between habitat unit size and fish abundance, more precise population estimates can be developed using the surface area information. Thus, managers should consider collecting surface area data at the same time fish abundance estimates are made.

7.1.2.2. Rivers and Lakes

Methods for quantitatively estimating total fish abundance in large rivers and lakes are not well-developed (Welcomme, 1975; Gardiner, 1984; Gray *et al.*, 1986; Slaney and Martin, 1987; Bohlin *et al.*, 1989), primarily because effective, nonselective sampling methods are lacking (Hendricks *et al.*, 1980; Beamesderfer and Rieman, 1988; Cone *et al.*, 1988; Johnson *et al.*, 1988; Vadas and Orth, 1993). As a result, estimates of total abundance often are foregone in favor of indices such as catch per unit effort (CPUE; Buynak and Mitchell, 1993; Burkhardt and Gutreuter, 1995; Simonson and Lyons,

1995) or abundance within restricted habitat zones such as nearshore areas or coves (Dauble and Gray, 1980; King *et al.*, 1981; Johnson *et al.*, 1988; Rider *et al.*, 1994). These methods assume that the calculated index is proportional to total population size, allowing trends through time to be detected. Unfortunately, violating this assumption is easy, but detecting the violation is not (Hillborn and Walters, 1992). Given this situation, suggestions for estimating population abundance in deep-water habitats must be tentative.

Many aspects of the multistage sampling design described for small streams apply to rivers and lakes as well. At the first stage, the lakes or rivers of interest are defined and, if appropriate, stratified into groups that should be similar in terms of fish abundance, based on pilot studies or professional judgment. Within each stratum, a sample is selected (see Chapter 2 for sampling designs and guidelines for allocating samples among strata).

The second stage of the design involves sampling within a lake or stream randomly selected during the first stage. Again, it is probable that stratification is justified. For lakes, stratification is usually by water depth (e.g., Espegren and Bergersen, 1990), most simply a division between near-shore, pelagic, and profundal zones. For rivers, stratification could be by habitat type (pool, riffle, run), water depth (near-shore versus midchannel), or both. Sometimes the target species inhabits primarily one zone (some species are usually found near shore, others are pelagic). If so, most or all of the sampling can be restricted to the appropriate areas. However, if the body of water is nonrandomly sampled, the survey data are only an index of abundance and therefore subject to the attending problems (see Chapters 1 and 3). Further, the study is open to "black hole arguments" (Hilborn and Walters, 1992), wherein opponents of resulting management decisions argue that the population did not decline, but simply moved into the areas not sampled. Managers deciding to sample only particular habitats should have proof that the species of interest does not occur outside those areas.

Second-stage sampling involves dividing each stratum into smaller units, and selecting a sample of units for measurement. The number of units sampled in each stratum can be equal (e.g., Espegren and Bergersen, 1990) or proportional to the size (area or volume) of the strata. The size of the sampling unit will depend in part on the sampling gear being used. For example, with active gear such as toxicants, electrofishing, trawls, or hydroacoustics, sample unit size can be specified by the investigator as a specific volume or area of water. In contrast, it is more difficult to define the area sampled by passive gear such as traps or gillnets, where fish must move to the gear to be captured. Therefore, active methods are more appropriate when sampling data from within a specified area will be used to extrapolate to other areas. However, passive gear can be useful for mark–recapture and presence–absence studies.

A third stage may be required if complete counts within second-stage units are not realistic. If this is the case, there are various approaches to estimating fish abundance in lakes and streams, which we will discuss in more detail later in this chapter.

Although there is no proven method for estimating total fish abundance in deep-water habitats, some specific suggestions will be made as a starting point for those designing such a study. There are two basic sampling techniques for estimating fish populations in deep water. The first involves using active gear to make complete counts of fish within specified areas and extrapolating to unmeasured areas. The simplest design would be to use an area-based approach in which the water surface is divided into a grid of sampling units equal to the size of the area for which it is feasible to make a complete count. For example, if a trawl were used, the grid would be based on the width of the net opening and length of the transect to be towed. Hilborn and Walters (1992) suggest that a grid-based systematic sample is best as long as sampling involves more than at least a few grid cells within each stratum. Based on Cochran's (1977) conclusions regarding systematic versus random sampling, they feel that the systematic grid sample will generate more precise abundance estimates because the grid will cut across gradients in fish density. Note that if fish are not evenly distributed from surface to bottom, the grid should be three-dimensional. When the area to be sampled is large, devising a grid-based sampling scheme is difficult, and it may be easier to sample based on randomly generated Cartesian starting points (two- or three-dimensional), rather than a grid system. Another alternative is to define a series of equally spaced transects over the entire water surface from which a random or systematic sample is selected for measurement.

There are three basic methods for expanding from measured areas to unmeasured ones (see Foote and Stefánsson, 1993). For simplicity, assume that we are working within one stratum. The simplest method is to calculate average density (per unit area or volume) in the measured areas, and to apply this density to unmeasured ones. The other two methods use sample data to generate a "contour map" of density in the target area, and numerical integration over that area to estimate total abundance and its variance. The two methods differ in their mathematical approaches. In one, a generalized linear model is fit to the data, producing a "response surface" for integration. In the other, some type of smoothing and interpolation technique (usually kriging) is used. Unfortunately, Foote and Stefánsson (1993) noted that statistically rigorous development of these techniques is lacking, so we cannot make definite recommendations about a preferred technique.

However, conceptually and mathematically, the first method is more straightforward, and Espegren and Bergersen (1990) and Brandt *et al.* (1991) have shown that it can be applied effectively in large and small lakes.

The second general method for making fish abundance estimates in rivers and lakes is to use mark–recapture or removal techniques. These are best applied either when a large fraction of the population (>0.1–0.3; Nichols *et al.*, 1981; White *et al.*, 1982) can be captured during each sampling occasion, or when the population to be estimated is very large (>5000) so that even low sampling efficiency (<0.1) provides a large number of marked fish for release (Hightower and Gilbert, 1984). Although the method has worked well in large systems (Burton and Weisberg, 1994), it also can produce very biased estimates when basic assumptions are violated (Cone *et al.*, 1988). Chapter 3 provides details on methods and assumptions. It is difficult to apply these methods within strata because individual fish may move among strata, violating the assumption of closure. Therefore, they are most suitable when stratification is not used (i.e., the entire body of water can be sampled), or when strata are so large relative to distances moved by the organism that the closure assumption is met. Caution should be exercised when making a judgment about strata size because many fish species are more mobile than currently recognized (Riley *et al.*, 1992; Gowan *et al.*, 1994).

Note that mark–recapture and removal techniques also could be used in conjunction with a grid-based sampling scheme (with consequent extrapolation to unmeasured grid cells), but that grid-cell size must be large so that closure can be assumed. Since mark–recapture estimates within cells have an associated variance, this approach is essentially a three-stage design (bodies of water within the state, grid cells within each body of water, and a sample within each grid cell). For example, in a large river, the grid may be long reaches of equal length from which a subsample is selected (randomly or systematically) and mark–recapture or removal estimates made.

7.1.2.3. General Comments

In order to adequately plan any study, the sample-size and sample-allocation formulas provided in Appendix C should be used. With multistage sampling designs, you will see that different sources of variation are explicitly considered in the formulas. For example, a two-stage design could include among-unit variation due to differences in abundance among locations within a stream, plus enumeration variation resulting from the need to estimate abundance (rather than making census or complete count) in specific locations. A review of the literature on multistage fish sampling shows that among unit variation is almost universally high (Bohlin *et al.*, 1982; Armour *et al.*, 1983; Hankin and Reeves, 1988; Cyr *et al.*, 1992; Hilborn and Walters, 1992; Foote and Stefánsson, 1993). That is, fish tend to be clumped in their distribution so that some locations contain no fish, whereas others contain very many. As a result, it is usually better to trade intensive (i.e., precise) sampling at a smaller number of locations for less intensive sampling at more locations. This is contrary to the way many managers and researchers typically work, but is necessary if fish populations over large spatial scales are to be measured precisely.

Another common approach in fisheries monitoring is the use of index or "representative" sites (King *et al.*, 1981; Bovee, 1982). These sites are typically a nonrandom sample based on the qualitative judgment that conditions within the sites accurately represent conditions over some larger area. Data from these sites are used in two ways. First, total abundance is estimated by extrapolating to unmeasured areas. Because sites are not selected using a statistically valid method, this extrapolation is inappropriate; experience has shown that the resulting abundance estimates can be substantially biased (Hankin and Reeves, 1988). In our view, representative sites should never be used for estimating abundance beyond the boundaries of the site.

A second way index sites are used is for detecting trends through time by periodically measuring abundance. The assumption is that trends at the index site reflect trends in the population as a whole. As noted above, this assumption is easy to violate (Hilborn and Walters, 1992). However, some authors (e.g., Bohlin *et al.*, 1989) have advocated use of fixed sites in monitoring studies, with the caveat that those sites are initially selected in a statistically valid manner. This recommendation is based on an assumed positive correlation for abundance estimates through time. That is, sites tending to hold more fish than average at one time tend to hold more than average at other times as well, even if the average is changing. A strong correlation between periods reduces the variance of the estimated change in abundance through time by reducing among-unit variation, compared to sampling new sites each period. Bohlin *et al.* (1989) provide empirical data to suggest that strong correlations are common, at least for stream fish, and show that sample sizes required to detect a given population change can be substantially lowered using index sites.

We would like to add a couple of notes of caution with respect to using the same sites from one time period to the next. First, selected sites must continue to be representative of the nonselected sites; otherwise, estimates of net change will be biased. For instance, managers should avoid treating some parts of a stream (or adjacent terrestrial habitats that affect the stream) differently because of the presence or absence of monitoring sites (see the discussion of permanent versus temporary plots in Chapter 6 for a possible remedy). Moreover, it appears that fish movement is common in many systems (Riley *et al.*, 1992; Gowan *et al.*, 1994), and these movements can rapidly change the number of fish within a given stream section (Gowan and Fausch, 1996). Consequently, the timing of estimates (relative to stream flow, temperature, and other factors influencing habitat quality) could greatly influence the number of fish present in the site. Each count should be conducted within a similar time period and under similar physical conditions from year to year in order to minimize variation among counts and increase the probability of detecting a trend.

A sometimes unappreciated source of variation in abundance can result from fish behavior, for example from diel migrations between different habitats (e.g., Naud and Magnan, 1988). If these behaviors are unknown to the investigator, sampling at somewhat different times of the day could lead to substantially different estimates of total abundance, and over time could increase total variance and thereby reduce power to detect abundance trends. These problems have recently received attention with regard to hydroacoustic sampling (Appenzeller and Leggett, 1992; Fréon *et al.*, 1993; Luecke and Wurtsbaugh, 1993), but certainly affect other situations as well (Magnan, 1991). As with any research, population estimation requires a good understanding of the natural history of the organisms being studied. A proper emphasis on rigorous statistical design complements the talents of a knowledgeable biologist, but does not replace them.

A final comment has to do with sample size. Given high sampling variability in fish abundance, and the cost of sampling, it is probable that required sample sizes calculated using formulas provided in Appendix C will be unrealistically large (Armour *et al.*, 1983). In these circumstances, the best approach may be to calculate the precision that can be achieved given budget and time constraints. If this precision is unacceptably low (i.e., variances are large), the scope of the study should be revised until reasonable precision can be achieved with the money available. The point is that unreliable results are not worth the money.

7.2. FISH COLLECTION METHODS

7.2.1. Ichthyocides

Ichthyocides (primarily rotenone, antimycin, or sodium cyanide; Lennon, 1970) are probably the least selective sampling method available (Lambou and Stern, 1958; Hendricks *et al.*, 1980). Rotenone and sodium cyanide are fast acting (several minutes to 2 h) compared to antimycin (10–24 h) and therefore preferred for the types of sampling discussed above (Hendricks *et al.*, 1980). Sodium cyanide is least expensive and can be used in concentrations that anesthetize but do not kill, but it is toxic to humans, special handling is required, and threats to domestic water supplies may cause public concern (Platts *et al.*, 1983; Wiley, 1984). For these reasons, rotenone is the most commonly used ichthyocide. Methods for applying and detoxifying rotenone are presented in Davies and Shelton (1983). Although generally nonselective in terms of lethality (at appropriate concentrations), rotenone-killed fish still must be collected, and this can introduce bias due to differential retrieval efficiency among habitats, species, and fish sizes (Tarzwell, 1942; Boccardy and Cooper, 1963; Shireman *et al.*, 1981; Bayley and Austen, 1990). In streams and large bodies of water, block nets are required to keep killed fish within the sampling area while collections are being made, sometimes over a period of days (Wiley, 1984). Some investigators (Carlander and Lewis, 1948; Davies and Shelton, 1983; Jensen, 1992) suggested that marked fish be used to estimate retrieval efficiency for population estimation. Jensen (1992) provided a thorough statistical treatment of the issue. Johnson *et al.* (1988) described a quadrat rotenone technique and two-stage sampling design that does not require fish marking. Bayley and Austen (1990) developed a multiple-regression model to estimate sampling efficiency of rotenone that may be widely applicable to warmwater lakes.

7.2.2 UNDERWATER OBSERVATION

Enumeration using direct underwater observation is a common technique for estimating fish abundance, especially in flowing waters. Typically, divers using snorkel or scuba gear move upstream through designated stream sections identifying and counting fish [see Helfman (1983) and Thurow (1994) for details on methods]. One diver can be used in small streams, whereas multiple divers swimming in prescribed lanes are required for larger rivers. Although some investigators imply that counts can be treated as a complete census or that consistency between divers indicates unbiased estimates (Platts *et al.*, 1983), there is ample evidence that accuracy of diver counts is strongly influenced by water depth, temperature and clarity, habitat complexity, fish abundance and behavior (including species differences), time of day, and diver technique and experience (Northcote and Wilkie, 1963; Goldstein, 1978; Schill and Griffith, 1984; Slaney and Martin, 1987; Heggenes *et al.*, 1990; Graham, 1992; Hillman *et al.*, 1992; Rodgers *et al.*, 1992). Thus, unadjusted diver counts should not be used as an unbiased measure of fish abundance.

Despite potential biases in diver counts, the method offers several important advantages. In clear streams it is particularly useful in water too deep, fast, turbulent, low in conductivity, or remote for other methods such as electrofishing or seining to be practical (Schill and Griffith, 1984; Slaney and Martin, 1987; Thurow, 1994). More importantly, snorkeling requires much less equipment, time, and manpower (i.e., cost) compared to other methods, making it particularly effective when fish abundance must be estimated over large spatial scales. In lakes, counts are made using scuba equipment or underwater video (Bergstedt and Anderson, 1990). However, although direct observation could be successfully applied to lakes, statistical methods to generate valid estimates of fish abundance using the distance sampling techniques of Buckland et al. (1993) have not been applied in most studies [but see Ensign et al. (1995)].

7.2.3. ACTIVE GEAR

A variety of active gear can be used to collect quantitative samples from a known volume of water for extrapolation to larger areas. Typical gear includes [see Hayes (1983) for details] bottom, midwater, or surface trawls (nets with rigid openings towed through the water by boat), seines (nets deployed by hand or boat that encircle fish), pushnets (mounted on the bow of a boat), pop nets (deployed on the bottom and quickly raised to the surface by inflating a bladder on the net frame), and drop nets (dropped from the water surface). All active gear is known to be variably selective in different habitats, water temperatures, and turbidities, and for different species and sizes of fish, so complete counts should not be assumed (Hayes, 1983; Vadas and Orth, 1993). Efforts to quantify gear selectivity (e.g., Parsely et al., 1989; Tinsely et al., 1989; Pierce et al. 1990; Millar, 1992; see the bibliography by Dahm, 1987) show that simple generalities are not possible and that site-specific studies are required in each instance. In fact, Hayes (1983) suggested that gear performance can be so variable that continuous monitoring during a study may be necessary to ensure unbiased results. Overall, active gear is a valuable sampling tool, but the potential for bias is high.

7.2.4. PASSIVE GEAR

Passive gear includes any device requiring that fish be moving for capture, including gill, trammel, hoop, fyke, and trap nets, minnow or pot traps, and fish weirs [see Hubert (1983) for details]. Because it is usually impossible to quantify the area or volume being sampled, passive gear is best suited for presence–absence or mark–recapture studies where quantitative samples are not needed. Moreover, passive gears are extremely selective, but quantifying this selectivity often is difficult (Hubert, 1983). Catch can be substantially influenced by choice of gear location and sample timing (He and Lodge, 1990; Magnan, 1991), spatial distribution and behavior of the target species (Allen *et al.*, 1960), relative abundance in multispecies communities (He and Lodge, 1990), gear spacing (Ryan and Kerekes, 1989), and fish size (Mattson, 1994). Perhaps more effort has been put into estimating selectivity of gill nets than any other gear (Winters and Wheeler, 1990; Helser *et al.*, 1991; Henderson and Wong, 1991; Borgstrøm, 1992; Borgstrøm and Plahte, 1992; Allison *et al.*, 1994; Helser *et al.*, 1994; Machiels *et al.*, 1994; Mattson, 1994; Acosta and Appeldoorn, 1995; Castro and Lawing, 1995; for a review of earlier work, see Hamley, 1975), but widely applicable results remain elusive. Other passive gears are likely to be equally selective, but detailed studies are lacking (Hubert, 1983).

7.2.5. ELECTROFISHING

Electrofishing is possibly the most popular method for sampling fish populations (Bohlin, 1981), but, surprisingly, there is not a complete physiological model for how it works (Kolz, 1989). What is known is that all aquatic organisms are immobilized (electronarcosis) when the voltage gradient along the body from nose to tail exceeds a certain value (Bohlin *et al.*, 1989). Higher voltages result in death, lower ones in a fright response (Reynolds, 1983). The goal is to maximize the size of the stun zone relative to death and fright zones. Details on the electrical and physiological theory behind electrofishing are provided in Northrop (1967), Reynolds (1983), Bohlin *et al.* (1989), Kolz (1989), and Snyder (1993).

Electrofishing gear consists of three major components: a power source (a generator, usually producing alternating current, or a battery), a transformer to convert current from the power source to different voltages or to direct current, and electrodes placed in the water to create an electrical field. In general, direct current (DC) is preferred over alternating current (AC) because it produces an "attraction" zone within which fish actively swim toward the anode (galvanotaxis), is usually less injurious to fish, and is less dangerous for operators (Hendricks *et al.*, 1980). Pulsed DC requires less voltage than unpulsed DC to achieve comparable stun zones (Reynolds, 1983), but may cause more injuries than unpulsed DC (Snyder, 1993). Despite advantages of DC, AC produces larger stun and death zones and may be preferable when capture efficiency takes priority over minimizing fish injury. Alternating current most often is used in boat-mounted systems for lakes and larger rivers (Heidinger et al., (1983).

Water conductivity is probably the single most important factor influencing electrofishing efficiency (Bohlin et al., 1989). At low conductivities (<75 S/cm) typical of small streams and oligotrophic lakes, small generators or batteries provide sufficient power to produce voltages needed to stun fish, allowing use of backpack-mounted systems. At higher conductivities, larger generators are necessary, requiring either shore-based configurations that are difficult to move, or boat-mounted gear.

Reynolds (1983) provided general designs for backpack and boat-mounted electrofishing gear. Numerous variations have been proposed. Electric seines (Angermeier et al., 1991) are effective in small and medium streams, and may provide better estimates of species richness and relative abundance compared to backpack electrofishers (Bayley et al., 1989). Aadland and Cook (1992) described an electric bottom trawl useful for sampling benthic fish in deep, turbid streams. Electric samplers prepositioned on the stream bottom offer promise for sampling benthic fish (Weddle and Kessler, 1993) and for use in stratified designs requiring sampling from quadrats of known size (Bain et al., 1985; Fisher and Brown, 1993).

As with all sampling gear, electrofishing is known to be selective, with fish length probably being the most important physiological variable (Bohlin et al., 1989). Capture efficiency is usually higher for large fish (Borgstrøm and Skaala, 1993) due to the greater voltage gradient produced across their bodies (Regis et al., 1981). Size selectivity can be reduced by using higher voltages (Stewart, 1975), but increased injury and mortality can result (Snyder, 1993). Species differ in internal conductivity and behavioral response to electric fields, and these also can affect capture efficiency (Sternin et al., 1972; Reynolds, 1983). Finally, characteristics of the habitat being sampled are extremely important, and thus capture efficiency can differ by species due to differing habitat preferences. Fish are difficult to see and net in turbid water, so the attraction produced by DC is especially important under these conditions. Similarly, DC usually works better than AC when fish are hidden in cover. Substrate size and composition affect strength and shape of the electric field, with soft-bottom substrates generally reducing capture efficiency more than coarse ones (Reynolds, 1983). Different results can be obtained in day versus night (Paragamian, 1989).

An emerging drawback of electrofishing is the potential for fish injury (Hauck, 1949; Sharber and Carothers, 1988; Hollender and Carline, 1994; Sharber et al., 1994). Based on this, Snyder (1995) recommended that "where electrofishing injury is a problem and cannot be adequately reduced, the technique must be abandoned or severely limited." However, Schill and Beland (1995) noted that compared to other sampling techniques (e.g., ichthyocides), electrofishing is relatively benign, and that injury or mortality imposed when sampling a relatively small fraction of a population is unlikely to result in significant risk to the resource. Besides direct injury or mortality, electrofishing is known to cause behavioral changes that tend to reduce catchability for several hours to 1 day (Cross and Stott, 1975; Mesa and Schreck, 1989), effects that could violate assumptions of certain techniques for estimating abundance, such as capture–recapture.

7.2.6 HYDROACOUSTICS

Hydroacoustic monitoring is an increasingly popular and sophisticated method to enumerate fish [see Thorne (1983) for a general overview]. Although in theory hydroacoustic surveys should be less biased than methods requiring capture or direct observation of fish, the potential for error still is great. Fish counts can be affected by miscalibration of the gear, errors in signal processing, prevailing conditions that produce false echoes (e.g., air bubbles or plankton), extinction of the signal by dense schools of fish, avoidance of the boat by fish, or by "dead zones" near surface, bottom, and shallow waters in which fish cannot be detected (Foote and Stefánsson, 1993). In particular, the near-bottom dead zone can substantially affect population estimates conducted in day versus night, or during a full moon, because many pelagic fishes remain near bottom during high-light conditions and are undetectable. For this reason, hydroacoustic sampling is usually most effective at night (Burczynski and Johnson, 1986; Fréon *et al.*, 1993; Luecke and Wurtsbaugh, 1993). Perhaps most troubling, even when hydroacoustics detect fish, identification of species, age, and size composition usually requires additional sampling using selective gear described in the preceding section (Thorne, 1983). Fortunately, increasingly sophisticated equipment will continue to reduce these problems (Mathisen, 1992; Guillard *et al.*, 1994) and is likely to make hydroacoustics the most valuable method for accurately estimating fish abundance in large waters.

The major advantage of hydroacoustics is that large samples can be collected relatively quickly and at low cost relative to other gear. In addition, split- and dual-beam transducers allow identification of both target strength (a metric of fish size; Love, 1971) and target location in three dimensions (Thorne, 1983). Locating fish in three-dimensional space facilitates a stratified sampling scheme wherein estimated fish densities within a depth stratum are expanded to absolute abundance based on total water volume within strata (e.g., Luecke and Wurtsbaugh, 1993). Alternately, linear models or interpolation (kriging) techniques can be used to generate a contour map of fish density that then is integrated to estimate total abundance

(Foote and Stefánsson, 1993). However, there is still much debate over proper selection of transect lines and statistical treatment of resulting data (Baroudy and Elliott, 1993; Foote and Stefánsson, 1993). In the simplest case, equally spaced transects are defined across the body of water in question, a random sample of lines selected (Appenzeller and Leggett, 1992), and survey data expanded based on area or volume. However, the efficiency of this approach (both statistically and logistically) is site specific and alternative designs should be considered (Baroudy and Elliott, 1993).

7.2.7. Comparisons among Gears

Because gear selectivity long has been recognized by fishery biologists, there are a number of studies that specifically compare efficiency of various gears under different conditions. To help managers select the appropriate gear for their applications, a sampling of such studies is summarized in Table 7.1.

7.3. ESTIMATING POPULATIONS

7.3.1. Complete Counts

Obtaining a census of fish over a large area is not possible, partly because all sampling gear is selective to some degree. One exception may be small ponds or controlled streams that can be completely dewatered and fish enumerated directly (e.g., Cone *et al.*, 1988; Espegren and Bergersen, 1990). Alternately, some collection methods are very effective over small spatial scales, and researchers often assume a census (complete count) within sampled areas and estimate total abundance by extrapolating these counts based on surface area or volume. Some examples follow.

Quadrat rotenoning techniques [Shireman *et al.*, 1981; Johnson *et al.*, 1988; Jensen, 1992; see also Metzger and Shafland (1986) for a discussion of sampling with explosives as an alternative to rotenone] are commonly used in reservoirs to obtain quantitative samples from a known area or volume of water. Johnson *et al.* (1988) used a two-stage sampling design to estimate age-0 gizzard shad abundance in a small pond, relying on standardized, prepositioned bottom nets to collect quantitative (second stage) subsamples of dead fish from within enclosed 0.15-ha (first-stage) quadrats to which rotenone was applied. This approach obviates the need for marked fish and allows use of multistage sampling formulas for calculating abundance and variance estimates, as well as study planning. Jensen

Table 7.1

Summary of Studies Comparing Different Gears in Different Habitats

Gears compared	Habitat	Preferred gear	Source
Electrofishing, seines	Warmwater streams	Electrofishing	Wiley and Tsai (1983)
Electrofishing, seines	Coolwater river	Varies	Dauble and Gray (1980)
Electrofishing, drop net, pop net	Warmwater lakes	Pop net and drop net in vegetation, any gear otherwise	Dewey (1992)
Electrofishing, visual counts from underwater or stream banks	Coldwater streams	Electrofishing in fast, shallow water, visual counts otherwise	Heggenes *et al.* (1990)
Electrofishing, stream bank counts	Coldwater streams	Either for fry; electrofishing for adults	Bozek and Rahel (1991)
Electrofishing, rotenone	Coldwater streams	Rotenone	Boccardy and Cooper (1963)
Electrofishing, rotenone	Warmwater lakes	Electrofishing	Rider *et al.* (1994)
Electrofishing, rotenone	Warmwater river	Rotenone	Jacobs and Swink (1982)
Electrofishing, rotenone, explosives	Warmwater streams	Electrofishing	Leyher and Maughan (1984)
Seines, diving[a]	Coldwater streams	Seines	Rodgers *et al.* (1992)
Seines, diving	Warmwater streams	Diving	Goldstein (1978)
Seines, fyke nets, gill nets	Warmwater lakes	Varies	Weaver *et al.* (1993)
Rotenone, detonating cord	Warmwater ponds	Rotenone	Bayley and Austen (1988)
Rotenone, diving	Warmwater lakes	Either	Dibble (1991)

Preferred gear is based on recommendations of the investigators, but in our view should not be taken as definitive. Readers must review the original publications before drawing conclusions.

[a] The main purpose of the study by Rodgers *et al.* (1992) was to compare abundance estimates made by mark–recapture, removal, and visual counts. Mark–recapture and removal estimates were based on data obtained from seining, and these estimates were closer to the known abundance of juvenile salmon in the test sections than were snorkel counts.

(1992) provided additional statistical details on multistage sampling using quadrat rotenone techniques. As with most quadrat techniques, his methods require fish marking prior to rotenoning so that estimates of collection efficiency are obtained.

Electrofishing has been suggested for quadrat sampling to estimate fish abundance. Bain *et al.* (1985) tested prepositioned, 23- or 5.7-m² electrofishing frames set on stream bottoms to estimate fish density. Based on tests in which known numbers of fish were held inside the frame prior to activation, Bain *et al.* (1985) concluded that the sampler provided a complete census. They suggest that fish abundance could be estimated using a stratified (by habitat) sampling design, but did not specifically test this approach. Other investigators used area-sampling techniques to estimate fish density for extrapolation to unmeasured habitat, but typically complete censuses were not assumed. Instead, mark–recapture or removal estimators were used to measure quadrats (Pajos and Weise, 1994).

Visual counts often are used in clearwater streams to estimate fish abundance. Recall that in the multistage sampling designs summarized above, it was almost universally true that variation due to patchy fish distributions among sampling units was much higher than that due to estimation errors within units. Based on this, Hankin and Reeves (1988) developed statistical methods estimating stream fish populations with diver counts. They used regression techniques to relate diver counts within sampling units to more rigorous estimates obtained by electrofishing, then used adjusted diver counts from a much larger number of habitat units to estimate total abundance of two salmonid species in a 9.6-km reach of a small stream in Oregon. As expected, uncertainty related to diver counts was a small part of the total variance of the abundance estimates. Despite high spatial variation in fish abundance, overall coefficients of variation (CVs) on the abundance estimates were only 8.4–10.8% because a large number of habitat units could be sampled, i.e., about 18% of all units in the 9.6-km reach. The methods of Hankin and Reeves (1988) are likely to be the most efficient and cost-effective way to estimate fish abundance in small streams and larger rivers suitable for diving.

Underwater observation also has proven effective in lakes using techniques adapted from work on coral reef communities (Sale and Douglas, 1981; Bohnsack and Bannerot, 1986). Keast and Harker (1977) used strip transects to estimate densities of centrarchids in three habitat zones (weed, gravel, and sand bottoms) in an Ontario lake. Because successive counts over the same 1.5-m-wide strip transects (transect length ~15–55 m) were generally consistent (CV = 1–26% depending on habitat type), Keast and Harker (1977) assumed that counts were a complete census. Graham (1992) used diver counts on point transects to estimate centrarchid densities near artificial structures in a 3885-ha impoundment in Virginia. Graham (1992) did not assume a complete census at each point transect, but used an *ad hoc* method to adjust counts based on estimated "fish visibility" reported by divers. Dibble (1991) used strip transects to estimate relative abundance of a mixture of 21 warm water species in an Arkansas lake. Although care was taken to randomly select transect locations, diver counts were assumed to be complete counts. Subsequent rotenone sampling indicated that diver counts were biased by fish size and species, with larger, near-shore species overrepresented. Bergstedt and Anderson (1990) showed that distance sampling (see Buckland *et al.*, 1993) could be used successfully underwater, and we strongly recommend this approach over the *ad hoc* methods just described.

Active gear, although very selective, can yield good results in the right circumstances. For example, Espegren and Bergersen (1990) used a rising net ("pop net") to estimate fish abundance in a small trout pond. After stratifying the lake into four depth categories, they randomly sampled 40 locations in each using a net that was positioned on the bottom, allowed to sit undisturbed for 10–12 h, and then retrieved by inflating a bladder on the frame that raised it quickly (ca. 5 s) to the surface. All fish captured within the 92.9-m^2 net were counted and returned to the lake. Confidence intervals on population estimates extrapolated from these samples, based on surface area or volume, included the true population size determined when the lake was drained. Although Espegren and Bergersen (1990) assumed that the net was 100% efficient under experimental conditions, they confirmed this in a small-scale pilot study, and reported results from Larson *et al.* (1986) indicating that fish were neither attracted nor repelled from a slowly rising pop net. Cyr *et al.* (1992) showed that estimates of larval fish abundance in a 250-km-long reach of the Hudson River could be estimated with trawl data collected in a stratified random sampling design. They also provided details on estimating required sample sizes to achieve given precision, based on literature and empirical data for the relationship between mean abundance and sampling variance.

Hydroacoustics is becoming an extremely important tool for estimating fish abundance in large bodies of water. In a dramatic example, Brandt *et al.* (1991) estimated total abundance of pelagic fishes in Lake Michigan at 43.4 ± 10.1 billion during the spring of 1987. This precise estimate was obtained using a stratified sampling design (although transects were not randomly selected) requiring only 10 days of boat time. Because dual-beam equipment was used, fish could be located in three dimensions and estimates based on either surface area or volume. Estimates from the two methods differed by only 9–11%. Brandt *et al.* (1991) provided a good discussion of potential sampling biases, particularly the need to estimate species composition from trawl data that were known to be selective for certain species. Burczynski *et al.* (1987) achieved good precision (CVs = 2.8–13.9%) using a stratified systematic design to estimate pelagic fish abundance in a large North–South Dakota reservoir, as did Burczynski and Johnson (1986) for a British Columbia lake. Burczynski and Johnson (1986) summarized statistical methods to expand acoustic estimates of fish density to lakewide abundance estimates.

7.3.2. PRESENCE–ABSENCE

Sampling to assess spatial distribution (presence–absence) of a species can be accomplished using any gear described thus far, including passive gear (such as traps, gill nets, or trammel nets) that requires fish to move to the gear. Once again, all gear is selective so it is especially important in distribution surveys to employ a variety of gear-types to ensure detection of the species interest (Bain, 1992; Vadas and Orth, 1993; Weaver *et al.,* 1993).

Key points in presence–absence sampling are to decide how large an area to inspect, and how intensively to sample. These issues were studied for streams by Angermeier and Schlosser (1989), Lyons (1992), and Paller (1995), but their general conclusions should hold for rivers (Lyons, 1992) and lakes as well. The number of species detected increases with size of area sampled because total effort is greater (i.e., more individuals are examined) and additional habitats are encountered. Similarly, greater sampling intensity within a specified area (e.g., more traps set) yields more species because more individuals are captured. In both cases, an asymptote is reached wherein sampling additional area, or sampling more intensively, yields no new species. Naturally, there are trade-offs between sample area and intensity so, for example, the asymptote for area is reached more quickly if sampling is more intensive.

In theory, presence–absence sampling is designed so that the asymptote is achieved using the least-cost combination of sampling intensity and sample area. If the species of interest is not detected, it is assumed not present. In practice, the most efficient gear available should be used, and sampling continued until some predefined criteria for stopping is achieved. For example, in small and medium streams (4.9–17.2 mean width at base flow) with a diverse warmwater fauna (ca. 30 species), Lyons (1992) continued sampling until no new species were encountered in 50 m of stream length. Although this distance seems short, data from Lyons (1992) indicated that it was necessary to sample a linear distance equal to 35 times the mean stream width to achieve 95% of the asymptotic species richness. Paller (1995), also working in warmwater streams with about 30 species, found that 35–158 stream widths were necessary. These relatively long distances were required despite use of sampling gear (electrofishing) capable of capturing 67–90% of individuals in the sampled reaches (Paller, 1995).

In presence–absence work, it is tempting to define sample area based on inclusion of all meso- and microhabitats, under the assumption that all species will be represented. This is a poor assumption because, especially for rare species, a large amount of suitable habitat may need to be searched before the species is encountered (Paller, 1995). Moreover, because fish distributions tend to be clumped, intensive sampling in a small area is not a substitute for sampling over a large area. For these reasons, it is better in most cases to sample a large area with low effort than a small area with high effort.

7.3.3. Indices of Relative Abundance

By far the most common index to abundance in fisheries work is catch per unit effort (CPUE). The assumption is that the number of fish caught per unit of effort expended (often time) is proportional to stock size. However, experience from commercial fisheries shows that CPUE can remain high in the face of a rapidly declining stock, or decline even if the stock is relatively stable (Hilborn and Walters, 1992).

The lack of a linear relationship between CPUE and stock size is due, in part, to characteristics of the fishery. For example, if stocks are declining, CPUE will stay high if efficient gear is increasingly concentrated in areas where fish density remains high. Further, CPUE for any particular gear can depend on season, water temperature and turbidity, and fish behavior, size, and species (Hubert, 1983). Thus, for CPUE to have any value in stock assessment, managers must standardize sampling protocols through time. For this reason, we will not consider CPUE methods that rely on data from commercial or sport fisheries [e.g., creel surveys; see Demory and Golden (1983) and Malvestuto (1983) for methods to estimate CPUE from commercial and recreational fisheries data]. Also, some authors use the term CPUE in reference to removal estimators, such as Leslie or DeLury methods [see Seber (1982) and Ricker (1975)], in which CPUE or its logarithm is plotted against cumulative effort to estimate absolute abundance. Here, we use CPUE in the strict sense of an index of abundance, and summarize two of the more common methods for CPUE sampling, electrofishing and passive gear [see Stevens *et al.* (1985) and Wilson and Weisberg (1993) for examples where CPUE is estimated using active gear].

Catch per effort indices often are used to test for population trends in large rivers and lakes because estimating absolute abundance is extremely difficult. Electrofishing is a popular CPUE sampling method because, in theory, effort can be standardized by fishing at a given voltage or amperage for a given period of time (Hendricks *et al.*, 1980). Burkhardt and Gutreuter (1995) argued that improved consistency can be gained by standardizing power (wattage), a function of the difference in conductivities between fish and water. However, in an analysis of data from 278 electrofishing collections in a variety of habitats in the Mississippi and Illinois Rivers, Burkhardt and Gutreuter (1995) concluded that variation in power explained only 12.1–14.9% of the variation in catch. Unfortunately, characteristics of the fish, habitat, operating conditions, and crew also influence catch rates (Reynolds, 1983; Hardin and Conner, 1992; Hill and Willis, 1994), so it is difficult in practice to achieve equal effort among samples. Despite this difficulty, electrofishing CPUE has proven a useful index of abundance in some circumstances. Serns (1982, 1983), Hall (1986), Coble (1992), Buynak and Mitchell (1993), McInerny and Degan (1993), and Simonson and Lyons (1995) demonstrated significant relationships between CPUE and absolute abundance (usually estimated by mark–recapture or removal) for various species in lakes and streams throughout the United States [see Parkinson *et al.* (1988) for a discussion of sample size requirements to detect changes in CPUE through time].

Passive gear offers another option for estimating CPUE, with the advantages of ease of use and low cost (Hubert, 1983). Although selective, if gear is standardized and fished at similar times under similar conditions, reasonable results can be obtained (Welcomme, 1975; LeCren *et al.*, 1977; Ryan, 1984; Hamley and Howley, 1985; Kelso *et al.*, 1986). However, Ryan and Kerekes (1989) showed that CPUE calculated from experimental gill net sampling was substantially influenced by fishing intensity (nets set per hectare of lake surface), demonstrating that small variations in study design could qualitatively alter conclusions about relative fish abundance. Moreover, Hilborn and Walters (1992) warn that passive gear is subject to saturation (traps or nets becoming full), such that CPUE may remain constant over a broad range of fish densities. For example, Bernard *et al.* (1993) demonstrated an asymptotic relationship between hoop-trap CPUE and abundance of burbot in 15 lakes ranging in size from 130 to 6519 ha.

7.3.4. CAPTURE–RECAPTURE AND REMOVAL ESTIMATORS

Capture–recapture and removal methods are popular for estimating fish abundance in lakes and streams. Summaries of these approaches are provided in Chapter 3 and White *et al.* (1982).

7.3.4.1. Capture–Recapture

The most common capture–recapture method is Chapman's modification of the Petersen[1] estimator (see Ricker, 1975), in which abundance is estimated from the number of fish marked and released during an initial sampling phase, and the ratio of marked to unmarked fish in a follow-up phase. However, this estimator assumes (1) no recruitment into the size class of interest (i.e., closure), (2) no individual heterogeneity in capture probability, (3) no effect of marking on capture probability or survival relative to unmarked fish, and (4) no marks lost or missed. In particular, the final three assumptions are difficult to meet in fisheries applications because all sampling gear is selective, making some individuals readily catchable and some essentially immune from capture, and because handling and permanently marking fish without affecting behavior or survival may be impossible (Arnason and Mills, 1987; Cone *et al.*, 1988). For example, Mesa and Schreck (1989) showed that stream salmonids captured by electrofishing were less susceptible to recapture for a period of about 24 h due to physiological and behavioral changes. Jacobs and Swink (1982) reported similar results for warmwater species in two Kentucky rivers.

Whenever possible, multiple mark–recapture methods described in Chapter 3 should be used because they have fewer assumptions, and can account for differences in behavior following initial capture [see Bozeman *et al.* (1985) for an example]. If the Chapman estimator is applied, different gear should be used during capture and recapture phases to help eliminate the effects of gear bias (Jacobs and Swink, 1982; Seber, 1982), but even a combination of gear may not solve the problem (Beamesderfer and Rieman, 1988). Despite these obstacles, innovative designs can yield good results. Burton and Weisberg (1994) generated estimates of larval fish abundance in a 67-km reach of the Delaware River using marked hatchery fish released randomly and recaptured with trawls. Rodgers *et al.* (1992) reported that mark–recapture techniques tested in salmonid streams in Oregon produced accurate population estimates when seines were used as the method of capture. Palmisano and Burger (1988) estimated salmon abundance in a turbid Alaskan river.

Capture–recapture methods can be applied when the assumption of population closure is not met. These Jolly–Seber estimators are best suited to estimate survival between sampling periods (Burnham *et al.*, 1987), but can be used for estimating abundance as well. For example, Hightower and Gilbert (1984) used open-population models to estimate fish abundance in a 1120-ha reservoir in South Carolina. However, precision of the estimates

---

[1] This is designated the Lincoln–Petersen estimator in terrestrial literature and elsewhere in this book.

was low (CV = 27–65%), in part because there are more parameters to estimate in open compared to closed population models (Cormack, 1979). Moreover, Cone *et al.* (1988) showed that assumptions of the model could be easily violated, the violations were largely undetectable, and that resulting abundance estimates could be in error by 40% or more. These results were especially troubling because sampling efficiency in the small lake being studied was quite high during each period with 0.2–0.3 of the population captured. One of the main problems encountered by Cone *et al.* (1988) was heterogeneity in capture probability among fish, due in large part to nonrandom sampling. Fishing effort was concentrated along the shoreline, but a large proportion of the population apparently remained near the center of the lake and was never available for capture. These results, and those of Bernard *et al.* (1993), show that in any capture–recapture study, care must be taken to ensure either that marked fish are randomly mixed within the total population, or that the recapture effort is conducted randomly within the study area. Note that individual heterogeneity in capture probability can be modeled using the closed-population models of White *et al.* (1982), but even these cannot detect when a segment of the population is fully unavailable for capture. An appropriate sampling design that includes all habitats helps reduce this possibility.

7.3.4.2. Removal

Removal estimators typically applied in fisheries come in two forms, both of which assume population closure and require relatively high capture efficiencies (>0.3) to be effective (Bohlin, 1982). The first form is regression-based techniques such as DeLury and Leslie estimators (Ricker, 1975) in which CPUE is regressed against cumulative catch. The x-axis intercept is the population estimate. Although widely applied, these estimators have a number of assumptions that are difficult to meet, especially equal capture probability for all individuals and sampling periods (Miller and Mohn, 1993). Because all gear is selective, the more vulnerable animals tend to be captured first, resulting in reduced capture probabilities during each successive sample (Mahon, 1980). The consequence is that abundance is underestimated (Mahon, 1980; Kelso and Shuter, 1989). Moreover, regression estimators will give larger confidence intervals compared to maximum likelihood estimators. Although some successful applications have been reported (Maceina *et al.*, 1993, 1995; Rider *et al.*, 1994), we recommend against CPUE methods.

The second form of removal estimator uses maximum-likelihood techniques based on numbers captured during successive sampling periods [the "generalized" removal estimator of White *et al.* (1982)]. The main advantage of this approach is that constant capture probability between samples is not assumed. However, the method requires that changes in capture probability occur in some predicable way as sampling proceeds, usually decreasing as the more easily captured animals are removed during initial samples. Capture probability should not fluctuate randomly from sample to sample.

Despite relaxation of the assumption of equal capture probability among sample periods, the generalized removal estimator still tends to produce underestimates of abundance, particularly if only two samples are taken (Riley and Fausch, 1992), or if capture probability or abundance of the target species is low (Bohlin, 1982; Riley *et al.*, 1993). However, good results can be obtained, especially using electrofishing in small streams when capture probability is high, at least three samples are made over a short period of time, and population closure is achieved using block nets (e.g., Gowan and Fausch, 1996).

7.3.5. DISTANCE SAMPLING

Line transect techniques have not typically been applied to freshwater fisheries applications [but see Ensign *et al.* (1995)], but are used to estimate marine mammal populations (Anganuzzi and Buckland, 1989; Buckland *et al.*, 1992), and fish abundance on coral reefs (Sale and Douglas, 1981). Studies using line or point transects in freshwater have either assumed completed counts (Keast and Harker, 1977; Dibble, 1991) or used *ad hoc* methods to estimate detection rates (Graham, 1992). Certainly, statistically rigorous distance sampling methods described by Buckland *et al.* (1993) could be applied. The main problem may be the assumption that the animal's position is fixed at the time of detection (or at least that its movements are slow relative to the observer's). Even for territorial fish species tending to hold a fixed position, disturbance due to divers is likely to elicit a fright response and invalidate the assumption of fixed position. However, this may be accounted for by grouping observations into appropriate distance categories, i.e., categories that are large enough to account for any fish movement but small enough to be usable. Buckland *et al.* (1993) described techniques for applying distance sampling methods to mobile organisms, but they are suited only for animals moving more or less unidirectionally past an observer. Buckland *et al.* (1993) also presented analytical techniques for line transect data taken in three dimensions, an extension with obvious applications in fisheries work.

Bergstedt and Anderson (1990) experimentally tested the performance of distance sampling methods for underwater applications. They used an underwater video camera, mounted on a sled and towed along 91.4-m-long transects, to estimate density of construction bricks placed randomly on the bottom of a shallow (3–8 m) bay in Lake Huron. Density estimates (95.5 bricks/ha) matched well with known density (89.9/ha) and there was no evidence of consistent bias. However, note that the assumption of no movement was met and that distances were measured in two and not three dimensions. O'Connell and Carlile (1994) also tested underwater video gear for estimating fish abundance, concluding that line transect methods could be applied, but they did not do so. Although underwater distance sampling holds promise, and certainly should be better than *ad hoc* methods employed in the past, more studies are needed to show that the technique will work under typical conditions encountered during fish surveys.

7.4. EXAMPLE

7.4.1. BACKGROUND

We want to estimate the number of Colorado River cutthroat trout in Cattle Creek, 1 of the 18 streams identified by Martinez (1988) as containing a genetically pure population of this subspecies. A total of 8.0 km of stream is accessible to cutthroat trout. Our general approach will be to select a sample of habitat units (pools, riffles, and runs) and make population estimates in each (a complete census is impractical). Thus, a two-stage design is most appropriate because there are two components of variation: (1) variation among habitat units within a stream (among-unit variation), and (2) variation within each habitat unit due to enumeration variation. Further, because fish density is often different on average in pools, riffles, and runs, we will stratify by habitat type. Finally, because streams are linear systems, we will use a systematic sampling scheme in which every $k_{s(h)}$th habitat unit of each type is sampled. The subscript h denotes habitat type.

We will use equations appropriate for a two-stage stratified random sample with equal-sized sampling units (see Appendix C for details). Although our sampling units are not equal in size (some pools are larger than others, for example), a more complicated sampling scheme involving proportional sampling is probably not warranted. Proportional sampling would be used only if fish density (number per unit area) was equal in all units within a stratum. Because fish density usually varies widely from one habitat unit to another, proportional sampling would not improve the precision of our estimates (Hankin, 1984; Hankin and Reeves, 1988). The basic approach is to estimate the average number of fish in habitat units

Example 219

of each type, $\overline{N}_h$. This average is multiplied by the total number of units of type h, providing an estimate $(\hat{N}_h)$ of the total number of fish in type h. The $\hat{N}_h$ are summed to provide an estimate of the total number of fish in the 8.0-km reach, $\hat{N}$.

The variance of $\hat{N}$ is equal to the sum of the variances of the $\hat{N}_h$, i.e., $\mathrm{V\hat{a}r}(\hat{N}) = \sum_{h=1}^{H} \mathrm{V\hat{a}r}(N_h)$, and H is the number of strata (in this case, habitat types). The variance estimator of $\hat{N}_h$ is provided here for convenience:

$$\mathrm{V\hat{a}r}(\hat{N}_h) = U_h^2 \left[\left(1 - \frac{u_h}{U_h} \right) \frac{\hat{S}_{\hat{N}_{hi}}^2}{u_h} + \frac{\overline{\mathrm{V\hat{a}r}(\hat{N}_{hi}/N_{hi})}}{U_h} \right].$$

As indicated by this equation, to calculate $\mathrm{V\hat{a}r}(\hat{N}_h)$ we need the total number of habitat units of each type (U_h), the number of units sampled of each type (u_h), an estimate of the among-unit variance of abundance in each habitat type (the first term in the brackets; see Appendix C for the formula to calculate this estimate), and an estimate of the average within-unit or enumeration variance of the abundance estimates in each habitat type $(\overline{\mathrm{V\hat{a}r}(\hat{N}_{hi}/N_{hi})};$ see Appendix C for the formula to calculate this estimate).

Hankin and Reeves (1988) described methods to obtain these values for a small stream. We will start at the downstream end of the reach inhabited by Colorado River cutthroat, and move upstream measuring fish abundance in every $k_{s(h)}$th pool, riffle, and run. Using pools as an example, a random number between 1 and $k_{s(\mathrm{pool})}$ will be generated, that pool measured, and then every $k_{s(\mathrm{pool})}$th pool thereafter. This same process will be repeated for riffles and runs. At the end of the process, we will know U_h and u_h. The abundance estimates for measured units in each habitat type are used to calculate $\hat{S}_{\hat{N}_{hi}}^2$ and the among-unit variance. The major question is how to estimate within-unit or enumeration variance, $\mathrm{V\hat{a}r}(\hat{N}_{hi}/N_{hi})$. This depends on the technique used to estimate abundance within units.

In theory, the best technique to estimate abundance within a unit would be one that quickly provides a precise, unbiased estimate. Unfortunately, no technique is fast, unbiased, and precise. For example, multiple-pass removal electrofishing can provide unbiased and precise estimates of abundance, but it is relatively slow and labor intensive. In contrast, snorkel counts can quickly be made within a unit, but they are usually biased and imprecise. Given this trade-off, the question is whether it is best to measure fewer units very precisely, or more units less precisely. The answer depends on the relative magnitude of among-unit versus within-unit variance.

Typically, fish abundance is very variable among habitat units, even if those units are the same habitat type. For this reason, and because time and money are always limited, it is usually better to measure a larger number of habitat units with less precision per unit, rather than a smaller number of units with high precision. Thus, Hankin and Reeves (1988) suggest that snorkel counts are the preferred method. A relatively quick count is made by divers in a large number of units, and in a small subsample of those units, a more intensive technique (electrofishing) also is used to gather very precise abundance estimates. The ratio of diver counts to the precise estimates is calculated. Diver counts in the units not sampled by electrofishing are multiplied by this ratio to estimate the true number of fish in those units, $\hat{N}_{hi}$ (i indicates a single habitat unit). Hankin and Reeves (1988) provide detailed methods and formulas for estimating the variance of these within-unit estimates, $\hat{\mathrm{Var}}(\hat{N}_{hi}/N_{hi})$ (see their Eq. 6; they use the notation $\hat{V}(\hat{y}_i)$ instead of $\hat{\mathrm{Var}}(\hat{N}_{hi}/N_{hi})$). The average of these values for each habitat-type, $\overline{\hat{\mathrm{Var}}(\hat{N}_{hi}/N_{hi})}$, is the value we need for our formula for $\hat{\mathrm{Var}}(\hat{N})$.

To streamline this example, we will not present the formulas derived by Hankin and Reeves (1988) to estimate $\hat{\mathrm{Var}}(\hat{N}_{hi}/N_{hi})$ from snorkel counts. Instead, we will make the assumption that sufficient time and money are available to use multiple-pass removal electrofishing in all sampled habitat units. Thus, $\hat{\mathrm{Var}}(\hat{N}_{hi}/N_{hi})$ can be obtained directly from, for example, program CAPTURE.

7.4.2. Pilot Study

We start our sampling program with a pilot study to get rough estimates of among and within-unit variation, and to determine U_h. We randomly select 4 pools, 3 runs, and 3 riffles from the 0.8-km reach, and make three-pass removal electrofishing estimates in each. The entire reach is walked and the total number of pools (80), runs (75), and riffles (90) is counted.

Pools	Electrofishing estimate $\hat{N}_{hi}$	Variance $\hat{\mathrm{Var}}(\hat{N}_{hi}/N_{hi})$
1	47	6
2	76	10
3	30	3
4	68	11
Average	$55.25 = \overline{N}_h$	$7.5 = \overline{\hat{\mathrm{Var}}(\hat{N}_{hi}/N_{hi})}$
$\hat{S}^2_{\hat{N}_{hi}}$	432.92	
$\hat{N}_h$	$80 \times 55.25 = 4420$	

Example 221

Runs	Electrofishing estimate $\hat{N}_{hi}$	Variance $\hat{V}ar(\hat{N}_{hi}/N_{hi})$
1	37	8
2	25	10
3	15	9
Average	$25.67 = \overline{N}_h$	$9.0 = \overline{\hat{V}ar(\hat{N}_{hi}/N_{hi})}$
$\hat{S}^2_{\hat{N}_{hi}}$	121.33	
$\hat{N}_h$	$75 \times 25.67 = 1925$	

Riffles	Electrofishing estimate $\hat{N}_{hi}$	Variance $\hat{V}ar(\hat{N}_{hi}/N_{hi})$
1	15	10
2	6	2
3	5	3
Average	$8.67 = \overline{N}_h$	$5.0 = \overline{\hat{V}ar(\hat{N}_{hi}/N_{hi})}$
$\hat{S}^2_{\hat{N}_{hi}}$	30.33	
$\hat{N}_h$	$8.67 \times 90 = 780$	

Then, $\hat{V}ar(\hat{N}_h)$ is calculated for each habitat type:

$$\hat{V}ar(\hat{N}_{pool}) = 80^2 \left(1 - \frac{4}{80}\right) \frac{432.92}{4} + 80(7.5) = 658{,}638.4,$$

$$\hat{V}ar(\hat{N}_{run}) = 75^2 \left(1 - \frac{3}{75}\right) \frac{121.33}{3} + 75(9.0) = 219{,}069.0$$

and

$$\hat{V}ar(\hat{N}_{riffle}) = 90^2 \left(1 - \frac{3}{90}\right) \frac{30.33}{3} + 90(5.0) = 76{,}611.3.$$

Next, the total population estimate, $\hat{N}$, for the 8.0-km reach is calculated as the sum of the individual estimates for pools, runs, and riffles: $4420 + 1925 + 780 = 7125$. The variance of this sum is the sum of the $\hat{V}ar(\hat{N}_h)$: $658{,}638.4 + 219{,}069.0 + 76{,}611.3 = 954{,}318.7$. Now that we have the estimate and its variance, 95% confidence intervals can be put on the estimate in the usual way: $\hat{N} \pm 1.96\sqrt{\hat{V}ar(\hat{N})} = 7125 \pm 1.96\sqrt{954{,}318.7}$. Thus, the lower confidence interval is 5210 and the upper interval is 9040. This interval is wide because we only measured a few sampling units. The next step is to use these data to design a study to achieve the precision we desire.

7.4.3. SAMPLE SIZE CALCULATIONS

The sample size required to meet a desired level of precision can be calculated using the appropriate formula in Appendix C (reproduced here for convenience)

$$\left[\frac{eN}{z_{1-\alpha/2}}\right]^2 = \sum_{h=1}^{H}\left[\frac{U_h^2\left(1 - \frac{u_h}{U_h}\right)S_{N_{hi}}^2 + \overline{\text{Var}(\hat{N}_{hi}/N_{hi})}}{u_h}\right],$$

where $S_{N_{hi}}^2 = \hat{S}_{N_{hi}}^2 - \overline{\text{Var}(\hat{N}_{hi}/N_{hi})}$. Assume we want a relative error of 10% ($e = 0.1$) and a 95% confidence level ($\alpha = 0.05$; $z = 1.96$). Using the estimates we obtained during the pilot study:

$$\left[\frac{0.1 \times 7125}{1.96}\right]^2 = \left[\frac{80^2\left(1 - \frac{u_{\text{pool}}}{80}\right)(432.92 - 7.5) + 7.5}{u_{\text{pool}}}\right]$$

$$+ \left[\frac{75^2\left(1 - \frac{u_{\text{run}}}{75}\right)(121.33 - 9.0) + 9.0}{u_{\text{run}}}\right]$$

$$+ \left[\frac{90^2\left(1 - \frac{u_{\text{riffle}}}{90}\right)(30.33 - 5.0) + 5.0}{u_{\text{riffle}}}\right].$$

This equation reduces to

$$132,147.09 = \frac{2,722,695.5 - 34,033.6 \times u_{\text{pool}}}{u_{\text{pool}}} + \frac{631,865.25 - 8424.75 \times u_{\text{run}}}{u_{\text{run}}}$$

$$+ \frac{205,178 - 2279.7 \times u_{\text{riffle}}}{u_{\text{riffle}}},$$

and this further reduces to

$$176,885.14 = \frac{2,722,269.5}{u_{\text{pool}}} + \frac{631,865.25}{u_{\text{run}}} + \frac{205,178}{u_{\text{riffle}}}.$$

Next, we solve this equation for the three unknowns. Naturally, there are many different combinations of u_{pool}, u_{run}, and u_{riffle} that satisfy the equation, but we are interested in the solution that minimizes the total number of units to be sampled (i.e., we want $u_{pool} + u_{run} + u_{riffle}$ to be the smallest value possible). This can be done by trial and error using a spreadsheet program (some spreadsheet programs have built-in routines to find the optimum solution automatically). Using a spreadsheet routine, we found an optimum solution to be 28 pools, 13 runs, and 8 riffles, or 49 units total.

Because we are planning to use a systematic sampling protocol, the final step is to calculate $k_{s(h)}$ for each habitat type. Because there are 80 pools and we want to sample 28 of them, $k_{s(pool)} = 80/28 = 2.9$. Rounding up, we will measure every 3rd pool. For runs, $k_{s(run)} = 75/13 = 5.8$; we will measure every 6th run. Finally, $k_{s(riffle)} = 90/8 = 11.2$; we will measure every 11th riffle.

Obviously, this study is extremely labor intensive, and, in fact, it may be more expensive than the budget allows. If so, it may be necessary to relax the precision requirements. For example, if we increase our allowable error to 20% ($e = 0.2$), the total number of units to be sampled would be on the order of 18 rather than 49. This would represent a dramatic cost savings. Overall, managers can use the approach outlined in this example to plan a study that achieves the required precision with acceptable cost.

7.5. DICHOTOMOUS KEY TO ENUMERATION METHODS

The dichotomous key that follows should be used in conjunction with information presented earlier in this chapter (sections are in parentheses) and in previous chapters on sampling design, enumeration methods, and guidelines for planning surveys. In particular, readers should keep in mind that for methods with unbiased estimators, assumptions must be met if resulting estimates are to be considered valid. This key does not address violations of assumptions.

1. (a) Goal is to determine species distribution or presence–absence .....
   .... Distribution and species lists (7.3.2)
   (b) Goal is to determine population abundance.................................2
2. (a) A complete count of individuals in all randomly chosen stream reaches or lake quadrats is possible with available funding...........
   .... Complete counts using hydroacoustics, ichthyocides, rising nets, visual counts, etc. (7.3.1)
   (b) A complete count is either impossible or too costly in all randomly chosen plots or surveyed area...3
3. (a) Complete counts are possible on a portion of selected stream reaches or lake quadrats and are used to correct index counts over

all selected reaches or quadrats .... Double-sampling
methods such as Hankin and Reeves (1988) (7.1.2.1 and 7.3.1)
    (b) Complete counts are not possible for any reaches or quadrats....4
4. (a) Individuals cannot be marked as groups or individually 5
    (b) Individuals can be marked as groups (batch marked) or individually
....8
5. (a) Individuals cannot be caught..6
    (b) Individuals can be caught...7
6. (a) Perpendicular distances from either a line or a point to an individual
(or distance category) can be obtained, and every individual on
the line or point can be detected............Distance sampling (7.3.5)
    (b) Not as above ....
.... Indices of relative abundance; visual observations (7.2.1.2)
7. (a) Sampling gear is efficient so that capture probabilities are high
(>0.3) within a stream reach or lake quadrat, captured individuals
can be removed from reaches or quadrats after each sampling
period, and at least three successive sampling periods are possible
.... Removal methods (7.3.4.2)
    (b) Not as above .... Indices of relative abundance; CPUE (7.3.3)
8. (a) Individuals are batch marked.............. Petersen estimator (7.3.4.1)
    (b) Individuals are individually marked...9
9. (a) Sampling gear is efficient so that capture probabilities are high
(0.3) within a stream reach or lake quadrat, or the population is
large enough so that low capture probabilities still yield many
(>1000) marked individuals in the population.............................10
    (b) Sampling gear is inefficient and the population within a stream
reach or lake quadrat is small (<500) ....
.... Indices of relative abundance; CPUE (7.3.3)
10. (a) Captured individuals are removed from stream reaches or lake
quadrats after each sampling period and held until all sampling
periods are complete (or individuals captured during previous
sampling periods can be identified from unique marks), and at least
three successive sampling periods are possible................................
.... Removal methods (7.3.4.2)
    (b) Captured individuals are not removed from reaches or quadrats;
marked individuals are rereleased into reaches or quadrats after
each sampling period, and at least three successive sampling periods
are possible............................Capture–recapture methods (7.3.4.1)

LITERATURE CITED

Aadland, L. P., and Cook, C. M. (1992). An electric trawl for sampling bottom-dwelling fishes
in deep turbid streams. *N. Am. J. Fish. Mgmt.* **12:** 652–656.

Acosta, A. R., and Appeldoorn, R. S. (1995). Catching efficiency and selectivity of gill-nets and trammel nets in coral reefs from southwestern Puerto Rico. *Fish. Res.* **22:** 175–196.

Allen, G. H., Delacy, A. D., and Gotshall, D. W. (1960). Quantitative sampling of marine fishes—A problem in fish behavior and fishing gear. *Waste Disposal in the Marine Environment* (E. A. Pearson, Ed.), pp. 448–511. Pergammon Press, New York.

Allison, E. H., Davies, A., Ngatunga, B. P., and Thompson, A. B. (1994). A method for studying pelagic fish communities in deep lakes using drifting gillnets. *Fish Res.* **20:** 87–91.

Anganuzzi, A. A., and Buckland, S. T. (1989). Reducing bias in estimated trends from dolphin abundance indices derived from tuna vessel data. *Rep. Int. Whaling Comm.* **39:** 323–334.

Angermeier, P. L., and Schlosser, I. J. (1989). Species-area relationships for stream fishes. *Ecology* **70:** 1450–1462.

Angermeier, P. L., Smogor, R. A., and Steele, S. D. (1991). An electric seine for collecting fish in streams. *N. Am. J. Fish. Mgmt.* **11:** 352–357.

Appenzeller, A. R., and Leggett, W. C. (1992). Bias in hydroacoustic estimates of fish abundance due to acoustic shadowing: Evidence from day-night surveys of vertically migrating fish. *Can. J. Fish. Aquat. Sci.* **49:** 2179–2189.

Armour, C. L., Burnham, K. P., and Platts, W. S. (1983). *Field Methods and Statistical Analyses for Monitoring Small Salmonid Streams.* U.S.D.I. Fish. Wildl. Serv., Washington, DC. [FSW/OBS-83/33].

Arnason, A. N., and Mills, K. H. (1987) Detection of handling mortality and its effects on Jolly-Seber estimates for mark-recapture experiments. *Can. J. Fish. Aquat. Sci.* **44**(Suppl. 1): 64–73.

Bain, M. B. (1992). Study designs and sampling techniques for community-level assessment of large rivers. *N. Am. Benthol. Soc. Tech. Info. Workshop* **5:** 63–74.

Bain, M. B., Finn, J. T., and Booke, H. E. (1985). A quantitative method for sampling riverine microhabitats by electrofishing. *N. Am. J. Fish. Mgmt.* **5:** 489–493.

Baroudy, E., and Elliott, J. M. (1993). The effect of large-scale spatial variation of pelagic fish on hydroacoustic estimates of their population density in Windermere (northwest England). *Ecol. Fresh. Fish* **2:** 160–166.

Bayley, P. B., and Austen, D. J. (1988). Comparison of detonating cord and rotenone for sampling fish in warmwater impoundments. *N. Am. J. Fish. Mgmt.* **8:** 310–316.

Bayley, P. B., and Austen, D. J. (1990). Modeling the sampling efficiency of rotenone in impoundments and ponds. *N. Am. J. Fish. Mgmt.* **10:** 202–208.

Bayley, P. B., Larimore, R. W., and Dowling, D. C. (1989). Electric seine as a fish-sampling gear in streams. *Trans. Am. Fish. Soc.* **118:** 447–453.

Beamesderfer, R. C., and Rieman, B. E. (1988). Size selectivity and bias in estimates of population statistics of smallmouth bass, walleye, and northern squawfish in a Columbia River reservoir. *N. Am. J. Fish. Mgmt.* **8:** 505–510.

Bergstedt, R. A., and Anderson, D. R. (1990). Evaluation of line transect sampling based on remotely sensed data from underwater video. *Trans. Am. Fish. Soc.* **119:** 86–91.

Bernard, D. R., Parker, J. F., and Lafferty, R. (1993). Stock assessment of burbot populations in small and moderate-size lakes. *N. Am. J. Fish. Mgmt.* **13:** 657–675.

Bisson, P. A., Nielsen, J. L., Palmason, R. A., and Grove, L. E. (1982). A system of naming habitat types in small streams, with examples of habitat utilization by salmonids during low streamflow. In *Acquisition and Utilization of Aquatic Habitat Inventory Information* (N. B. Armantrout, Ed.), pp. 62–73. Proc. West. Div. Am. Fish. Soc., Am. Fish. Soc., Bethesda, MD.

Boccardy, J. A., and Cooper, E. L. (1963). The use of rotenone and electrofishing in surveying small streams. *Trans. Am. Fish. Soc.* **92:** 307–310.

Bohlin, T. (1981). Methods of estimating total stock, smolt output and survival of salmonids using electrofishing. *Inst. Fresh. Res. Drott. Rep.* **59:** 5–14.

Bohlin, T. (1982). The validity of the removal method for small populations—Consequences for electrofishing practice. *Inst. Fresh. Res. Drott. Rep.* **60:** 15–18.

Bohlin, T., Dellefors, C., and Faremo, U. (1982). Electro-fishing for salmonids in small streams—Aspects of the sampling design. *Inst. Fresh. Res. Drott. Rep.* **60:** 19–24.

Bohlin, T., Hamrin, S., Heffberget, T. G., Rasmussen, G., and Saltveit, S. J. (1989). Electrofishing—Theory and practice with special emphasis on salmonids. *Hydrobiologia* **173:** 9–43.

Bohnsack, J. A., and Bannerot, S. P. (1986). *A Stationary Visual Census Technique for Quantitatively Assessing Community Structure of Coral Reef Fishes.* [NOAA Tech. Rep. NMFS 41].

Borgstrøm, R. (1992). Effect of population density on gillnet catchability in four allopatric populations of brown trout (*Salmo trutta*). *Can. J. Fish. Aquat. Sci.* **49:** 1539–1545.

Borgstrøm, R., and Plahte, E. (1992). Gillnet selectivity and a model for capture probabilities for a stunted brown trout (*Salmo trutta*) population. *Can. J. Fish. Aquat. Sci.* **49:** 1546–1554.

Borgstrøm, R., and Skaala, O. (1993). Size-dependent catchability of brown trout and Atlantic salmon parr by electrofishing in a low conductivity stream. *Nordic. J. Freshw. Res.* **68:** 14–21.

Bovee, K. D. (1982). *A Guide to Stream Habitat Analysis Using the Instream Flow Incremental Methodology.* Instream Flow Information Paper 12. U.S.D.I. Fish Wildl. Serv., Washington, DC. [FWS/OBS-82/26].

Bozek, M. A., and Rahel, F. J. (1991). Comparison of streamside visual counts to electrofishing estimates of Colorado River cutthroat trout fry and adults. *N. Am. J. Fish. Mgmt.* **11:** 38–42.

Bozeman, E. L., Helfman, G. S., and Richardson, T. (1985). Population size and home range of American eels in a Georgia tidal creek. *Trans. Am. Fish. Soc.* **114:** 821–825.

Brandt, S. B., Mason, D. M., Patrick, E. V., Argyle, R. L., Wells, L., Unger, P. A., and Stewart, D. J. (1991). Acoustic measures of the abundance and size of pelagic planktivores in Lake Michigan. *Can. J. Fish. Aquat. Sci.* **48:** 894–908.

Buckland, S. T., Anderson, D. R., Burnham, K. P., and Laake, J. L. (1993). *Distance Sampling: Estimating Abundance of Biological Populations.* Chapman and Hall, London.

Buckland, S. T., Cattanack, K. L., and Gunnlaugsson, T. (1992). Fin whale abundance in the North Atlantic, estimated from Icelandic and Faroese NASS-87 and NASS-89 data. *Rep. Int. Whaling Comm.* **42:** 645–651.

Burczynski, J. J., and Johnson, R. L. (1986) Application of dual-beam acoustic survey techniques to limnetic populations of juvenile sockeye salmon (Oncorhynchus nerka). *Can. J. Fish. Aquat. Sci.* **43:** 1776–1788.

Burczynski, J. J., Michaletz, P. H., and Marrone, G. M. (1987). Hydroacoustic assessment of the abundance and distribution of rainbow smelt in Lake Oahe. *N. Am. J. Fish. Mgmt.* **7:** 106–116.

Burkhardt, R. W., and Gutreuter, S. (1995). Improving electrofishing catch consistency by standardizing power. *N. Am. J. Fish. Mgmt.* **15:** 375–381.

Burnham, K. P., Anderson, D. R., White, G. C., Brownie, C., and Pollock, K. H. (1987). Design and analysis of methods for fish survival experiments based on release-recapture. *Am. Fish. Soc. Monogr.* **5.**

Burton, W. H., and Weisberg, S. B. (1994). Estimating abundance of age-0 striped bass in the Delaware River using marked hatchery fish. *N. Am. J. Fish. Mgmt.* **14:** 347–354.

Buynak, G. L., and Mitchell, B. (1993). Electrofishing catch per effort as a predictor of largemouth bass abundance and angler catch in Taylorsville Lake, Kentucky. *N. Am. J. Fish. Mgmt.* **13:** 630–633.

Carlander, K. D., and Lewis, W. M. (1948). Some precautions in estimating fish populations. *Prog. Fish Cult.* **10:** 134–137.

Castro, M., and Lawing, W. (1995). A study of sampling strategies for estimating growth parameters in fish populations. *Fish. Res.* **22:** 59–75.

Coble, D. W. (1992). Predicting population density of largemouth bass from electrofishing catch per effort. *N. Am. J. Fish. Mgmt.* **12:** 650–652.

Cochran, W. G. (1977). *Sampling Techniques,* third ed. Wiley, New York.

Cone, R. S., Robson, D. S., and Krueger, C. C. (1988). Failure of statistical tests to detect assumption violations in the mark-recapture population estimations of brook trout in Adirondack ponds. *N. Am. J. Fish. Mgmt.* **8:** 489–496.

Cormack, R. M. (1979). Models for capture–recapture. In *Sampling Biological Populations* (R. M. Cormack, G. P. Patil, and D. S. Robson, Eds.), pp. 217–255. Intern. Co-op. Publ. House, Fairland, MD.

Cross, D. B., and Stott, B. (1975). The effect of electrical fishing on the subsequent capture of fish. *J. Fish Biol.* **7:** 349–357.

Cyr, H., Downing, J. A., Lalonde, S., Baines, S. B., and Pace, M. L. (1992). Sampling larval fish populations: Choice of sample number and size. *Trans. Am. Fish. Soc.* **121:** 356–368.

Dahm, E. (1987). *Bibliography of Existing Literature on Selectivity of Inland Water Fishing Gear Published by European Authors.* [Eur. Inland Fish. Adv. Comm. Food Agric. Org. Occas. Pap. 18].

Dauble, D. D., and Gray, R. H. (1980). Comparison of small seine and a backpack electro-shocker to evaluate nearshore fish populations in rivers. *Prog. Fish-Cult.* **42:** 93–95.

Davies, W. D., and Shelton, W. L. (1983). Sampling with toxicants. In *Fisheries Techniques* (L. A. Nielsen and D. L. Johnson, Eds.), pp. 199–214. American Fisheries Society, Bethesda, MD.

Demory, R. L., and Golden, J. T. (1983). Sampling the commercial catch. In *Fisheries Techniques* (L. A. Nielsen and D. L. Johnson, Eds.), pp. 421–430. American Fisheries Society, Bethesda, MD.

Dewey, M. R. (1992). Effectiveness of a drop net, a pop net, and an electrofishing frame for collecting quantitative samples of juvenile fishes in vegetation. *N. Am. J. Fish. Mgmt.* **12:** 808–813.

Dibble, E. D. (1991). A comparison of diving and rotenone methods for determining relative abundance of fish. *Trans. Am. Fish. Soc.* **120:** 666–668.

Ensign, W. E., Angermeier, P. L., and Dolloff, C. A. (1995). Use of line transect methods to estimate abundance of benthic stream fishes. *Can. J. Fish. Aquat. Sci.* **52:** 213–222.

Espegren, G. D., and Bergersen, E. P. (1990). Quantitative sampling of fish populations with a mobile rising net. *N. Am. J. Fish. Mgmt.* **10:** 469–478.

Fisher, W. L., and Brown, M. E. (1993). A prepositioned areal electrofishing apparatus for sampling stream habitats. *N. Am. J. Fish. Mgmt.* **13:** 807–816.

Foote, K. G., and Stefánsson, G. (1993). Definition of the problem of estimating fish abundance over an area from acoustic line-transect measurements of density. *ICES J. Mar. Sci.* **50:** 369–381.

Fréon, P., Soria, M., Mullon, C., and Gerlotto, F. (1993). Diurnal variation in fish density estimate during acoustic surveys in relation to spatial distribution and avoidance reaction. *Aquat. Liv. Resour.* **6:** 221–234.

Gardiner, W. R. (1984). Estimating population densities of salmonids in deep water in streams. *J. Fish. Biol.* **24:** 41–49.

Goldstein, R. M. (1978). Quantitative comparison of seining and underwater observation for stream fishery surveys. *Prog. Fish-Cult.* **40:** 108–111.

Gowan, C., and Fausch, K. D. (1996). Long-term demographic responses of trout populations to habitat manipulation in six Colorado streams. *Ecol. Appl.* **6:** 931–946.

Gowan, C., Young, M. K., Fausch, K. D., and Riley, S. C. (1994). Restricted movement in resident stream salmonids: A paradigm lost? *Can. J. Fish. Aquat. Sci.* **51:** 2626–2637.

Graham, R. J. (1992). Visually estimating fish density at artificial structures in Lake Anna, Virginia. *N. Am. J. Fish. Mgmt.* **12:** 204–212.

Gray, R. H., Page, T. L., Neitzel, D. A., and Dauble, D. D. (1986). Assessing population effects from entrainment of fish at a large volume water intake. *J. Environ. Sci. Health* **A21:** 191–209.

Guillard, J., Boet, P., Gerdeaux, D., and Roux, P. (1994). Application of mobile acoustic techniques fish surveys in shallow water: The River Seine. *Reg. Riv. Res. Mgmt.* **9:** 121–126.

Hall, T. J. (1986). Electrofishing catch per hour as an indicator of largemouth bass density in Ohio impoundments. *N. Am. J. Fish. Mgmt.* **6:** 397–400.

Hamley, J. M. (1975). Review of gillnet selectivity. *J. Fish. Res. Board Can.* **32:** 1943–1969.

Hamley, J. M., and Howley, T. P. (1985). Factors affecting variability of trap-net catches. *Can. J. Fish. Aquat. Sci.* **42:** 1079–1087.

Hankin, D. G. (1984). Multistage sampling designs in fisheries research: Applications in small streams. *Can. J. Fish. Aquat. Sci.* **14:** 1575–1591.

Hankin, D. G., and Reeves, G. H. (1988). Estimating total fish abundance and total habitat area in small streams based on visual estimation methods. *Can. J. Fish. Aquat. Sci.* **45:** 834–844.

Hardin, S., and Connor, L. L. (1992). Variability of electrofishing crew efficiency, and sampling requirements for estimating reliable catch rates. *N. Am. J. Fish. Mgmt.* **12:** 612–617.

Hauck, F. R. (1949). Some harmful effects of the electric shocker on large rainbow trout. *Trans. Am. Fish. Soc.* **77:** 61–64.

Hayes, M. L. (1983). Active fish capture methods. In *Fisheries Techniques* (L. A. Nielsen and D. L. Johnson, Eds.), pp. 123–146. American Fisheries Society, Bethesda, MD.

He, X., and Lodge, D. M. (1990). Using minnow traps to estimate fish population size: The importance of spatial distribution and relative species abundance. *Hydrobiologia* **190:** 9–14.

Heggenes, J., Brabrand, Å., and Saltveit, S. J. (1990). Comparison of three methods for studies of stream habitat use by young brown trout and Atlantic salmon. *Trans. Am. Fish Soc.* **119:** 101–111.

Heidinger, R. C., Helms, D. R., Hiebert, T. I., and Howe, P. H. (1983). Operational comparisons of three electrofishing systems. *N. Am. J. Fish. Mgmt.* **3:** 254–257.

Helfman, G. S. (1983). Underwater methods. In *Fisheries Techniques* (L. A. Nielsen and D. L. Johnson, Eds.), pp. 349–370. American Fisheries Society, Bethesda, MD.

Helser, T. E., Condrey, R. E., and Geaghan, J. P. (1991). A new method for estimating gillnet selectivity, with an example for spotted seatrout, *Cynocion nebulosus. Can. J. Fish. Aquat. Sci.* **48:** 487–492.

Helser, T. E., Geaghan, J. P., and Condrey, R. E. (1994). Estimating size composition and associated variances of a fish population from gillnet selectivity, with an example for spotted seatrout (*Cynocion nebulosus*). *Fish. Res.* **19:** 65–86.

Henderson, B. A., and Wong, J. L. (1991). A method for estimating gillnet selectivity of walleye (*Stizostedion vitreum vitreum*) in multimesh multifilament gill nets in Lake Erie, and its application. *Can. J. Fish. Aquat. Sci.* **48:** 2420–2428.

Hendricks, M. L., Hocutt, C. H., and Stauffer, J. R., Jr. (1980) Monitoring fish in lotic habitats. In *Biological Monitoring of Fish* (J. R. Stauffer, Jr., Ed.), pp. 205–321. Lexington Books, Lexington, MA.

Hightower, J. E., and Gilbert, R. J. (1984). Using the Jolly-Sever model to estimate population size, mortality, and recruitment for a reservoir fish population. *Trans. Am. Fish. Soc.* **113:** 633–641.

Hilborn, R., and Walters, C. J. (1992). *Quantitative Fisheries Stock Assessment, Choice, Dynamics and Uncertainty.* Chapman and Hall, New York.

Hill, T. D., and Willis, D. W. (1994). Influence of water conductivity on pulsed AC and pulsed DC electrofishing catch rates for largemouth bass. *N. Am. J. Fish. Mgmt.* **14:** 202–207.

Hillman, T. W., Mullan, J. W., and Griffith, J. S. (1992). Accuracy of underwater counts of juvenile chinook salmon, coho salmon, and steelhead. *N. Am. J. Fish. Mgmt.* **12:** 598–603.

Hollender, B. A., and Carline, R. F. (1994). Injury to wild brook trout by backpack electrofishing. *N. Am. J. Fish. Mgmt.* **14:** 643–649.

Hubert, W. A. (1983). Passive capture techniques. In *Fisheries Techniques* (L. A. Nielsen and D. L. Johnson, Eds.), pp. 95–122. American Fisheries Society, Bethesda, MD.

Jacobs, K., and Swink, W. D. (1982). Estimations of fish population size and sampling efficiency of electrofishing and rotenone in two Kentucky tailwaters. *N. Am. J. Fish. Mgmt.* **2:** 239–248.

Jensen, A. L. (1992). Integrated area sampling and mark-recapture experiments for sampling fish populations. *Biometrics* **48:** 1201–1205.

Johnson, B. M., Stein, R. A., and Carline, R. F. (1988). Use of a quadrat rotenone technique and bioenergetics modeling to evaluate prey availability to stocked piscivores. *Trans. Am. Fish. Soc.* **117:** 127–141.

Keast, A., and Harker, J. (1977). Strip counts as a means of determining densities and habitat utilization patterns in lake fishes. *Environ. Biol. Fish.* **1:** 181–188.

Kelso, J. R., Minns, C. K., Gray, J. E., and Jones, M. L. (1986). Acidification of surface waters in eastern Canada and its relationship to aquatic biota. *Can. Spec. Publ. Fish. Aquat. Sci.* **87.**

Kelso, J. R. and Shuter, B. J. (1989). Validity of the removal method for fish population estimation in a small lake. *N. Am. J. Fish. Mgmt.* **9:** 471–476.

King, T. A., Williams, J. C., Davies, W. D., and Shelton, W. L. (1981). Fixed versus random sampling of fishes in a large reservoir. *Trans. Am. Fish. Soc.* **110:** 563–568.

Kolz, A. L. (1989). A power transfer theory for electrofishing. In *Electrofishing: A Power Related Phenomenon* (A. L. Kolz and J. B. Reynolds, Eds.), pp. 1–11. U. S. D. I. Fish Wildl. Serv., Washington, DC. [Fish Wildl. Tech. Rep. 22].

Lambou, V. W., and Stern, H., Jr. (1958). An evaluation of some of the factors affecting the validity of rotenone sampling data. *Proc. Ann. Conf. Southeast. Assoc. Game Fish Comm.* **11:** 91–98.

Larson, E. W., Johnson, D. L., and Lynch, W. E., Jr. (1986). A buoyant pop net for accurately sampling fish at artificial habitat structure. *Trans. Am. Fish. Soc.* **115:** 351–355.

LeCren, E. D., Kipling, C., and McCormack, J. C. (1977). A study of the numbers, biomass and year-class strengths of perch (*Perca fluviatilis* L.) in Windermere from 1941 to 1966. *J. Anim. Ecol.* **46:** 281–307.

Lennon, R. E. (1970). Control of freshwater fish with chemicals. *Proc. Vertebrate Pest Conf.* **4:** 129–137.

Leyher, W. G., and Maughan, O. E. (1984). Comparison efficiencies of three sampling techniques for estimating fish populations in small streams. *Prog. Fish-Cult.* **46:** 180–184.

Love, R. H. (1971). Dorsal-aspect target strength of an individual fish. *J. Acoustic Soc. Am.* **49:** 816–823.

Luecke, C., and Wurtsbaugh, W. A. (1993). Effects of moonlight and daylight on hydroacoustic estimates of pelagic fish abundance. *Trans. Am. Fish. Soc.* **122:** 112–120.

Lyons, J. (1992). The length of stream to sample with a towed electrofishing unit when fish species richness is estimated. *N. Am. J. Fish. Mgmt.* **12:** 198–203.

Machiels, M. A. M., Klinge, M., Lanters, R., and van Densen, W. L. T. (1994). Effects of snood length and hanging ratio on efficiency and selectivity of bottom-set gillnets for pikeperch, *Stizostedion lucioperca* L., and bream, *Abramis brama. Fish. Res.* **19:** 231–239.

Maciena, M. J., Rider, S. J., and Lowery, D. R. (1993). Use of a catch-depletion method to estimate population density of age-0 largemouth bass in submersed vegetation. *N. Am. J. Fish. Mgmt.* **13:** 847–851.

Maciena, M. J., Wrenn, W. B., and Lowery, D. R. (1995). Estimating harvestable largemouth bass abundance in a reservoir with an electrofishing catch depletion technique. *N. Am. J. Fish. Mgmt.* **15:** 103–109.

Magnan, P. (1991). Unrecognized behavior and sampling limitations can bias field data. *Environ. Biol. Fish.* **3:** 403–406.

Mahon, R. (1980). Accuracy of catch-effort methods for estimating fish density and biomass in streams. *Biol. Fish.* **5:** 343–363.

Malvestuto, S. P. (1983). Sampling the recreational fishery. In *Fisheries Techniques* (L. A. Nielsen and D. L. Johnson, Eds.), pp. 397–420. American Fisheries Society, Bethesda, MD.

Martinez, A. M. (1988). Identification and status of Colorado River cutthroat trout in Colorado. *Am. Fish. Soc. Symp.* **4:** 81–89.

Mathisen, O. A. (1992). Hydroacoustics as a tool in fisheries research—"where are we and what are the future expectations?" *Fish. Res.* **14:** 91–93.

Mattson, N. S. (1994). Direct estimates of multi-mesh gillnet selectivity to *Oreochromis shiranus chilwae*. *J. Fish Biol.* **45:** 997–1012.

McInerny, M. C., and Degan, D. J. (1993). Electrofishing catch rates as an index of largemouth bass population density in two large reservoirs. *N. Am. J. Fish. Mgmt.* **13:** 223–228.

Mesa, M. G., and Schreck, C. B. (1989). Electrofishing mark-recapture and depletion methodologies evoke behavioral and physiological changes in cutthroat trout. *Trans. Am. Fish. Soc.* **118:** 644–658.

Metzger, R. J., and Shafland, P. L. (1986). Use of detonating cord for sampling fish. *N. Am. J. Fish. Mgmt.* **6:** 113–118.

Millar, R. B. (1992). Estimating the size-selectivity of fishing gear by conditioning on the total catch. *J. Am. Stat. Assoc.* **87:** 962–968.

Miller, R. J., and Mohn, R. K. (1993). Critique of the Leslie method for estimating sizes of crab and lobster populations. *N. Am. J. Fish. Mgmt.* **13:** 676–685.

Naud, M., and Magnan, P. (1988). Diel onshore-offshore migration in northern redbelly dace, *Phoxinus eos* (Cope), in relation to prey distribution in a small oligotrophic lake. *Can. J. Zool.* **66:** 1249–1253.

Nichols, J. D., Noon, B. R., Stokes, S. L., and Hines, J. E. (1981). Remarks on the use of mark-recapture methodology in estimating avian population size. *Stud. Avian Biol.* **6:** 121–136.

Northcote, T. C., and Wilkie, D. W. (1963). Underwater census of stream fish populations. *Trans. Am. Fish. Soc.* **92:** 146–151.

Northrop, R. B. (1967). Electrofishing. *IEEPD Trans. Biomed. Eng.,* **14:** 191–200.

O'Connell, V. M., and Carlile, D. W. (1994). Comparison of a remotely operated vehicle and a submersible for estimating abundance of demersal shelf rockfishes in the eastern Gulf of Alaska. *N. Am. J. Fish. Mgmt.* **14:** 196–201.

Pajos, T. A., and Weise, J. G. (1994). Estimating populations of larval sea lamprey with electrofishing sampling methods. *N. Am. J. Fish. Mgmt.* **14:** 580–587.

Paller, M. H. (1995). Relationships among number of fish species sampled, reach length surveyed, and sampling effort in South Carolina coastal plain streams. *N. Am. J. Fish. Mgmt.* **15:** 110–120.

Palmisano, A. M., and Burger, C. V. (1988). Use of a portable electric barrier to estimate chinook salmon escapement in a turbid Alaskan river. *N. Am. J. Fish. Mgmt.* **8:** 475–480.

Paragamian, V. L. (1989). A comparison of day and night electrofishing: Size structure and catch per unit effort for smallmouth bass. *N. Am. J. Fish. Mgmt.* **9:** 500–503.

Parkinson, E. A., Berkowitz, J., and Bull, C. J. (1988). Sample size requirements for detecting changes in some fisheries statistics from small trout lakes. *N. Am. J. Fish. Mgmt.* **8:** 181–190.

Parsely, M. J., Palmer, D. E., and Burkhardt, R. W. (1989). Variation in capture efficiency of a beach seine for small fishes. *N. Am. J. Fish. Mgmt.* **9:** 239–244.

Pierce, C. L., Rasmussen, J. B., and Leggett, W. C. (1990). Sampling littoral fish with a seine: Corrections for variable capture efficiency. *Can. J. Fish. Aquat. Sci.* **47:** 1004–1010.

Platts, W. S., Megahan, W. F., and Minshall, G. W. (1983). *Methods for Evaluating Stream, Riparian, and Biotic Conditions.* [U.S.D.A. For. Serv. Intermountain For. Range Exp. Sta. Gen. Tech. Rept. INT-138].

Regis, J., Pattee, E., and Lebreton, J. D. (1981). A new method for evaluating the efficiency of electric fishing. *Arch. Hydrobiol.* **93:** 68–82.

Reynolds, J. B. (1983). Electrofishing. In *Fisheries Techniques* (L. A. Nielsen and D. L. Johnson, Eds.), pp. 147–164. American Fisheries Society, Bethesda, MD.

Ricker, W. E. (1975). Computation and interpretation of biological statistics of fish populations. *Bull. Fish. Res. Board Can.* **191.**

Rider, S. J., Maciena, M. J., and Lowery, D. R. (1994). Comparisons of cove rotenone and electrofishing catch-depletion estimates to determine abundance of age-0 largemouth bass in unvegetated and vegetated areas. *J. Fresh. Ecol.* **9:** 19–27.

Riley, S. C., and Fausch, K. D. (1992). Underestimation of trout population size by maximum likelihood removal estimates in small streams. *N. Am. J. Fish. Mgmt.* **12:** 768–776.

Riley, S. C., Fausch, K. D., and Gowan, C. (1992). Movement of brook trout (*Salvelinus fontinalis*) in four small subalpine streams in northern Colorado. *Ecol. Freshwater Fish* **1:** 112–122.

Riley, S. C., Haedrich, R. L., and Gibson, R. J. (1993). Negative bias in removal estimates of Atlantic salmon parr relative to stream size. *J. Fresh. Ecol.* **8:** 97–101.

Rodgers, J. D., Solazzi, M. F., Johnson, S. L., and Buckman, M. A. (1992). Comparison of three techniques to estimate juvenile coho salmon populations in small streams. *N. Am. J. Fish. Mgmt.* **12:** 79–86.

Ryan, P. M. (1984). Fyke net catches as indices of the abundance of brook trout, *Salvelinus fontinalis,* and Atlantic salmon, *Salmo salar. Can. J. Fish. Aquat. Sci.* **41:** 377–380.

Ryan, P. M., and Kerekes, J. J. (1989). Correction of relative fish abundance estimates from catch data for variable fishing intensity during lake surveys. *Can. J. Fish. Aquat. Sci.* **46:** 1022–1025.

Sale, P. F., and Douglas, W. A. (1981). Precision and accuracy of visual census techniques for fish assemblages on coral patch reefs. *Environ. Biol. Fish.* **6:** 333–339.

Schill, D. J., and Beland, K. F. (1995). Electrofishing injury studies, a call for population perspective. *Fisheries* **20**(6): 28–29.

Schill, D. J., and Griffith, J. S. (1984). Use of underwater observations to estimate cutthroat trout abundance in the Yellowstone River. *N. Am. J. Fish. Mgmt.* **4:** 479–487.

Seber, G. A. F. (1982). *The Estimation of Animal Abundance and Related Parameters,* second ed. Griffin, London.

Serns, S. L. (1982). Relationship of walleye fingerling density and electrofishing catch per effort in northern Wisconsin lakes. *N. Am. J. Fish. Mgmt.* **2:** 38–44.

Serns, S. L. (1983). Relationship between electrofishing catch per effort and density of walleye yearlings. *N. Am. J. Fish. Mgmt.* **3:** 451–452.

Sharber, N. G., and Carothers, S. W. (1988). Influence of electrofishing pulse shape on spinal injuries in adult rainbow trout. *N. Am. J. Fish. Mgmt.* **8:** 117–122.

Sharber, N. G., Carothers, S. W., Sharber, J. P., de Vos, J. C., Jr., and House, D. A. (1994). Reducing electrofishing-induced injury of rainbow trout. *N. Am. J. Fish. Mgmt.* **14:** 340–346.

Shireman, J. V., Colle, D. E., and DuRant, D. F. (1981). Efficiency of rotenone sampling with large and small block nets in vegetated and open-water habitats. *Trans. Am. Fish. Soc.* **110:** 77–80.

Simonson, T. D., and Lyons, J. (1995). Comparison of catch per effort and removal procedures for sampling stream fish assemblages. *N. Am. J. Fish. Mgmt.* **15:** 419–427.

Slaney, P. A., and Martin, A. D. (1987). Accuracy of underwater census of trout populations in a large stream in British Columbia. *N. Am. J. Fish. Mgmt.* **7:** 117–122.

Snyder, D. E. (1993). *Impacts of Electrofishing on Fish.* Report to the U.S.D.I. Bur. Reclam. and Glen Canyon Environ. Studies Team, Salt Lake City, UT/Flagstaff, AZ.

Snyder, D. E. (1995). Impacts of electrofishing on fish. *Fisheries* **20**(1): 26–27.

Sternin, V. G., Nikonorov, I. V., and Bumeister, Y. K. (1972). *Electrical Fishing, Theory and Practice.* Israel Program for Scientific Translations, Jerusalem, 1976. [English translation]

Stevens, D. E., Kohlhorst, D. W., Miller, L. W., and Kelly, D. W. (1985). The decline of striped bass in the Sacramento-San Joaquin Estuary, California. *Trans. Am. Fish. Soc.* **114:** 12–30.

Stewart, P. E. M. (1975). Catch selectivity by electrical fishing systems. *Const. Int. Explor. Mer.* **36:** 106–109.

Tarzwell, C. M. (1942). Fish populations in the back-waters of Wheeler reservoir and suggestions for their management. *Trans. Am. Fish. Soc.* **71:** 201–214.

Thorne, R. E. (1983). Hydroacoustics. In *Fisheries Techniques* (L. A. Nielsen and D. L. Johnson, Eds.), pp. 239–260. American Fisheries Society, Bethesda, MD.

Thurow, R. F. (1994). *Underwater Methods for Study of Salmonids in the Intermountain West.* [U.S.D.A. For. Serv. Intermountain Res. Sta. Gen. Tech. Rep. INT-GTR-307]

Tinsely, V. R., Nielsen, L. A., and Wahl, D. H. (1989). Pushnet sampling as a supplement to seine sampling in rivers. *Fish. Res.* **7:** 201–206.

Vadas, R. L., Jr., and Orth, D. J. (1993). A new technique for estimating the abundance and habitat use of stream fishes. *J. Fresh. Ecol.* **8:** 305–317.

Weaver, M. J., Magnuson, J. J., and Clayton, M. K. (1993). Analyses for differentiating littoral fish assemblages with catch data from multiple sampling gears. *Trans. Am. Fish. Soc.* **122:** 1111–1119.

Weddle, G. K., and Kessler, R. K. (1993). A square-metre electrofishing sampler for benthic riffle fishes. *J. N. Am. Benthol. Soc.* **12:** 291–301.

Welcomme, R. L. (Ed.) (1975). *Symposium on the Methodology for the Survey, Monitoring, and Appraisal of Fishery Resources in Lakes and Large Rivers.* Food Agric. Organiz. United Nations, Rome, Italy. [European Inland Fish. Advisory Comm. Tech. Paper 23 (Suppl. 1)]

White, G. C., Anderson, D. R., Burnham, K. P., and Otis, D. L. (1982). Capture-recapture and removal methods for sampling closed populations. Los Alamos Natl. Lab., Los Alamos, NM. [LA-8787-NERP]

Wiley, M. L., and Tsai, C-F. (1983). The relative efficiencies of electrofishing vs. seines in Piedmont streams of Maryland. *N. Am. J. Fish. Mgmt.* **3:** 243–253.

Wiley, R. W. (1984). A review of sodium cyanide for use in sampling stream fishes. *N. Am. J. Fish. Mgmt.* **4:** 249–256.

Wilson, H. T., and Weisberg, S. B. (1993). Design considerations for striped bass beach seine surveys. *N. Am. J. Fish. Mgmt.* **13:** 376–382.

Winters, G. H., and Wheeler, J. P. (1990). Direct and indirect estimation of gillnet selection curves of Atlantic herring (*Clupea harengus harengus*). *Can. J. Fish. Aquat. Sci.* **47:** 460–470.

Chapter 8

Amphibians and Reptiles

8.1. **Complete Counts**
    8.1.1. Entire Sampling Frame
    8.1.2. Portion of Sampling Frame
8.2. **Incomplete Counts**
    8.2.1. Index Methods
    8.2.2. Adjusting for Incomplete
        Detectability

8.3. **Recommendations**
    8.3.1. General Comments
    8.3.2. Dichotomous Key to
        Enumeration Methods
Literature Cited

Historically, herpetofauna have received much less attention from wildlife managers and researchers than other vertebrate species. In fact, many fish and wildlife biologists may not receive formal training in herpetology (Jones, 1986). We encourage those interested in a detailed overview of ecology and counting techniques of amphibians and reptiles to read: Heyer *et al.* (1994) for amphibians; Turner (1977) for reptiles; and Scott (1982), Jones (1986), Karns (1986), and Zug (1993) for herpetofauna in general. Hammerson (1982) described ecology and distribution of amphibians and reptiles in Colorado.

This chapter provides an overview of some potential methods for surveying populations of amphibians and reptiles. Availability of funds and species of interest will largely dictate which techniques can be employed; index methods may be all that can be realistically applied if many species must be monitored concurrently. However, methods that account for incomplete detectability of individuals must be used in cases that require unbiased estimates at a specified level of precision for detecting a certain percentage change over time. We discuss the shortcomings of both index and abundance/density estimation methods, and provide a dichotomous key at the end of this chapter as a general guide for biologists in selecting an enumeration method. Finally, we wish to reiterate that knowledge of the ecology of the species of interest is vital to designing a survey. Herpetofauna are particularly sensitive to weather changes, particularly temperature and precipitation (Zug, 1993). Hence, surveys should be scheduled at appropriate times and conditions to maximize the number of individuals that can be detected.

8.1. COMPLETE COUNTS

Both highly trained observers and complete detectability of individuals are required for complete counts to be possible. Observers must be knowledgeable in techniques used for both searching and species identification. Further, areas to be searched must be small enough to allow a census. Area size will largely depend upon the density of the species of interest and the searching method employed. Counts of any kind should be conducted when individuals are most visible; for temperate terrestrial salamanders, this would occur immediately after warm rains during spring (Jaeger and Inger, 1994). Lizards and snakes, however, may move 2 or 3 days after rains depending on ambient temperature (Vogt and Hine, 1982).

8.1.1. ENTIRE SAMPLING FRAME

A true census of a target population of herpetofauna is extremely unlikely unless the entire sampling frame is contained within a small geographical area or perhaps a small discrete habitat feature (e.g., small prairie pond). Therefore, we will concentrate on methods for obtaining complete counts within a portion of a sampling frame.

8.1.2. PORTION OF SAMPLING FRAME

Although complete counts often are not possible over large areas, a few amphibians and reptiles may be censused within small areas. Jaeger and Inger (1994) suggested that complete counts may be possible for forest-floor or streamside dwelling amphibians, where the biologist can search the habitat "up close." If so, this situation would correspond well with a two-stage cluster sampling design in which an area of interest is divided into quadrats and each quadrat is composed of equal-sized subquadrats that are small enough to permit complete counts. Thus, the quadrats themselves are too large to census, but a random selection of subquadrats within randomly chosen quadrats is not. For instance, 1×1-m subquadrats could be used and all stones, litter, and woody debris removed quickly to count all amphibians therein. The removed materials should be replaced if the subquadrats are to be censused at a later time (Jaeger and Inger, 1994). Quadrat size will depend on sizes of both subquadrats (size depends on how large an area can be completely counted) and the sampling frame.

Complete counts on small plots of ground have been used in monitoring salamander densities in temperate forest over time (Jaeger, 1980; Mathis, 1990). Bury and Luckenbach (1977) censused 1-ha subquadrats within 25- to 100-ha quadrats for tortoises and set up permanent stations to monitor their populations.

Patch sampling (Jaeger 1994) is a method in which only specific micro-habitats or features (e.g., fallen logs, stones, and pools within tree hollows) known to be frequented by herpetofauna are searched within a defined area. All surrounding habitat is ignored. Corn and Bury (1990) used this technique to search a predetermined number of downed logs for salamanders in Pacific Northwest forests, but correctly noted that resulting density estimates could only be applied to other logs within their study area (assuming a random sample) and not to the habitat between. Thus, this method is of little use for obtaining overall density estimates unless all or essentially all of the target population is contained within these features. Also, patches must be discrete, operationally definable (i.e., specific dimensions given), and able to be censused (Jaeger, 1994).

Another potential drawback of patch sampling is that all sampling units or features within a defined area probably cannot be mapped. In the log example, every downed log would have to be located, put in the appropriate size category, and assigned a unique number if the usual procedure of random selection was to be employed. To circumvent this problem, Corn and Bury (1990) suggested taking a systematic sample in which every 1-in-k_s log encountered would be censused. We add that both the starting point and the direction of travel should be randomly selected. This selection procedure could be very time-consuming and inefficient if the patches are not readily visible and easy to locate.

Bart *et al.* (1984) located seemingly all calling rails in strip transects over the course of several counts and calculated their detectability during a given count based on these results. They then determined how many counts were needed to detect all, or nearly all, calling yellow rails within a given strip. Zimmerman (1994) suggested this approach is relevant for use with many rainforest frogs because they too are secretive and uncommon, and call during the evening. However, there are difficulties with realistically meeting the assumptions underlying this method. First, frogs are usually smaller and therefore relatively more difficult to locate than rails. Also, frogs generally occur at higher local densities than rails, so identifying and counting every individual frog in a strip seems extremely unlikely. Moreover, the assumptions of equal detectability of individuals during each sampling occasion and constant detectability across sampling occasions will rarely be met [see further discussion of Bart *et al.*'s (1984) approach in the next chapter]. Therefore, we generally recommend against the use of this approach, and strongly recommend a rigorous evaluation of the validity of underlying assumptions if its use is being considered.

Complete counts of herpetofauna at any scale are very much the exception rather than the rule. Thus, biologists should evaluate each situation as to whether a complete count is realistic based on the ecology of the species of interest and the habitat structure and composition in which it occurs.

8.2. INCOMPLETE COUNTS

Complete counts of herpetofauna on any spatial scale are only plausible in a relatively limited number of situations. Therefore, biologists must rely on incomplete counts to meet their monitoring objectives. These objectives will dictate which class of methods, either indices or methods properly adjusted for incomplete detectability, will be most appropriate.

8.2.1. INDEX METHODS

Index methods are used to document either occurrence or relative status of species. Their advantage is in their ease of use and comparatively low costs. Unfortunately, they are inadequate for inferential monitoring programs (i.e., those in which detection of a specified change in numbers is desired). However, depending on available monies and numbers of species that require monitoring, index methods may be the only feasible alternative in a number of situations.

8.2.1.1. Presence–Absence

Collecting presence–absence data is an option when little or no faunal information has been collected in an area and inadequate funds are available for financing a full-fledged monitoring program. Consequently, methods as presented in this section should be used for compiling the most basic distribution and occurrence information and should not be used as "stand alone" techniques in a monitoring program. These techniques may be expanded to relative index methods when they are standardized and all individuals encountered are recorded.

Visual Searches

One method for assessing distribution and compiling a species list is to systematically search an area and record all species of herpetofauna observed. This effort is further enhanced by focusing on certain habitats (e.g., ponds, riparian zones) and microhabitats (e.g., woody debris) where some herpetofauna are known to occur, and by conducting searches at the time of day and year when target species are most active. In addition, many nocturnal amphibians and reptiles not normally encountered during the day may be seen on or near roads at night (Jones, 1986). Searches can be conducted over large areas by driving and surveying roads at night (Campbell and Christman, 1982; Shaffer and Juterbock, 1994). All available types of habitat within an area should be searched in some way, whether on foot, in a vehicle, or a combination of both. Crump and Scott (1994) described visual search methodology in some detail.

Visual detections may be of animal sign as well. Lillywhite (1982) successfully identified a variety of snake species by their tracks on fine-textured soils, especially on dirt roads. This technique may be applicable to medium to large lizards as well if they leave tracks that are distinguishable among species. Occupied burrows of fossorial species like the desert tortoise also could be used as indicators of occurrence.

Audio Searches

Encounters are not limited to observation only; many male frogs in their breeding phase use species-specific calls to attract females and discourage intrusion by rival males (Wells, 1977). Personnel trained to identify frog species by their calls should survey spring breeding pools and similar habitat. Audio surveys are much more efficient for recording presence of calling frogs than visually based methods because both arboreal and ground-dwelling species can be surveyed simultaneously (Zimmerman, 1994).

Trapping and Netting

Drift fences and pitfall traps are commonly used to trap amphibians and reptiles in order to assess species distribution and compile species lists (Friend, 1984; Bury and Corn, 1987). Material, such as aluminum flashing and plastic sheeting (Dodd and Scott, 1994), of some predefined length and height is erected to serve as a barrier to redirect ground-traveling individuals into an open container or containers buried to the rim in the ground (Fig. 8.1). Hard plastic buckets with 3-liter (Shields 1985), 8-liter (Raphael, 1988; Corn and Bury, 1990), or 19-liter (Dodd and Scott, 1994) capacity make the most effective containers (pitfalls) because they are resistant to collapse and moisture-induced deterioration (Dodd and Scott, 1994). Funnel traps may be used in place of pitfall traps. Details regarding the configuration, construction, and maintenance of drift fences and associated traps are provided in Gibbons and Semlitsch (1981), Jones (1986), Corn (1994), and Dodd and Scott (1994).

Lagler (1943) and Petokas and Alexander (1979) used floating pitfall traps to catch turtles in lakes and ponds. These types of traps are most effective in trapping species, such as the painted turtle, that frequently bask on logs and other structures. Funnel traps also have been adapted for aquatic use. Lagler (1943) and Moulton (1954) used them for both larval and adult salamanders and frogs. Feuer (1980) compared hoop nets and chicken wire traps in their usefulness in capturing turtles.

Many aquatic species may be collected through the use of seines, dip nets, electroshockers, and enclosure samples (Jones, 1986; Shaffer *et al.*,

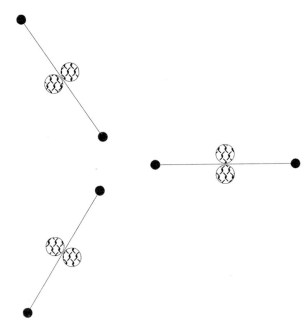

Figure 8.1 Configuration of three 5-m drift fences and pitfall traps suggested by Corn (1994). (⊗) 19-liter trap, (●) 8-liter trap, (—) fence.

1994). Seines can be reasonably effective in capturing larval and adult amphibians when aquatic vegetation is sparse and substrates are more or less uniform in structure, but decrease in effectiveness as vegetation increases in density and substrates in structural complexity (Shaffer *et al.*, 1994). An alternative approach is to spread a seine across a stream and "beat the brush" upstream by turning over stones, wading through and disturbing aquatic vegetation, and stirring up the substrate (Jones, 1986). Dip nets are generally used to capture individuals in areas too narrow, shallow, or structurally complex to drag a seine. Dip nets also are commonly used in conjunction with electroshocking activities.

Electroshocking is a technique that uses equipment to generate an electrical field through an aqueous medium in order to stun individuals for easy capture. A direct current (DC) field is preferable to an alternating current (AC) field because a DC field causes individuals to involuntarily approach its source (a condition called electrotaxis) and therefore increases capture success; is less apt to cause permanent physical harm to stunned individuals, and is potentially less dangerous to operators (Williams *et al.*, 1981). Electroshocking has been widely used for years by biologists to capture fish but has seldom been used for herpetofauna. Gunning and Lewis (1957) subjected snakes and turtles to an AC field and observed no ill effects. Williams *et al.* (1981) successfully used a DC electroshocker to capture hellbenders in streams of northwestern Pennsylvania. They concluded that electroshocking in conjunction with either dipnetting or seining (netting method depending on stream habitat) was much more successful in capturing hellbenders than seining alone, search and seizure, or raking the substrate.

The effectiveness of electroshocking for capturing herpetofauna is greatly lessened in stream sections or water bodies that are turbid (unlike fish, stunned amphibians and reptiles sink rather than float), have low electrical conductivity, and are deep and wide. In addition, individuals with small surface to volume ratios (i.e., large body size) are usually not stunned by electroshockers (Jones, 1986). Shaffer *et al.* (1994) recommended against using electroshocking methods for collecting larval amphibians because of equipment costs, potential dangers to operators, and lack of improved capture efficiently compared to other methods.

Enclosure samplers, such as box samplers, are used to capture larval and adult amphibians in shallow water environments, although dredges may be used in deeper water. A box sampler, which is open on the top and bottom, is placed in the water and forced down into the underlying substrate (Shaffer *et al.*, 1994). Individuals trapped within the enclosure then are dip-netted out through the top, preferably with a dip net configured to match the width of the enclosure. This method is essentially limited to shallow water areas with few rocks and debris in the underlying sandy or muddy substrate (Shaffer *et al.*, 1994).

8.2.1.2. Indices of Relative Abundance or Density

Many of the methods mentioned in the preceding section may be used to obtain indices of relative abundance or density when survey areas are well delineated and procedures have been standardized for habitat type, effort expended as measured in units of time, timing of survey, weather, observers, and sample unit selection (i.e., some type of probability sampling design is used). Such standardization is necessary if comparisons of indices across space and/or time are to be of any potential use. That is, all counts should be conducted under conditions as similar as possible. Survey designs for choosing sampling units follow directly from information given in the chapter on sampling designs, where index counts within plots are treated as complete counts for a single survey (i.e., no enumeration variance) and as incomplete counts for repeated surveys (i.e., estimate of enumeration variance available).

As with all comparisons of constant-proportion indices, we assume that changes in relative abundance are directly proportional to changes in true abundance. In other words, we assume that any biases in counting individuals will be constant across similar habitats and conditions. If this assumption does not hold, which is frequently the case, surveys results could lead to incorrect conclusions regarding changes in population levels.

Visual Searches

Techniques for conducting visual searches can generally be described in terms of the type and amount of area covered. Methods are presented in general order of smallest to largest spatial scale of the sampling units, but terms may overlap considerably depending upon how they are spatially defined.

Patch sampling (Jaeger, 1994), which was described earlier in this chapter, applies here when patches are too large or structurally complex to census. Any indices obtained from searching patches can only be applied to other patches (assuming a random selection of patches) and not to the surrounding area. The advantage of this technique is that it targets specific microhabitats that may be used almost exclusively by certain species.

Quadrat sampling consists of searching a regularly shaped plot of ground or volume of water. The term quadrat may refer to a plot of any size, but usually is applied to a square or rectangular plot that contains multiple patches but is much smaller than the entire area of interest. Quadrats should be small enough so that they can be adequately searched. Conversely, subquadrats could be randomly chosen from randomly selected quadrats and searched, or a set of equally spaced strip transects (i.e., small scale) could be randomly placed within a selected quadrat.

Strip transects are basically quadrats that are long and narrow. The width depends on how much area on either side of the center line can be effectively searched by an observer. The length will depend on the mode of travel of the observer, i.e., whether the search is on foot or from a vehicle. Lengths of road may be thought of both as strip transects, because of their configuration, and as a patch sample, because they are a specific feature of an area (space beyond the immediate roadside is excluded). Campbell and Christman (1982) offered suggestions for designing road cruising surveys. We recommend against relying on road surveys to obtain indices of abundance because they probably are not representative of surrounding areas (Figs. 8.2 and 8.3).

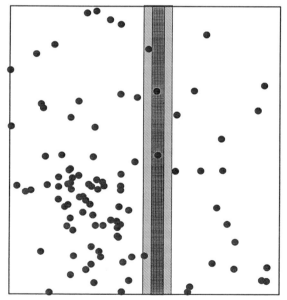

Figure 8.2 A road (gray area) count within an undefined sampling frame. Only those individuals (i.e., black dots) either on the road or within the hatched area are capable of being detected. Inferences can only be made properly within the counted area; expanding the count to the entire area of interest assumes densities outside the counted area are the same as within it.

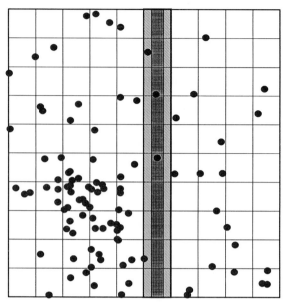

Figure 8.3 Same as previous figure except sampling frame is well-defined. Plots containing the road may or may not be randomly selected. Because the probability of selecting each plot is known, valid inferences may be drawn about the entire frame.

Audio Searches

Surveys of calling frogs are analogous in many ways to songbird surveys; hence, some procedures for obtaining indices of relative abundance for singing birds are applicable to calling frogs (see Chapter 9). In particular, areas could be stratified by frog breeding sites and surveyed from randomly placed points or transects. Zimmerman (1994) suggested that placing transects along existing trails is appropriate if trails were not constructed with respect to frog breeding sites. For example, a trail following a stream would not satisfy this requirement. However, limiting transects to existing trails is just another form of road survey and, as such, removes most "non-trailed" areas from the sampling frame. We strongly urge placement of transects or points to be random irrespective of existing trail systems.

Trapping and Netting

Setting out arrays of drift fences/pitfall traps is a common technique of obtaining indices of relative abundance for amphibians and reptiles. This method varies in effectiveness depending on species. Herpetofauna that are strong jumpers or climbers, or are large relative to trap size, are much more difficult, if not impossible, to trap than those that are not (Gibbons and Semlitsch, 1981; Franz and Ashton, 1989; Dodd, 1991; Corn, 1994). These arrays should be randomly placed within the area of interest; stratification by habitat may well increase potential for capture.

Drift fences/pitfall traps also have been used to trap amphibians moving to and from their breeding ponds (Dodd and Scott, 1994; Franz and Ashton, 1989; Gibbons and Semlitsch, 1981; Storm and Pimentel, 1954). If the breeding pond is small enough, it may be completely encircled by fence with pitfall traps spaced apart on both sides. Larger ponds may not be realistically fenced off because of the amount of fencing needed. Important considerations for placement of partial fence/trap arrays include distance from body of water and placement in relation to movement corridors (Gibbons and Semlitsch, 1981). Areas around a body of water should be stratified by its potential as a movement corridor with more fencing/traps placed within areas of potentially high rates of travel across them. Within a given stratum discrete lengths of fence/traps could be equally spaced, with the leading fence/trap unit randomly placed. Drift fences/pitfall traps also can be used to trap certain species of turtles traveling between their terrestrial wintering area and summer aquatic habitat (Gibbons and Semlitsch, 1981).

A relatively new method for "capturing" amphibians and reptiles is through artificial cover such as cover boards. Cover boards (e.g., plywood sheets of prespecified dimensions) attract certain herpetofauna seeking refuge under them, especially following a rainstorm (Fellers and Drost, 1994). These individuals then are captured by hand or some other means when boards are overturned and checked. Artificial cover should be arranged in a standard configuration, like a grid array, so that capture comparisons can be more easily drawn across areas. Each array formation should be randomly placed within a plot or study area. Fellers and Drost (1994) provided detailed descriptions of research designs, methods, and materials needed for use of this technique. Fitch (1992) used artificial cover to survey snake populations in Kansas. The artificial cover approach is only useful when applied to those species that frequently seek cover under surface objects. This method may be applicable for use in trapping web designs (Anderson *et al.*, 1983) in lieu of other trapping methods; however, the large size of cover items may be impractical for use in trapping webs (Fellers and Drost, 1994).

Netting, trapping, and capturing methods previously discussed for use in aquatic habitats are applicable here except that sampling units must be delineated and methods standardized. Shaffer *et al.* (1994) gave a detailed discussion of these techniques.

Techniques that obtain indices of relative abundance or density are attractive because they are cheaper and easier to implement than either complete counts (at whatever spatial scale) or incomplete counts adjusted for incomplete detectability of individuals. However, index techniques suffer from the fact that they often are subject to numerous sources of bias, the magnitudes of which are unknown, but have no mechanism to correct for these biases. To assume that these various sources of bias somehow cancel out or are directly proportional across space and time is likely nothing more than wishful thinking. These biases could very easily mask changes in numbers or point to nonexistent trends. Therefore, we strongly recommend that biologists employ, if at all possible, methods that account for incomplete detectability or, in limited cases when feasible, complete counts at some level to obtain abundance estimates when conducting an inferential monitoring program.

8.2.2. ADJUSTING FOR INCOMPLETE DETECTABILITY

When appropriately applied, procedures based on capture–recapture and related methods, distance sampling, and encounter probabilities of tracks may be used at various spatial scales to obtain, on average, unbiased estimates of abundance and density. The key words here are "when appropriately applied." Methods discussed in this section are no better than indices, and are much more expensive to perform, if underlying assumptions are not satisfied. Thus, biologists should critically evaluate assumptions before these methods are implemented. In the following, we discuss potential applications of these techniques to amphibians and reptiles.

8.2.2.1. Capture–Recapture and Related Methods

Closed Population Models

Capture–recapture methods and their models have been commonly used to obtain abundance estimates for herpetofauna (Donnelly and Guyer, 1994). The two-sample Lincoln–Petersen estimator, in particular, has been applied to amphibians (Inger and Greenburg, 1966) and reptiles (Parker, 1976; Freedman and Catling, 1978; Martens, 1995). Yet, Lindeman (1990) observed that capture–recapture studies of turtle populations commonly ignored, and therefore did not test, underlying model assumptions required for obtaining valid abundance estimates. Further, the two model assumptions that were ignored most often were population closure and equal catchability, especially in studies using the Lincoln–Petersen estimator. This led Lindeman to comment, "As a result, many estimates of population size in turtle studies are of dubious value" (Lindeman, 1990, p. 78). Turtle studies are not unique among herpetofaunal studies in this respect (e.g., see review by Turner, 1977, pp. 202–216).

Capture–recapture studies assuming a closed population are best applied over a short period of time, perhaps up to a week, on discrete habitats that can be entirely sampled and within which the target species is more or less contained (e.g., small ponds and marshes). A common violation of the closure assumption occurs when capture activities are conducted over a period of weeks or more, or when capture data from different seasons or years are combined. Failure to adequately select and delineate a self-contained or nearly self-contained area also may violate this assumption.

An ill-defined sampling unit or study area is subject to "edge effect" in mobile populations (Fig. 8.4). That is, home range-related or other movements of some animals may carry them outside of the sampling unit or study area. Consequently, some animals whose home ranges overlap only a part of the unit may be captured, which results in an overestimation of density (White *et al.*, 1982, p. 120). A trapping web design (Anderson *et al.*, 1983; Wilson and Anderson, 1985; Parmenter *et al.*, 1989) may be an

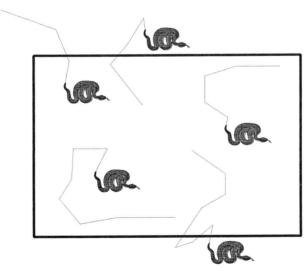

Figure 8.4 Hypothetical movement patterns of five snakes within a selected plot. Movements of only two of the five are entirely contained within the plot boundaries. Radio telemetry could be used to obtain an estimate of these movements and hence provide a correction for the capture–recapture estimator used.

option in situations in which sampling units do not accurately represent areas actually being trapped. White (1996) developed program NORE-MARK for mark–resight studies of radio-marked animals, including an estimator for populations subject to movements on and off sampling units or study areas. This approach requires a representative portion of animals within a selected plot or study area to be captured and equipped with radio transmitters. This estimator assumes sampling without replacement and an equal sighting probability for each animal within each sampling occasion [but this probability may vary across occasions; White (1996)].

A methodology employing radio-telemetry data to adjust abundance estimates has potential for use in herpetofaunal studies. Radio-telemetry technology has previously been used in studies of frogs and toads (van Nuland and Claus, 1981; Smits, 1984; Sinsch, 1989), salamanders (Stouffer et al., 1983), turtles (Obbard and Brooks, 1981; Brown et al., 1990; Lovich, 1990), and snakes (Madsen, 1984; Weatherhead and Hoysak, 1989; Plummer, 1990). Richards et al. (1994) discussed radio-telemetry applications for amphibians in general.

Regarding the assumption of equal catchability, capture probabilities may vary among individuals due to capture method, age, sex, marking technique used, and trap location (i.e., individuals with greater overlap of their activity areas with the trapping area are more likely to be captured). These factors are not mutually exclusive; some capture methods, like pitfall traps, may be more effective at trapping smaller subadults than larger adults (Gibbons et al., 1977). Conversely, van Gelder and Rijsdijk (1987) reported higher netting rates for larger male toads than smaller ones. Marking method also could affect the subsequent behavior of individuals. Clark (1972) reported that recapture rates of toads decreased with the number of toes clipped.

Unequal capture probabilities can be accounted for in some models in program CAPTURE (White et al., 1982; Rexstad and Burnham, 1991). Szaro et al. (1988) used a model allowing unequal catchability to estimate numbers of garter snakes in riparian habitat in New Mexico. However, these models require at least three capture occasions (four in the case of M_{bh}) to appropriately fit capture data. An alternative approach is to stratify the data into classes of similar capture rates, such as age and sex, and fit data within each stratum. This, however, may reduce the sample sizes to a point below which adequate abundance estimates cannot be computed.

Using the same capture technique over multiple occasions may yield a biased abundance estimate (Seber, 1982) because of trapping selectivity, behavioral response, and capture heterogeneity. These latter two factors may be modeled using program CAPTURE, but trapping selectivity may only be addressed via use of different capture methods. Behavioral response to a capture method also can be avoided by using a mark–resight approach in conjunction with program NOREMARK (White, 1996). Unfortunately, practical and logistical difficulties will probably limit the number of different capture techniques that can be realistically implemented. However, in certain situations, different capture techniques could possibly be used over more than two sampling occasions (Example 8.1). For example, a three-sample capture–recapture study of frogs in a small pond may use electro-shocking and dipnetting for the initial capture, seining for the second capture, and resighting of marked/unmarked individuals for the third "capture." Unfortunately, such an approach may not be feasible in a number of instances due to habitat structure precluding the use of certain methods, low capture rates of some methods under certain conditions, and/or capture methods that lead to violations of model assumptions. If the same physical capture technique must be used, it should be applied in a manner that attempts to minimize potential bias. For instance, seining parallel to a shoreline in one direction for the initial capture and seining in the opposite direction for the second capture may lessen the magnitude of bias. That is, because of habitat structure some individuals may be much less available for capture in one seining direction than another. This probably will not remove the bias associated with using the same capture technique, but may lower it.

Rosenberg *et al.* (1995) applied jackknife estimators to capture data from a small population characterized by low and heterogeneous capture probabilities. They reported that reliable abundance estimates could be obtained using low-order jackknife estimators, an *ad hoc* adjustment to variance estimators (which they provided) to improve confidence interval coverage, and ≥12 capture occasions. We refer readers to their paper for details of this approach.

Example 8.1. A Capture–Recapture Pilot Study To Estimate Frog Abundance

A capture–recapture pilot study was conducted on a northern leopard frog population in 16 ponds in and around Arapaho National Wildlife Refuge in northcentral Colorado. Capture efforts, and therefore inferences, were limited to wadeable portions of ponds. Ponds were placed in 1 of 3 strata based on size. Five ($U_1 = 5$) were roughly 15 ha, 7 ($U_2 = 7$) were approximately 5 ha, and 4 ($U_3 = 4$) were about 2 ha. Two ponds from each stratum ($u_1 = u_2 = u_3 = 2$) were selected for collection of preliminary capture data. Captured individuals were marked with a fluorescent dye (Taylor and Deegan, 1982; Nishikawa and Service, 1988) in unique combinations of colors (9) and numbers (Donnelly *et al.*, 1994). Initial capture was done via a combination of electroshocking, dip-netting, and physical displacement of individuals among littoral vegetation. Seining and littoral displacement were used in the second and third capture events except that seining was done in the opposite direction (parallel to the shore) within each. The final "capture" was performed using a spotlight and portable ultraviolet lamp to detect frogs at night. Program CAPTURE chose model M_{th} as an adequate fit to capture data. Abundance estimates ($\hat{N}_{hi}$) and associated variance estimates [$\hat{\mathrm{Var}}(\hat{N}_{hi}/N_{hi})$, in parentheses] for the two ponds in each stratum were: stratum I, 852 (10,404) and 594 (8648); stratum II, 171 (960) and 251 (1764); and stratum III, 67 (196) and 95 (290). These results, in conjunction with formulas in Appendix C, section C.2.2.2 (stratified random sample), were used to produce the following table.

Stratum	u_h	U_h	$\hat{S}^2_{N_{hi}}$	$\hat{\mathrm{Var}}(\hat{N}_{hi}/N_{hi})$	$\hat{N}_h$	$\hat{\mathrm{Var}}(\hat{N}_h)$
I	2	5	33,282	9526	3615	297,245
II	2	7	3,200	1362	1477	65,534
III	2	4	392	243	324	2,540

Overall estimates for abundance and precision were: $\hat{N} = 3615 + 1477 + 324 = 5416$; $\hat{\mathrm{Var}}(\hat{N}) = 297,245 + 65,534 + 2540 = 365,319$; $\hat{\mathrm{SE}}(\hat{N}) = 604.4$; and $\hat{\mathrm{CV}}(\hat{N}) = 604.4/5416 = 0.112$. An optimizer function in a spreadsheet was used to obtain sample sizes needed for each stratum to satisfy $e = 0.1$ and $\alpha = 0.05$ based on the sample size formula in Appendix C, section C.2.2.2 (stratified random sample). These sample sizes were $u_1 = 4$, $u_2 = 4$, and $u_3 = 2$. Total cost of the survey was computed using $c_{11} = \$500$, $c_{12} = \$600$, $c_{13} = \$900$, $c_{21} = \$200$, $c_{22} = \$400$, and $c_{23} = \$1100$, and was calculated to be \$10,800.

Removal Models

Models based on nonselective removals and equal effort are special cases of capture–recapture models (i.e., M_b and M_{bh}) in which captured individuals are either physically removed or removed via marking from the population of interest. Bury et al. (1977) and Bury (1982) used removal estimators for obtaining abundance estimates for herpetofauna. Hayek (1994) discussed removed methods particularly as they applied to amphibian populations. See White et al. (1982) and Lancia et al. (1994) for general discussions of removal models.

Open Population Models

The Jolly–Seber model (Jolly, 1965; Seber, 1965) may be used when capture events occur over long periods and/or data are combined over years. Yet, this model produces abundance estimates that have greater variability than those produced by closed models and does not account for individual differences in capture probabilities or behavioral responses. It also still requires geographic closure (i.e., no edge effect) for the individuals being studied. Although not commonly applied to herpetofauna, the Jolly–Seber model has been used for frogs and toads (Cooke and Oldham, 1995), turtles (Lindeman, 1990) and tortoises (Murray, 1994). Unfortunately, heterogeneity in capture probabilities are likely in most field situations, which leads to serious biases in population estimates based on open models unless

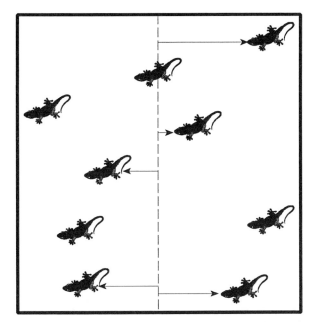

Figure 8.5   A line transect (dashed line) with perpendicular distances (arrows) to detected individuals.

capture probabilities are very high (Gilbert, 1973). Hence, we generally recommend using open population models for survival estimation only, i.e., when appropriate model assumptions are met.

8.2.2.2. Distance Sampling

Line transect methods based on perpendicular detection distances to observed individuals or objects (Fig. 8.5) have been applied only sparsely in herpetofaunal studies. We are not aware of any published studies using point transect methods. Nevertheless, distance sampling methods have the potential for broader use in studies of amphibian and reptile populations (Example 8.2). Buckland *et al.* (1993) presented an excellent review and discussion of distance sampling theory and applications.

Example 8.2. Line Transect Sampling To Estimate Snake Densities
   A study was undertaken to assess present density levels of racers in a 42-km^2 portion of Comanche National Grassland in southwestern Colorado. The project leader decided to use line transect sampling to obtain density estimates of snakes. The $\hat{CV}(\hat{D})$ required to detect

about a 20% change ($r = 0.20$) over 5 years ($n = 5$), with $\alpha = 0.05$ and $\beta = 0.10$ was [Eq. (6.7); program TRENDS]:

$$\hat{CV}(\hat{D}) \leq \sqrt{\frac{(0.2)^2(5)^3}{12(1.96 + 1.282)^2}} \leq 0.199.$$

An observer walked along a randomly placed, 500-m transect (L_0) and had eight sightings (n_0) of racers. Transect length (L) for the full survey was calculated [Eq. (3.4)] using a dispersion parameter (b) of 3,

$$L = \left(\frac{3}{[0.199]^2}\right)\left(\frac{500\ \text{m}}{8}\right) = 4735\ \text{m}.$$

Fifteen (l_i, $i = 1, \ldots, 15$) square-shaped 316-m (79 m per side) transects (Buckland et al., 1993, pp. 299–300) were randomly placed within the target area. Surveys were conducted during peak activity hours of racers. Sighting distances were successfully modeled in DISTANCE. Resulting estimates were $\hat{D} = 13.4$ racers/ha, $\hat{SE}(\hat{D}) = 2.57$, and $\hat{CV}(\hat{D}) = 0.192$. Total cost for the first year was computed using $c = $ cost per $l_i = \$150$ and $c_0 = \$5000$, and was $7250.

Lacki et al. (1994) used a simple curve function proposed by Eberhardt (1968) to fit distance data, and used the slope of the log-transformed curve data in a density formula to obtain density estimates of northern copperbelly water snakes in southern Indiana. Transects were the waterline perimeters of surveyed water bodies. In upland areas, transects were placed along contour lines within a certain elevational interval. Unfortunately, there are some problems with studies such as this. First of all, a fairly simple curve-fitting function was used to fit sighting data. Although Eberhardt (1968) was a landmark paper in its time, distance sampling theory and methods of analysis have progressed greatly since then. Program TRANSECT (Laake et al., 1979) has been available since at least 1979 and was recently replaced by program DISTANCE (Laake et al., 1993). Second, the curve-fitting model was not tested for validity, nor were other models considered. We highly recommend that sighting data be fitted to a number of different models, a global model tested for goodness of fit, and the AIC model selection procedure (Akaike, 1973; Burnham and Anderson, 1992) used to select among appropriate models. Program DISTANCE accomplishes these tasks. Finally, transects were placed nonrandomly and therefore inferences were limited to the area covered within the transects. Water snakes occurring outside the width of the perimeter transects around water bodies were excluded. Transects placed along contour lines are likely unrepresentative of the area of interest (Buckland et al., 1993).

Lohoefener (1990) used line transect methods to estimate density of gopher tortoise burrows in Alabama. He used a Fourier series to model detection distances. Dodd (1990) applied the same technique to salamander burrows. Both authors were confident that densities of burrows were accurately estimated. Both also pointed out that using burrow densities as direct indicators of population densities was problematic. One would need to know whether or not burrows were occupied as well as the number of occupants of each. Consequently, burrow densities alone can only be used as an index to population density.

Distance sampling methods should only be applied in situations in which key underlying model assumptions are met, such as detecting all individuals or objects on the line or point, recording individuals at their original locations, and accurately measuring detection distances (Buckland *et al.*, 1993). In addition, a minimum number of detections are needed to acquire reasonable density estimates (see Buckland *et al.*, 1993, pp. 295–312). Detecting all individuals on a line or point is particularly important. For instance, distance methods should be applied with caution to species that have at least some subterranean habits. One possible solution is to conduct surveys during times when essentially all individuals are active (Example 8.3); however, some species may only spend a small amount of time above ground. Gopher tortoises may spend several weeks underground during particular times of the year (Cox *et al.*, 1987). In these situations, proper inferences can only be made to the segment of the population above ground during the time of the survey unless individuals can be realistically located in burrows. In any event, we strongly urge that a pilot study be conducted to assess the applicability of either line or point transect methodology.

Example 8.3. Line Transect Sampling To Estimate Lizard Densities

A BLM biologist was instructed to design a monitoring program for a side-blotched lizard population in a 259-km² portion of the Sand Wash Basin Wild Horse Management Area in northwestern Colorado. She decided to use line transect methods in ground surveys to obtain estimates of lizard densities. The $\hat{CV}(\hat{D})$ required to detect about a 15% change ($r = 0.15$) over five sampling periods (evenly spaced surveys once every 2 years for 10 years; $n = 5$), with $\alpha = 0.05$ and $\beta = 0.10$, was [Eq. (6.7); program TRENDS]:

$$\hat{CV}(\hat{D}) \leq \sqrt{\frac{(0.15)^2(5)^3}{12(1.96 + 1.282)^2}} \leq 0.149.$$

A preliminary ground count was conducted along a randomly placed, 500-m transect (L_0) that was oriented against the prevailing topographic gradient. Length of transect (L) needed to satisfy the precision requirements was computed [Eq. (3.4)] using the number of sightings ($n_0 = 10$) during the preliminary survey and a dispersion parameter (b) set equal to 3 and was

$$L = \left(\frac{3}{[0.149]^2}\right)\left(\frac{500 \text{ m}}{10}\right) = 6756 \text{ m}.$$

A grid of 20 338-m lines was randomly placed within the target area. Lines were spaced 100 m apart and oriented against the topographic features. Four well-trained observers each walked five lines between 9 AM and 10 AM (time of peak lizard activity) and recorded distances to sighted lizards. The data were adequately fit by a model chosen by program DISTANCE. These results were: $\hat{D}$ = 22.1 lizards/ha, $\hat{SE}(\hat{D})$ = 3.0, and $\hat{CV}(\hat{D})$ = 0.136. The estimated dispersion parameter (b) was actually 2.50, so that line transect length for the next survey was shorter. Given c = cost per l_i = $100 and c_0 = $7500, the total cost of the survey for the first year was $9500.

8.2.2.3. Line-Intercept Sampling

Becker (1991) described a method for estimating abundance of individuals based on probability of intersecting their fresh tracks from randomly placed systematic transects across a rectangular area or plot. The horizontal length of a given set of tracks of a specific individual within a defined area represents its probability of being encountered during a survey. Resulting abundance estimates will be unbiased. A more thorough description of this technique was given in Chapter 3.

Although Becker (1991) developed his technique for use on wolverines and other furbearers in snow conditions, with appropriate modifications it also may be applicable to some reptile species living in areas with fine-textured soils. The target species would have to be large enough to leave recognizable tracks that are distinguishable from other species by an observer on foot. For example, Lillywhite (1982) successfully identified and followed tracks of three common species of snakes in semidesert chaparral. Further, tracks made prior to the survey must be discernible from those made during the survey period. This means that a track survey must either be scheduled immediately after a weather event (e.g., high winds or rain) that would obliterate existing tracks or be preceded by a preliminary foot survey in which all encountered tracks are marked on the transects. Also, surveys must be conducted at times when movements of individuals are minimal (e.g., early cool mornings for snakes). Other underlying assumptions of this method follow directly from Becker (1991, p. 731).

8.3. RECOMMENDATIONS

The key to success in any monitoring program is the ability to achieve the objectives within the cost limitations. Selection of counting techniques always should be viewed in terms of both cost and utility of results. Utility of results will depend on the level of precision and bias of population estimates. This section offers some general recommendations for choosing among enumeration methods already discussed in this chapter. A dichotomous key has been included for more detailed help. We emphasize that these are general recommendations only; biologists must rely on their knowledge of the species and survey situation to assess the validity of enumeration methods under consideration.

8.3.1. GENERAL COMMENTS

When attempting to monitor numerous species concurrently, unbiased abundance/density estimates will only be obtainable for relatively few of them. Costs and effort will vary even among index methods, with presence–absence methods being the easiest to conduct. We generally recommend defining a sampling frame and randomly choosing plots within it (or, similarly, within the boundaries of a defined area) to obtain information. However, road surveys or even less rigorous methods may have to suffice for some species when many species are being monitored, and funding and manpower are low. Index methods that obtain the most information at the lowest cost are obviously desirable. As such, index techniques such as pitfall trapping, audio surveys for frogs, and road counts will probably provide the most information for the highest number of species of amphibians and reptiles. Again, index information will only provide, at best, a rough guess at population trends, and should not be relied upon for monitoring species of special concern.

8.3.2. DICHOTOMOUS KEY TO ENUMERATION METHODS

The dichotomous key in this section should be used in conjunction with information already presented earlier in this chapter (sections are in parentheses) and in previous chapters on sampling design, enumeration methods, and guidelines for planning surveys. In particular, readers should keep in mind that assumptions underlying techniques that correct for incomplete detectability must be met if resulting estimates are to be considered valid.

1. (a) A complete count of individuals or burrows in all randomly chosen plots or surveyed areas is possible with available funding............2
   (b) A complete count is either impossible or too costly in all randomly chosen plots or surveyed areas...4
2. (a) All individuals can be counted within each plot during a single count .... Quadrat sampling (complete ground counts), patch sampling (8.12)
   (b) All burrows can be located within each plot 3

3. (a) Either number of individuals occupying burrows or relationship between number of occupied burrows and number of individuals is known ....Complete ground counts

   (b) Not as above 17

4. (a) Individuals can be caught and uniquely marked 5

   (b) Individuals cannot be caught, are too expensive to catch, or are not conducive to marking 10

5. (a) Individuals are completely contained (i.e., no edge effect) within either entire area of interest if survey is over the entire sampling frame or within a given plot if surveys are performed at the plot level..6

   (b) Not as above 8

6. (a) Time period of study is short enough to treat population as closed in terms of births/immigration and deaths/emigration 7

   (b) Not as above 9

7. (a) There are at least 100 individuals contained within the survey area (e.g., plot) and their capture probabilities are at least 0.3 (preferably 0.5 or above for the range of population sizes encountered in most herpetofaunal studies); lower capture probabilities are acceptable with larger population sizes (program CAPTURE has a simulation routine to check if estimated capture probabilities are adequate) or ≥ 12 sampling occasions ....Capture–recapture models for multiple sampling occasions (8.2.2.1); frogs in small ponds

   (b) Not as above 10

8. (a) Individuals can be equipped with radio transmitters in a cost-efficient manner, and readily and accurately located via radio-telemetry ....correction for immigration/emigration in Program NOREMARK (8.2.2.1)

   (b) Not as above 10

9. (a) Population is geographically closed and individuals are equally and easily catchable with no behavioral response to the capture technique ....Jolly–Seber open population model (8.2.2.1)

   (b) Not as above 10

10. (a) Perpendicular distances (or distance categories) from a line transect to an individual, a discrete group of individuals, or a burrow can be recorded in a cost-efficient manner..11

    (b) Perpendicular distances are either unobtainable or too costly to obtain..15

11. (a) Every individual or burrow on a line transect can be located....12

    (b) Not all individuals on a line transect can be detected because of vegetational structure, subterranean habits, undetected movements, or other reasons 15

12. (a) Element of interest is a burrow 13
    (b) Element of interest is an individual 14
13. (a) Either number of individuals per burrow or relationship between
        number of individuals and number of burrows is known............14
    (b) Not as above 15
14. (a) Adequate numbers of individuals, groups of individuals, or objects
        can be detected for accurate model selection in program
        DISTANCE and reasonably precise density estimates...................
        Distance sampling methods; line transects for snakes, lizards, and
        possibly turtles in open habitats, etc. (8.2.2.2)
    (b) Not as above 15
15. (a) Individuals leave identifiable tracks that can be followed from one
        end to the other without causing animal movement.....................16
    (b) Tracks not identifiable because of substrate or size of track, or
        tracks cannot be followed without causing further movement by
        individual 17
16. (a) Either fresh tracks are distinguishable from older tracks because
        of weather effects on substrate (e.g., rain) or older tracks are
        marked by an observer...Becker's (1991)
        line intercept estimator (8.2.2.3)
    (b) Not as above 17
17. (a) Only data on species occurrence required ....
        ....Presence–absence methods (8.2.1.1)
    (b) Uncorrected counts of all individuals/objects detected on a plot
        ....Relative index methods (8.2.1.2); e.g., pitfall trapping,
        seining, and aural surveys for frogs

LITERATURE CITED

Akaike, H. (1973). Information theory as an extension of the maximum likelihood principle. In *Second International Symposium on Information Theory* (B. N. Petrov and F. Csaki, Eds.), pp. 276–281. Akademiai, Budapest.

Anderson, D. R., Burnham, K. P., White, G. C., and Otis, D. L. (1983). Density estimation of small-mammal populations using a trapping web and distance sampling methods. *Ecology* **64:** 674–680.

Bart, J., Stehn, R. A., Herrick, J. A., Heaslip, N. A., Bookhout, T. A., and Stenzel, J. R. (1984). Survey methods for breeding yellow rails. *J. Wildl. Mgmt.* **48:** 1382–1386.

Becker, E. F. (1991). A terrestrial furbearer estimator based on probability sampling. *J. Wildl. Mgmt.* **55:** 730–737.

Brown, G. P., Brooks, R. J., and Layfield, J. A. (1990). Radiotelemetry of body temperatures of free-ranging snapping turtles (*Chelydra serpentina*) during summer. *Can. J. Zool.* **68:** 1659–1663.

Buckland, S. T., Anderson, D. R., Burnham, K. P., and Laake, J. L. (1993). *Distance Sampling: Estimating Abundance of Biological Populations.* Chapman and Hall, London.

Burnham, K. P., and Anderson, D. R. (1992). Data-based selection of an appropriate biological model: The key to modern data analysis. In *Wildlife 2001: Populations* (D. R. McCullough and R. H. Barrett, Eds.), pp. 16–30. Elsevier, London.

Bury, R. B. (1982). Structure and composition of Mojave reptile communities determined with a removal method. In *Herpetological Communities* (N. J. Scott, Jr., Ed.), pp. 135–142. U.S.D.I. Fish Wildl. Serv., Washington, DC. [Wildl. Res. Rep. 13].

Bury, R. B., and Corn, P. S. (1987). Evaluation of pitfall trapping in northwestern forests: Trap arrays with drift fences. *J. Wildl. Mgmt.* **51:** 112–119.

Bury, R. B., and Luckenbach, R. A. (1977). Censusing desert tortoise populations using a quadrat and grid location system. In *Desert Tortoise Council Symposium Proceedings* (M. Trotter and C. G. Jackson, Jr., Eds.), pp. 169–178. Desert Tortoise Counc., San Diego.

Bury, R. B., Luckenbach, R. A., and Busack, S. D. (1977). *Effects of Off-Road Vehicles on Vertebrates in the California Desert.* U.S.D.I. Fish Wildl. Serv., Washington, DC. [Wildl. Res. Rep. 8].

Campbell, H. W., and Christman, S. P. (1982). Field techniques for herpetofaunal community analysis. In *Herpetological Communities* (N. J. Scott, Jr., Ed.), pp. 193–200. U.S.D.I. Fish Wildl. Serv., Washington, DC. [Wildl. Res. Rep. 13].

Clark, R. D. (1972). The effect of toe clipping on survival in Fowler's toad (*Bufo woodhousei fowleri*). *Copeia* **1972:** 182–185.

Cooke, A. S., and Oldham, R. S. (1995). Establishment of populations of the common frog, *Rana temporaria,* and common toad, *Bufo bufo,* in a newly created reserve following translocation. *Herpetol. J.* **5:** 173–180.

Corn, P. S. (1994). Straight line drift fences and pitfall traps. In *Measuring and Monitoring Biological Diversity—Standard Methods for Amphibians* (W. R. Heyer, M. A. Donnelly, R. W. McDiarmid, L.-A. C. Hayek, and M. S. Foster, Eds.), pp. 109–117. Smithsonian Institution Press, Washington, DC.

Corn, P. S., and Bury, R. B. (1990). *Sampling Methods for Terrestrial Amphibians and Reptiles.* [U.S.D.A. For. Serv. Gen. Tech. Rep. PNW-GTR-256].

Cox, J., Inkley, D., and Kautz, R. (1987). *Ecology and Habitat Protection Needs of Gopher Tortoise (Gopherus polyphemus) Populations Found on Lands for Large-Scale Development in Florida.* Fla. Game Fresh Water Fish Comm., Tallahassee, FL. [Nongame Wildl. Prog. Tech. Rep. 4].

Crump, M. L., and Scott, N. J., Jr. (1994). Visual encounter surveys. In *Measuring and Monitoring Biological Diversity—Standard Methods for Amphibians* (W. R. Heyer, M. A. Donnelly, R. W. McDiarmid, L.-A. C. Hayek, and M. S. Foster, Eds.), pp. 84–92. Smithsonian Institution Press, Washington, DC.

Dodd, C. K., Jr. (1990). Line transect estimation of Red Hills salamander burrow density using a Fourier series. *Copeia* **1990:** 555–557.

Dodd, C. K., Jr. (1991). Drift fence-associated sampling bias of amphibians at a Florida sandhills temporary pond. *J. Herpetol.* **25:** 296–301.

Dodd, C. K., Jr., and Scott, D. E. (1994). Drift fences encircling breeding sites. In *Measuring and Monitoring Biological Diversity—Standard Methods for Amphibians* (W. R. Heyer, M. A. Donnelly, R. W. McDiarmid, L.-A. C. Hayek, and M. S. Foster, Eds.), pp. 125–130. Smithsonian Institution Press, Washington, DC.

Donnelly, M. A., and Guyer, C. (1994). Mark-recapture. In *Measuring and Monitoring Biological Diversity—Standard Methods for Amphibians* (W. R. Heyer, M. A. Donnelly, R. W. McDiarmid, L.-A. C. Hayek, and M. S. Foster, Eds.), pp. 183–205. Smithsonian Institution Press, Washington, DC.

Donnelly, M. A., Guyer, C., Juterbock, J. E., and Alford, R. A. (1994). Techniques for marking amphibians. In *Measuring and Monitoring Biological Diversity—Standard Methods for*

Amphibians (W. R. Heyer, M. A. Donnelly, R. W. McDiarmid, L.-A. C. Hayek, and M. S. Foster, Eds.), pp. 277–284. Smithsonian Institution Press, Washington, DC.

Eberhardt, L. L. (1968). A preliminary appraisal of line transects. *J. Wildl. Mgmt.* **32:** 82–88.

Fellers, G. M., and Drost, C. A. (1994). Sampling with artificial cover. In *Measuring and Monitoring Biological Diversity—Standard Methods for Amphibians* (W. R. Heyer, M. A. Donnelly, R. W. McDiarmid, L.-A. C. Hayek, and M. S. Foster, Eds.), pp. 147–150. Smithsonian Institution Press, Washington, DC.

Feuer, R. C. (1980). Underwater traps for aquatic turtles. *Herpetol. Rev.* **11:** 107–108.

Fitch, H. S. (1992). Methods of sampling snake populations and their relative success. *Herpetol. Rev.* **23:** 17–19.

Franz, R., and Ashton, R. E., Jr. (1989). *Behavior and Movements of Certain Small Sandhill Amphibians and Reptiles in Response to Drift Fences.* Fla. Game Fresh Water Fish Comm., Tallahassee, FL. [Nongame Wildl. Prog. Tech. Rep.]

Freedman, W., and Catling, P. M. (1978). Population size and structure of four sympatric species of snakes at Amherstburg, Ontario. *Can. Field-Nat.* **92:** 167–173.

Friend, G. T. (1984). Relative efficiency of two pitfall-drift fence systems for sampling small vertebrates. *Aust. Zool.* **21:** 423–433.

Gibbons, J. W., Coker, J. W., and Murphy, T. M., Jr. (1977). Selected aspects of the life history of the rainbow snake (*Farancia erytrogramma*). *Herpetologica* **33:** 276–281.

Gibbons, J. W., and Semlitsch, R. D. (1981). Terrestrial drift fences with pitfall traps: an effective technique for quantitative sampling of animal populations. *Brimleyana* **7:** 1–16.

Gilbert, R. O. (1973). Approximations of the bias in the Jolly-Seber capture-recapture model. *Biometrics* **29:** 501–526.

Gunning, G. E., and Lewis, W. M. (1957). An electrical shocker for the collection of amphibians and reptiles in the aquatic environment. *Copeia* **1957:** 52.

Hammerson, G. A. (1982). *Amphibians and Reptiles in Colorado.* Colo. Div. Wildl., Denver, CO. [Publ. DOW-M-I-27-82].

Hayek, L.-A. C. (1994). Removal sampling. In *Measuring and Monitoring Biological Diversity—Standard Methods for Amphibians* (W. R. Heyer, M. A. Donnelly, R. W. McDiarmid, L.-A. C. Hayek, and M. S. Foster, Eds.), pp. 201–205. Smithsonian Institution Press, Washington, DC.

Heyer, W. R., Donnelly, M. A., McDiarmid, R. W., Hayek, L.-A. C., and Foster, M. S. (Eds.) (1994). *Measuring and Monitoring Biological Diversity—Standard Methods for Amphibians.* Smithsonian Institution Press, Washington, DC.

Inger, R. F., and Greenburg, B. (1996). Ecological and competitive relations among three species of frogs (genus *Rana*). *Ecology* **47:** 746–759.

Jaeger, R. G. (1980). Density-dependent and density-independent causes of extinction of a salamander population. *Evolution* **34:** 617–621.

Jaeger, R. G. (1994). Patch sampling. In *Measuring and Monitoring Biological Diversity—Standard Methods for Amphibians* (W. R. Heyer, M. A. Donnelly, R. W. McDiarmid, L.-A. C. Hayek, and M. S. Foster, Eds.), pp. 107–109. Smithsonian Institution Press, Washington, DC.

Jaeger, R. G., and Inger, R. F. (1994). Quadrat sampling. In *Measuring and Monitoring Biological Diversity—Standard Methods for Amphibians* (W. R. Heyer, M. A. Donnelly, R. W. McDiarmid, L.-A. C. Hayek, and M. S. Foster, Eds.), pp. 97–102. Smithsonian Institution Press, Washington, DC.

Jolly, G. M. (1965). Explicit estimates from capture-recapture data with both death and immigration-stochastic model. *Biometrika* **52:** 225–247.

Jones, K. B. (1986). Amphibians and reptiles. In *Inventory and Monitoring of Wildlife Habitat*

(A. Y. Cooperrider, R. J. Boyd, and H. R. Stuart, Eds.), pp. 267–290. U.S.D.I. Bur. Land Manage. Serv. Ctr., Denver, CO.

Karns, D. R. (1986). *Field Herpetology: Methods for the Study of Amphibians and Reptiles in Minnesota.* Bell Mus. Nat. Hist., Univ. Minn., Minneapolis, MN. [Occas. Pap. 18].

Laake, J. L., Buckland, S. T., Anderson, D. R., and Burnham, K. P. (1993). *DISTANCE User's Guide,* v2.1. Colo. Coop. Fish Wildl. Res. Unit, Colo. State Univ., Fort Collins, CO.

Laake, J. L., Burnham, K. P., and Anderson, D. R. (1979). *User's Manual for Program TRANSECT.* Utah State Univ. Press, Logan, UT.

Lacki, M. J., Hummer, J. W., and Fitzgerald, J. L. (1994). Application of line transects for estimating population density of the endangered copperbelly water snake in Southern Indiana. *J. Herpetol.* **28:** 241–245.

Lagler, K. F. (1943). Methods of collecting freshwater turtles. *Copeia* **1943:** 21–25.

Lancia, R. A., Nichols, J. D., and Pollock, K. H. (1994). Estimating the number of animals in wildlife populations. In *Research and Management Techniques for Wildlife and Habitats* (T. A. Bookhout, Ed.), fifth ed., pp. 215–253. The Wildlife Society, Bethesda, MD.

Lillywhite, H. B. (1982). Tracking as an aid in ecological studies of snakes. In *Herpetological Communities* (N. J. Scott, Jr., Ed.), pp. 181–191. U.S.D.I. Fish Wildl. Serv., Washington, DC. [Wildl. Res. Rep. 13].

Lindeman, P. V. (1990). Closed and open model estimates of abundance and tests of model assumptions for two populations of the turtle, *Chrysemys picta. J. Herpetol.* **24:** 78–81.

Lohoefener, R. (1990). Line transect estimation of gopher tortoise burrow density using a Fourier series. In *Burrow Associates of the Gopher Tortoise* (C. K. Dodd, Jr., R. E. Ashton, Jr., R. Franz, and E. Wester, Eds.), pp. 44–69. Mus. Nat. Hist., Gainesville, FL [Proc. 8th Ann. Mtg. Gopher Tortoise Counc., FL].

Lovich, J. (1990). Spring movement patterns of two radio-tagged male spotted turtles. *Brimleyana* **16:** 67–71.

Madsen, T. (1984). Movements, home range size and habitat use of radio-tracked grass snakes (*Natrix Natrix*) in southern Sweden. *Copeia* **1874:** 707–713.

Martens, D. (1995). Population structure and abundance of grass snakes, *Natrix natrix,* in central Germany. *J. Herpetol.* **29:** 454–456.

Mathis, A. (1990). Territoriality in a terrestrial salamander: The influence of resource quality and body size. *Behaviour* **112:** 162–175.

Moulton, J. M. (1954). Notes on the natural history, collection, and maintenance of the salamander *Ambystoma maculatum. Copeia* **1954:** 64–65.

Murray, R. C. (1994). *Mark-Recapture Methods for Monitoring Sonoran Populations of the Desert Tortoise (Gopherus agassizi).* M.S. Thesis, Univ. Ariz., Tucson, AZ.

Nishikawa, K. C., and Service, P. M. (1988). A fluorescent marking technique for individual recognition of terrestrial salamanders. *J. Herpetol.* **22:** 351–353.

Obbard, M. E., and Brooks, R. J. (1981). A radio-telemetry and mark-recapture study of activity in the common snapping turtle, *Chelydra serpentina. Copeia* **1981:** 630–637.

Parker, W. S. (1976). Population estimates, age structure, and denning habits of whipsnakes, *Masticophis T. taeniatus,* in a northern Utah *Atriplex-Sarcobatus* community. *Herpetologica* **32:** 53–57.

Parmenter, R. P., MacMahon, J. A., and Anderson, D. R. (1989). Animal density estimation using a trapping web design: Field validation experiments. *Ecology* **70:** 169–179.

Petokas, P. J., and Alexander, M. M. (1979). A new trap for basking turtles. *Herpetol. Rev.* **10:** 90.

Plummer, M. V. (1990). Nesting movements, nesting behavior and nest sites of green snakes (*Opheodry aestivus*) revealed by radio telemetry. *Herpetologica* **46:** 190–195.

Raphael, M. G. (1988). Long-term trends in abundance of amphibians, reptiles, and mammals in Douglas-fir forests of northwestern California. In *Management of Amphibians, Reptiles and Small Mammals in North America* (R. C. Szaro, K. E. Severson, and D. R. Patton, Tech. Coords.), U.S.D.A. For. Serv., Fort Collins, CO. [Gen. Tech. Rep. RM-166].

Rexstad, E., and Burnham, K. (1991). *User's Guide for Interactive Program CAPTURE*. Colo. Coop. Fish Wildl. Res. Unit, Colo. State Univ., Fort Collins, CO.

Richards, S. J., Sinsch, U., and Alford, R. A. (1994). Radio tracking. In *Measuring and Monitoring Biological Diversity—Standard Methods for Amphibians* (W. R. Heyer, M. A. Donnelly, R. W. McDiarmid, L.-A. C. Hayek, and M. S. Foster, Eds.), pp. 155–158. Smithsonian Institution Press, Washington, DC.

Rosenberg, D. K., Overton, W. S., and Anthony, R. G. (1995). Estimation of animal abundance when capture probabilities are low and heterogeneous. *J. Wildl. Mgmt.* **59:** 252–261.

Scott, N. J., Jr. (Ed.) (1982). *Herpetological Communities*. U.S.D.I. Fish Wildl. Serv., Washington, DC. [Wildl. Res. Rep. 13].

Seber, G. A. F. (1965). A note on the multiple-recapture census. *Biometrika* **52:** 249–259.

Seber, G. A. F. (1982). *Estimation of Animal Abundance and Related Parameters,* second ed. Griffin, London.

Shaffer, H. B., Alford, R. A., Woodward, B. D., Richards, S. J., Altig, R. G., and Gascon, C. (1994). Quantitative sampling of amphibian larvae. In *Measuring and Monitoring Biological Diversity—Standard Methods for Amphibians* (W. R. Heyer, M. A. Donnelly, R. W. McDiarmid, L.-A. C. Hayek, and M. S. Foster, Eds.), pp. 130–141. Smithsonian Institution Press, Washington, DC.

Shaffer, H. B., and Juterbock, J. E. (1994). Night driving. In *Measuring and Monitoring Biological Diversity—Standard Methods for Amphibians* (W. R. Heyer, M. A. Donnelly, R. W. McDiarmid, L.-A. C. Hayek, and M. S. Foster, Eds.), pp. 163–166. Smithsonian Institution Press, Washington, DC.

Shields, M. A. (1985). Selective use of pitfall traps by southern leopard frogs. *Herpetol. Rev.* **16:** 14.

Sinsch, U. (1989). Behavioural thermoregulation of the Andean toad (*Bufo spinulosus*) at high altitudes. *Oecologia* **80:** 32–38.

Smits, A. W. (1984). Activity patterns and thermal biology of the toad *Bufo boreas halophilus. Copeia* **1984:** 689–696.

Storm, R. M., and Pimentel, R. A. (1954). A method for studying breeding populations. *Herpetologica* **10:** 161–166.

Stouffer, R. H., Jr., Gates, J. E., Hocutt, C. H., and Stauffer, J. R., Jr. (1983). Surgical implantation of a transmitter package for radio-tracking endangered hellbenders. *Wildl. Soc. Bull.* **11:** 384–386.

Szaro, R. C., Belfit, S. C., Aitkin, J. K., and Babb, R. D. (1988). The use of time fixed-area plots and a mark-recapture technique in assessing riparian garter snake populations. In *Management of Amphibians, Reptiles and Small Mammals in North America* (R. C. Szaro, K. E. Severson, and D. R. Patton, Tech. Coords.), pp. 239–246. U.S.D.A. For. Serv., Fort Collins, CO. [Gen. Tech. Rep. RM-166].

Taylor, J., and Deegan, L. (1982). A rapid method for mass marking of amphibians. *J. Herpetol.* **16:** 172–173.

Turner, F. B. (1977). The dynamics of populations of squamates, crocodilians and rhynchocephalians. In *Biology of Reptilia* (C. Gans and D. W. Tinkle, Eds.), Vol. 7, pp. 157–264. Academic Press, New York.

van Gelder, J. J., and Rijsdijk, G. (1987). Unequal catchability of male *Bufo bufo* within breeding populations. *Holarctic Ecol.* **10:** 90–94.

van Nuland, G. J., and Claus, P. F. H. (1981). The development of a radio tracking system for anuran species. *Amphibia-Reptilia* **2:** 107–116.

Vogt, R. C., and Hine, R. L. (1982) Evaluation of techniques for assessment of amphibian and reptile populations in Wisconsin. In *Herpetological Communities* (N. J. Scott, Jr., Ed.), pp. 201–217. U.S.D.I. Fish Wildl. Serv., Washington, DC. [Wildl. Res. Rep. 13].

Weatherhead, P. J., and Hoysak, D. J. (1989). Spatial and activity patterns of black rat snakes (*Elaphe obseleta*) from radiotelemetry and recapture data. *Can. J. Zool.* **67:** 463–468.

Wells, K. D. (1977). The social behaviour of anuran amphibians. *Anim. Behav.* **25:** 666–693.

White, G. C. (1996). NOREMARK: Population estimation from mark-resighting surveys. *Wildl. Soc. Bull.* **24:** 50–52.

White, G. C., Anderson, D. R., Burnham, K. P., and Otis, D. L. (1982) *Capture-Recapture and Removal Methods for Sampling Closed Populations.* Los Alamos Natl. Lab., Los Alamos, NM. [LA-8787-NERP].

Williams, R. D., Gates, J. E., and Hocutt, C. H. (1981). An evaluation of known and potential sampling techniques for hellbender, *Cryptobranchus alleganiensis. J. Herpetol.* **15:** 23–27.

Wilson, K. R., and Anderson, D. R. (1985). Evaluation of a density estimator based on a trapping web and distance sampling theory. *Ecology* **66:** 1185–1194.

Zimmerman, B. L. (1994). Audo strip transects. In *Measuring and Monitoring Biological Diversity—Standard Methods for Amphibians* (W. R. Heyer, M. A. Donnelly, R. W. McDiarmid, L.-A. C. Hayek, and M. S. Foster, Eds.), pp. 92–97. Smithsonian Institution Press, Washington, DC.

Zug, G. R. (1993). *Herpetology: An Introductory Biology of Amphibians and Reptiles.* Academic Press, San Diego.

Chapter 9

Birds

9.1. Complete Counts
   9.1.1. Entire Sampling Frame
   9.1.2. Portion of Sampling Frame
9.2. Incomplete Counts
   9.2.1. Index Methods
   9.2.2. Adjusting for Incomplete
       Detectability

9.3. Recommendations
   9.3.1. General Comments
   9.3.2. Dichotomous Key to
       Enumeration Methods
Literature Cited

There are a variety of techniques used to obtain distribution, abundance, and productivity estimates for avian species. No method can be universally applied to all species or populations because of the variety of both statistical and practical problems associated with different situations. Detailed information on survey methods may be obtained from reviews of counting techniques for songbirds (Verner, 1985; Wiens, 1989; Manuwal and Carey, 1991), raptors (Fuller and Mosher, 1981, 1987; Kochert, 1986), waterfowl (Eng, 1986a; Cowardin and Blohm, 1992), colonial waterbirds (Speich, 1986), upland gamebirds (Eng, 1986b), and all birds (Call, 1981; Bibby *et al.*, 1992). Andrews and Righter (1992) described distribution and habitat use of birds in Colorado.

In this chapter, we describe a variety of counting techniques for birds so that the reader may choose the technique that best fits his or her research objectives and requirements. We cover index methods for use in an index monitoring program (gathering general information on species status) as well as techniques for obtaining valid estimates of spatial distribution and abundance/density for use in an inferential monitoring program (i.e., one in which some average change in population can be reliably detected). This chapter concentrates on methods for counting breeding birds because these are the most common survey situations. In theory, similar methods could be applied to winter populations of birds, but a number of methods need an adequate number of observations to produce reliable estimates, which may or may not be the case in winter surveys. We conclude this chapter with a dichotomous key to enumeration methods to serve as a general guide for biologists.

9.1. COMPLETE COUNTS

A complete count or census of birds within a specified area is rarely possible in most field situations. Possible exceptions to this may include counts of birds in roosts, nesting colonies, or large, visible flocks. However, conditions affecting visibility and detectability of individuals or nests will determine if a true census is realistic. Note that the term "census" is frequently misused in place of survey (partial count of individuals or objects) in the ornithological literature.

9.1.1. ENTIRE SAMPLING FRAME

Complete counts of large birds (e.g., geese) have been attempted using remote sensing techniques such as single (Best and Fowler, 1981) and multispectral scanning (Strong et al., 1991) and aerial photography (Spinner, 1946; Kerbes, 1975; Haramis et al., 1985). The usefulness of these methods as "census" techniques depends on whether or not every individual or object can be counted. This seems unlikely given that aerial photography and related technology are subject to the same or similar factors affecting observability as aerial counts.

9.1.2. PORTION OF SAMPLING FRAME

9.1.2.1. Roost and Lek Counts

Large, highly visible birds such as bald eagles and turkey vultures may be counted while they are in communal roosts during winter (Keister, 1981). Smaller roosting birds also may be censused if they can be accurately counted.

Complete counts at roosts will only be possible if the following conditions are met: roosts are geographically discrete with no interroost movements of birds (Parr and Scott, 1978); there are few enough birds occupying a roost so that density of individuals does not affect count accuracy; individual birds are highly visible (e.g., not obscured by tree branches or poor weather visibility); species are readily discernable if more than one is sharing a roost (Parr and Scott, 1978); and counts occur while the roost is fully occupied. Because all of these conditions are rarely satisfied, roost counts are more appropriately used as an index rather than a complete count in most situations. A situation similar to roost counts is seen in lek counts of grouse (Cannon and Knopf, 1981; Martin and Knopf, 1981).

9.1.2.2. Total Mapping

Total mapping is a technique in which breeding territories of uniquely banded birds are carefully mapped out within a defined area and used as a complete count of the breeding population (Verner, 1985; Fig. 9.1). A combination of netting/trapping and audio/visual locations are used to better delineate territories. Nest sites for each breeding pair are located (Example 9.1). Playback of recorded songs may be used to enhance detectability of birds, especially hard to detect species (Verner, 1985). Territories overlapping outside areas may be recorded as the fraction contained within the plot (DeSante, 1986). Verner (1985, p. 292) described total mapping

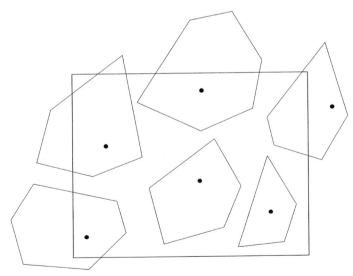

Figure 9.1 Representation of six breeding bird territories delineated using the total mapping method within a selected plot. Nests are black dots. Number of territories can be calculated by using the fraction of each territory or the number of nests contained within plot boundaries.

Example 9.1. Estimating Songbird Abundance via Total Mapping

A biologist wished to obtain an estimate of the number of breeding pairs of horned larks (hola), lark buntings (larb), western meadowlarks (weme), and Brewer's sparrows (brsp) in a 75-km^2 portion of the Pawnee National grassland in northcentral Colorado. The initial abundance estimate was to be within 20% ($e = 0.20$) of the true number of pairs about 90% ($\alpha = 0.10$) of the time.

The target area was stratified as follows: fifty 500 × 500-m plots were mapped with 1 side at the edge of the road (stratum 1; $U_1 = 50$) and 250 were at least 500 m away from the road (stratum 2; $U_2 = 250$). Bird territories within selected plots were delineated via the total mapping method; individual males were marked with unique combinations of color legbands within each species and nests were located. Three plots were randomly selected from each stratum ($u_1 = u_2 = 3$) and surveyed to provide necessary estimates to compute sample sizes (presumed optimum) and assess costs (optimum allocation). Results of the pilot study are shown below (see equations in Appendix C, section C.2.1.2: one-stage stratified sample).

Species	N_{1i}	N_{2i}	$\hat{N}$	$\hat{SE}(\hat{N})$	$\hat{S}^2_{N_{1i}}$	$\hat{S}^2_{N_{2i}}$	u	u_1	u_2
hola	8, 12, 7	15, 8, 13	3450	591.3	7	13	8	2	6
larb	27, 16, 23	8, 18, 13	4350	873.2	31	25	11	3	8
weme	2, 0, 4	4, 1, 4	850	304.5	4	3	27	6	21
brsp	4, 12, 11	7, 9, 17	3200	970.5	19	28	17	3	14

Total costs for each species was computed using an overhead cost (c_0) of $5000, a cost for stratum 1 (c_1) of $2000 and a cost for stratum 2 (c_2) of $3000. Costs for each species were $27,000, $35,000, $80,000, and $53,000, respectively. Available monies only allowed the biologist to survey horned larks. He assumed that the number of breeding territories has not changed since the pilot study so he uses results from it for stratum 1 and selects three new plots for stratum 2, which cuts his costs. He uses his computed $\hat{CV}(\hat{N})$ to ensure that he will be able to detect a 20% change in numbers over 5 years with $\alpha = \beta = 0.10$. Satisfied with his results, he computes the sample sizes needed for the next year of his study.

---

as "the only method suitable for studies in terrestrial communities requiring an absolute scale of density."

A major difficulty with the total mapping technique is accurately delineating all territories and finding all nests within a plot. Another drawback is the enormous amount of time and effort required to accomplish this technique. DeSante (1986) used total mapping to obtain "actual" number of breeding territories for birds within his study plots, which involved about 6 times the effort per unit area compared to using only locations of unbanded singing males (i.e., spot mapping). Further, total mapping estimates the size of the breeding population only; other members of the local population, like floaters, are not included in the estimate. However, in rare or less abundant species, total mapping may be the only option for obtaining reliable abundance estimates because there would be too few individuals for methods like capture–recapture (unless there are many capture occasions; Rosenberg *et al.*, 1995) or distance sampling to produce reliable results.

9.1.2.3. A Complete Count Obtained over Multiple Surveys

Bart *et al.* (1984) computed the detection probability of calling rails in a strip transect based on results from repeated surveys in Michigan. They assumed all calling rails were detected over the course of 4–10 surveys. The authors compared the proportion of rails detected for each survey and concluded that "there was no detectable tendency for all individuals to select the same night for remaining silent" (Bart *et al.*, 1984, p. 1386). Thus, rails were assumed to share the same detection probabilities so that the same proportion of birds was located during each sampling occasion. The detection probability (p) obtained from survey results was used to compute the proportion of rails detected over a given number of counts via the formula $(1 - [1 - p]^{n_s}) \times 100\%$, where n_s is the number of counts or surveys. For example, a constant detection probability of 0.50 over 5 counts would ensure that approximately 97% of the individuals would be detected.

There are a number of potential drawbacks to the approach of Bart *et al.* (1984). First, a complete count must be feasible over some number of sampling occasions. This may only be realistic for a relatively few species under favorable conditions. Second, detection probabilities of individuals would have to be constant for valid results. Although Bart *et al.* (1984) reported no detectable difference in calling rates of individuals, this seems unlikely due to the myriad factors potential affecting vocalization rates of birds (see section 9.2.1.2). The failure of their statistical test to detect a difference in proportions of calling rails seems more likely due to low power than an actual lack of difference in proportions. Third, this approach will likely be labor-intensive and therefore costly. Plots or strips will probably be large and surveyed on foot over multiple occasions. Target species that occur at low densities will require a lot of surveyed plots to obtain reasonable abundance estimates. Finally, individuals on neighboring plots could be heard on the surveyed plot (Bart *et al.*, 1984); an observer could spend a lot of time looking for an individual that turned out to be on an adjacent plot. The feasibility of this method should be thoroughly evaluated before it is implemented.

9.2. INCOMPLETE COUNTS

Complete counts of birds are rarely a viable option so biologists must use data from incomplete counts of birds in an attempt to meet their monitoring needs. Methods for conducting incomplete counts may be generally categorized as those that do not account for, or do not properly account for, incomplete and (very often) unequal detectability of individuals (i.e., index methods) and those that do. Index methods may further be classified as presence–absence and relative index techniques.

9.2.1. INDEX METHODS

Because of their relative ease of use and low cost, index methods are the most common population assessment technique used in studies of birds. Of these, relative index methods are more widespread. Methods that adjust for incomplete detectability sometimes are used as relative indices if their model assumptions are known to have been violated. In this section we outline the more commonly used indices employed in avian population studies.

9.2.1.1. Presence–Absence

Surveys for assessing the occurrence of bird species may be based on auditory signals (e.g., songs or calls of breeding birds, responses by birds to playback of tape recordings), sighted individuals (e.g., soaring raptors), or signs (e.g., regurgitated pellets of raptors, droppings of turkeys). Because the objective of these surveys is generally to obtain baseline information at a minimal cost, the method used in a given survey will depend on which is the most efficient and cost-effective.

The E.F.P. method (Echantillonnages Frequentiels Progressifs; Blondel, 1975; Blondel et al., 1981) is a technique for recording presence–absence of species on auditory cues recorded at various locations or points scattered throughout a study area. All species of birds detected within hearing distance of each point are recorded. Blondel et al. (1981) recommended a minimum of 12 points be surveyed within each biotype or habitat type. We recommend some random sampling scheme be used to determine locations of these points.

Bart and Klosiewski (1989) used data from the North American Breeding Bird Survey (Bystrak, 1981; Droege, 1990) to evaluate the ability of numbers of species recorded at survey stations (presence–absence data) to detect changes over time in numbers of individuals (density) recorded at the same stations. They concluded that presence–absence data were adequate for detecting a population change, but not for estimating magnitude of population change as represented by numbers of individuals. They assumed that changes in numbers of individuals recorded at survey stations were representative of true changes in the sample population (unbiased estimates of density were not available).

Temple and Temple (1976) used frequency of occurrence of bird species recorded in weekly checklists in an attempt to detect their population changes over time. They compared checklist data to other trend data for selected species and concluded that weekly checklists were useful in discerning long-term trends in bird populations. Again, they assumed that the trend data that they used as a benchmark accurately reflected true population trends.

Using presence–absence data to detect population trends is difficult because of problems associated with confirming absence of individuals and the general lack of a linear relationship between these data and population density. Although presence can be established from the detection of a single individual, confirmation of absence will likely require repeated surveying of a plot. If such intensive surveying is necessary, then available resources would be better applied to obtaining valid density estimates instead. Further, there is usually not a linear relationship between presence–absence data and density of birds even under the best situation (see Chapter 3). Finally, simply recording the presence of a species within some area may mask a decline in numbers (Fig. 9.2).

9.2.1.2. Index of Relative Abundance or Density

Numerous methods have been developed to collect count data on bird species to be used as indices of abundance or density and population change

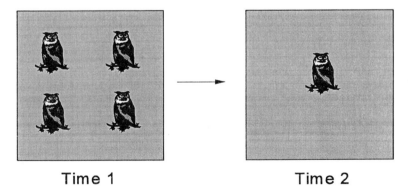

Time 1   **Time 2**

Figure 9.2 Number of great-horned owls existing on the same plot over successive time periods. Assuming complete detectability, presence–absence data would only demonstrate that owls were present during both sampling periods, but would not detect the 75% decline that had occurred.

over time. Most of these techniques rely on vocalizations associated with breeding behavior, whereas others are based on visibility of individuals associated with nesting, roosting, or migratory behavior. These methods are classified as indices because not all individuals or objects are detectable, and either estimators based on these methods are not adjusted for varying rates of detectability of individuals, or their adjustments do not have a proper statistical foundation.

Factors affecting the detection rates of birds have an important bearing on the validity of the proportionality assumption of relative index methods. These methods assume that a constant fraction of individuals is counted between areas at the same time, between areas over time, or within an area over time (Pendleton, 1995). If factors affecting detectability can be controlled through standardization of methods and designs so that a constant proportion of individuals is detected during each sampling occasion, index estimates may be reasonable for assessing population trends. If not, index estimates could easily lead to misleading conclusions.

Auditory and Visual Cues

Singing of male passerines during the breeding season is an obvious auditory cue on which to base counts. Consequently, many counting methods have been developed using songbirds as the target species. Similar techniques based on call counts or other sound cues may be applicable to nonpasserines such as mourning doves (Dolton, 1993), ruffed grouse (Petraborg *et al.*, 1953), pheasants (Greeley *et al.*, 1962; Gates, 1966), gray partridge (Rotella and Ratti, 1986), and various raptor species (Fuller and Mosher, 1987). Although primarily based on sound, these methods also may capitalize on sightings of individuals. Before outlining these techniques, we first will discuss a number of factors that could potentially cause both incomplete and unequal detectability of individuals in many survey situations. We offer this lengthy yet incomplete list of factors in hopes of dissuading those who may feel that uncorrected counts can be routinely "standardized" to yield useful results. In particular, note the number of factors that are impossible to standardize (e.g., density and breeding status of individuals).

Factors Affecting Detectability   Detection of auditory cues is influenced by how often a bird produces sound, how well this sound is carried to the observer, and how well an observer perceives the sound. Observability is influenced by morphological and behavioral traits of individuals, habitat structure and complexity, and environmental factors that may affect visual detection of birds (Fig. 9.3).

Figure 9.3 An example of how observability of birds can be influenced by vegetational structure. Trees on the left and far right have five birds perched in identical locations. Note that only three of the five birds in the tree on the far right are completely visible, whereas all five in the left tree are visible. A single additional tree was all it took to cause a decrease in observability.

Vocalization rates of birds may vary temporally, individually, and environmentally. Birds sing or call more frequently at some times than others during a 24-h period (Robbins, 1981a; Skirven, 1981; Morrell *et al.,* 1991), and these rates may vary during the course of a breeding season (Best and Petersen, 1982; Wilson and Bart, 1985). Peak times for vocalization often are asynchronous among species. Birds that are unmated (von Haartman, 1956; Sayre *et al.,* 1978) or have unsuccessfully bred (Diehl, 1981) vocalize more than their counterparts. In addition, singing rates may be related to density of individuals (Bart and Schoultz, 1984). Finally, inclement weather could reduce vocalization rates (Robbins, 1981b; Verner, 1985).

How well sound travels through air depends on its acoustical properties, habitat structure of the site, and weather. Calder (1990) reported that sound output in birds was correlated with body size. Larger birds generally had louder vocalizations than smaller birds and therefore were more detectable. Waide and Narins (1988) concluded that, in a tropical forest, birds producing low-frequency sounds were usually more detectable than those producing high-frequency songs. In general, habitats that are structurally complex, both horizontally and vertically, have a greater effect on sound transmission than less complex, more open habitats (Bibby and Buckland, 1987; Richards, 1981). Sound travel also can be affected by inclement weather (Robbins, 1981b; Verner, 1985) and temperature (Morrell *et al.,* 1991).

Observer-related factors may play a large role in sound detection and recognition. Hearing abilities vary among individuals (Cyr, 1981; Ramsey and Scott, 1981), and ability to detect high-frequency sounds often declines with age (Emlen and DeJong, 1992). Such factors as observer fatigue (Cyr, 1981) and noise from travel movement (Dawson, 1981) also may affect detectability. Moreover, song and call identification skills vary among observers; hence, Kepler and Scott (1981) recommended that observers be trained before surveys are conducted. Even with proper training and experience, observer efficiency for detecting birds is lowered with increasing avian densities (Bart and Schoultz, 1984; Scott and Ramsey, 1981). Observer effects can exist both among (Sauer *et al.*, 1994) and within (Kendall *et al.*, 1996) individuals over time.

Visibility of birds may be related to individual appearance and behavior, habitat structure, terrain, and weather. Large, colorful birds are more visible than smaller, drab ones. Solitary birds are less likely to be spotted than those that flock, and smaller flocks are generally less visible than larger ones (Buckland *et al.*, 1993). However, individuals are less easily counted in larger flocks than in smaller ones. Flocking behavior may vary seasonally, such as during migration (Fuller and Mosher, 1987) or winter (Balph *et al.*, 1979; Eng, 1986a; Sweeney and Fraser, 1986). Other seasonally based behavior, like territorial displays and incubation, also affect observability. Daily activity patterns and habitat usage (i.e., canopy dwelling versus ground dwelling) will influence visibility. Smith *et al.* (1995) discussed various factors affecting visibility of waterfowl in aerial surveys.

Habitat complexity and topographic relief may have a large impact on visibility of birds. Presence of certain features like highly visible song perches or hunting perches in open habitats may increase the number of observed birds. Birds residing within a clearcut area will be more detectable than those in an uncut forest. Further, inclement weather (e.g., rain) will decrease observability in terms of both visual acuity and bird activity. High winds may actually increase observability of soaring species (Fuller and Mosher, 1987). Schueck and Marzluff (1995) discussed the influence of weather on raptor behavior.

Ignoring unequal detectability rates of individuals within an area over time, among areas at the same time, or among areas over time could very easily lead to wrong conclusions about trends in abundance or density. That is, a perceived change or difference in bird density could simply be an artifact of unequal detectability, and the bias it has introduced, rather than a true change or difference. For instance, a particular bird species may have remained at the same densities on a surveyed area from year 1 to year 2, but may have experienced a change in detectability due to one or more of the factors previously discussed. A 20% change in detectability could be mistaken for a 20% change in density. Similarly, birds may occur at the same densities in two areas during the same year but may differ in their detectabilities at the two sites. Again, the difference in detectabilities may be misconstrued as a difference in densities (Pendleton, 1995).

Line Counts   Strip or belt transects are long, rectangular plots in which a count is conducted from the center line and all birds within some fixed width are recorded (Merikallio, 1958). This method was originally developed to be used as a complete count but in reality records only an unknown portion of birds within each strip.

In an attempt to address the detectability problem, Jarvinen and Vaisanen (1975, 1976) developed the Finnish line transect technique from the strip transect method by including birds detected outside the main strip in the estimation of density. Thus, bird detections are placed in either of two distance categories. They chose a linear model fit to empirical data that assumed bird detections decreased linearly from the observer. Emlen (1971, 1977) developed a somewhat similar approach, called the Emlen line transect, except he based estimates of bird densities on an empirically derived detection constant obtained from songbird detections at greater than two distance categories. Neither the Finnish nor Emlen line transect methods are based on rigorous statistical foundations and therefore should be considered indices rather than techniques that properly adjust for incomplete detectability. Modern line transect methods (Buckland *et al.*, 1993) should be used in place of these *ad hoc* approaches.

Point Counts   The IPA (indice ponctuel d'abondance) or point count method (Ferry and Frochot, 1970; Blondel *et al.*, 1981) records all birds detected within some distance from a point during a 20-min period. The distance may be either finite or infinite (i.e., all birds detected within hearing distance). Individual birds should not be recorded more than once within or among points. Although Ferry and Frochot (1970) recommend that points be placed in "representative" habitat as defined by the researcher, we urge that some type of random sampling design be used.

The point count method is the most widely used counting technique in avian population studies (Ralph *et al.*, 1995, p. iii). Ralph *et al.* (1995, pp. 161–168) presented guidelines for standardizing point count methodology and designs across studies. Ralph *et al.* (1993, pp. 33–34) provided standard data forms for recording both count and habitat information. Barker *et al.* (1993) presented a technique based on numerical methods for optimizing number of points required based on counting and traveling times. Several papers in Ralph *et al.* (1995) discuss effects of number of points and count duration per point on ability to detect birds.

Point counts also are used in the North American Breeding Bird Survey (BBS), which is a large-scale effort using volunteers to conduct bird counts at points along established routes (i.e., roads) throughout the United States and Canada during the late spring–early summer each year. Each route was initially selected randomly from existing road systems and consists of 50 points spaced 0.8 km apart. At each point an observer records all birds detected within a 0.4-km radius during a 3-min period (Droege, 1990). Sauer and Droege (1990) contains several papers describing techniques using BBS data to detect population trends. Bart and Schoultz (1984) concluded that effects of relative bias on detecting major changes in avian density are probably minimal in large surveys like the BBS. Conversely, Johnson (1995) stated that problems inherent with point counts may be lessened by large sample sizes, but this was not guaranteed. In any case, caution should be exercised in analyzing BBS data because of the high likelihood of unequal detection rates of birds across areas and its adverse effect on the proportionality assumption.

Standardizing point count methodology to control for sources of incomplete and unequal detectability of birds is not realistic. There are simply too many factors to account for, any number of which may be impossible to control. Wilson and Bart (1985) reported that random changes in singing rates of house wrens over a breeding season could lead to an error of up to 25% in the estimate of relative density. They cautioned that singing bird surveys should not be relied upon to produce reliable estimates of relative density. Bart and Schoultz (1984) reported that the percentage of individuals detected by observers decreased by as much as 50% as the number of singing birds at a point increased from 1 to 4. They warned that bird densities may be underestimated by as much as 25% for common species and 33% for abundant species in survey methods such as point counts. This decreasing detectability with increasing density violates the constant proportionality assumption and cannot be controlled for in survey design (Bart and Schoultz, 1984).

We cannot recommend using point counts for inferential monitoring programs because of the substantial problems associated with incomplete and unequal detectability. It is not realistic to think that a constant proportion of individuals is being counted in nearly any situation. White and Bennetts (1996) reported that more than 95% of the individuals within a surveyed area must be counted for the proportionality assumption to be met. Therefore, when unbiased counts are needed, we urge biologists to investigate every possibility for appropriately using some type of procedure that either provides a complete count (e.g., total mapping under ideal conditions) or accounts for incomplete detectability (described later in this chapter). Point counts should only be used when all other possibilities have been exhausted. If the proportionality assumption can be realistically satisfied, then analysis of point count data can be performed using the negative binomial distribution as described by White and Bennetts (1996).

Mapping   With spot-mapping (also called territory mapping), the observer plots locations of detected individuals recorded over several visits to an area divided into grids (Williams, 1936; Verner, 1985). Clusters of plotted locations are identified and defined as discrete territories for breeding pairs. Density of breeding pairs is based on numbers of clusters per unit area. Standardized guidelines for this procedure have been provided by the International Bird Census Committee (Robbins, 1970). Tomialojc (1980) proposed further modifications to increase the accuracy of this technique.

A number of studies have concluded that spot-mapping is a reliable method for obtaining a nearly accurate number of breeding territories for some species of songbirds (Hogstad, 1984; Morozov, 1994). However, the spot-mapping technique does not work well for detecting birds that are secretive, have large daily movements relative to plot size, or are nonterritorial "floaters" (Verner, 1985). Moreover, clusters of recorded locations may be interpreted differently by different observers (Best, 1975; Verner and Milne, 1990). Even trained observers may exhibit considerable variability in identifying species' territories (Verner and Milne, 1990). Color-banding birds would decrease this ambiguity, but requires much more time and effort. Verner (1985) discussed various assumptions and possible violations regarding this method.

Wiens (1969) developed a technique that mapped locations of songbird territories based on consecutively flushing a singing bird and noting its subsequent locations. This method will only be applicable only to open habitats where birds can be readily observed and followed. Wiens (1969) recommended at least 20 consecutive flushes per bird. This method is similar to spot-mapping except territories are mapped during one visit (Verner, 1985).

Playback of Taped Calls   Playback of taped calls to elicit responses of birds has been used to obtain information on both distribution and relative abundance for a variety of species. This method has been particularly useful in detecting secretive (e.g., rails; Glahn, 1974; Griese *et al.*, 1980; Marion *et al.*, 1981), nocturnal (e.g., owls; Johnson *et al.*, 1981; Palmer and Rawinski, 1986), and wide-ranging (e.g., hawks; Mosher *et al.*, 1990) birds exhibiting territorial behavior.

Factors affecting responses to taped calls have mostly been investigated for raptors. Morrell *et al.* (1991) reported that breeding season, lunar period, and weather affected response rates of great horned owls. Distance to nest (Kimmel and Yahner, 1990; Kennedy and Stahlecker, 1993) and breeding period (Kennedy and Stahlecker, 1993) affected detectability of northern goshawks. Other factors that may affect detectability include age and sex, density of territories, and habituation of individuals to taped broadcasts (Fuller and Mosher, 1987).

Broadcasting taped calls undoubtedly is useful in detecting rare, secretive, or nocturnal birds. Yet, because every target bird is not being detected, this method remains an index of relative abundance. Consequently, as it is presently employed, its usefulness in accurately detecting significant trends in populations is problematic. A possible exception may be in using taped broadcasts in repeated surveys of a given plot to ensure detectability of each nesting pair.

Visual Cues

Counts of Flocks, Roosts, or Colonies   Counts based primarily on visual cues capitalize on various avian behavioral activities such as colonial nesting, roosting, and other grouping behavior. Surveys of birds within nesting colonies have been conducted from the air, ground, or water. Anthony *et al.* (1995) compared aerial videography with complete ground counts in estimating numbers of black brant in a nesting colony. They concluded that aerial videography was a precise and efficient technique for counting visible, fixed objects such as large nesting birds. When corrected for detectability, these estimates may serve as unbiased abundance estimates.

Bennett (1989) used both helicopter and ground surveys to estimate numbers of sandhill cranes in marsh habitat and commented that helicopter counts were more efficient but counted fewer birds. Gibbs *et al.* (1988) concluded that sizes of great blue heron colonies in Maine were adequately estimated by aerial visual estimates and aerial photographic counts. Conversely, Dodd and Murphy (1995) felt that aerial methods alone were not sufficiently accurate and precise for estimating heron colony sizes in South Carolina. Both Dodd and Murphy (1995) and Rodgers *et al.* (1995) recommended using a combination of aerial and ground counts for surveying colonial waterbirds. Thomas (1993) described a statistical technique to calculate abundance estimates using index counts of nesting colonies conducted over time.

Counts at roosting colonies have been discussed earlier in this chapter. Survey techniques are similar for counting birds within roosts, nesting colonies, or breeding grounds (e.g., leks). Locating roosts has been accomplished through use of prior information (e.g., published reports), interviewing local residents, or aerial searches (Sykes, 1979; Fuller and Mosher, 1987). Daily, seasonal, and annual variation in numbers of birds using particular roosts could produce misleading results when counts are used to estimate population trends. For instance, birds counted at a particular roost in 1 year may move to another roost that may not be surveyed during the following year. A lower count at the first roost during the 2nd year may lead the biologist to mistakenly believe that the population had decreased when it had just redistributed. This also may be a problem when conducting counts of prairie grouse on lekking grounds (Cannon and Knopf, 1981).

Garner *et al.* (1995) used a commercially available thermal-infrared scanning system to survey several wildlife species, including wild turkeys. They suggested potential applications using computer-assisted analysis as an alternative to standard survey methods. They also pointed out that detectability would have to be assessed via ground counts.

Migrating raptors often are highly visible and occur in large concentrations along specific routes every fall and hence may be counted every year to provide an index of population trends over broad geographical areas. As such, this method is not well suited for assessing localized population trends so we will not discuss it further. Fuller and Mosher (1987) reviewed these techniques and discussed their potential shortcomings. Buckland *et al.* (1993, pp. 284–293) discussed application of distance sampling approaches to migrating individuals.

Nest Searching   A very labor-intensive method for acquiring both an index to abundance of breeding pairs and an estimate of productivity is via locating nests. Larger, more visible nests, such as those of raptors, are probably more suitable for generating indices of relative abundance than smaller, less visible songbird nests. Call (1978) provided guidelines for searching for raptor nests, whereas Martin and Geupel (1993) described a method for locating songbird nests. Cable (Labisky, 1957) and rope-dragging (Higgins *et al.*, 1969) have been used to locate nests of grassland and marsh nesting species, particularly waterfowl. Bromaghin and McDonald (1993) proposed a systematic nest-encounter design to be used in estimating success of nests located during nest searching. Nest searching is best used to supplement population assessment methods where productivity, causes of nest failure, and habitat use are of interest (Ralph *et al.*, 1993). This technique is limited to small areas because of the intensive effort involved. Perhaps in some situations a complete, or nearly complete (≥95%), count of active nests can be made, such as in conjunction with total mapping.

Trapping and Netting

A variety of capture techniques are used for trapping and netting birds. Bub (1991) gave a detailed discussion of the many techniques applied to various bird groups. A number of these methods are used for obtaining relative indices of abundance or density, particularly mist netting.

Mist netting is by far the most commonly used method for capturing both songbirds and smaller nonpasserines, such as most woodpeckers. MacArthur and MacArthur (1974) and Karr (1981) discussed a number of behavioral, temporal, and spatial factors influencing capture rates of birds in mist nets. They cautioned against using capture data for making inferences about bird densities because of the many factors affecting catchability (i.e., detectability) of birds. Even mesh size of nets (Pardieck and Waide, 1992), degree of exposure to sunlight, and wind velocity (Jenni *et al.*, 1996) can impact capture rates and which species are caught. Remsen and Good (1996) quantified via simulation the potential biases introduced by several factors affecting capture rates of birds in mist nets. They concluded that the bias produced by these factors precluded the use of mist net capture data for reliable comparisons of relative abundances of birds.

There are so many factors that could potentially impact catchability of birds in mist nets that its usefulness as a measure of abundance or density is dubious at best. Even attempts to "generalize cautiously" (Robinson, 1992, p. 416) from netting results is likely futile. Therefore, we cannot envision these types of netting data producing anything other than a poor index of abundance or density.

Combined Methods

Sometimes a combination of point counts and mist netting returns are used to compute indices of relative abundance or density (Holmes and Sturgess, 1975). Rappole *et al.* (1993) compared results from simultaneous point counts and mist netting and concluded both to be lacking. Although a combination of the two methods will no doubt result in larger count totals, violation of the constant proportion assumption is still very much an issue.

Nest Box Monitoring

Nest boxes have been used to monitor a variety of cavity-nesting birds (e.g., Zicus and Hennes, 1987; Hayward *et al.*, 1992; Petty *et al.*, 1994). Hayward *et al.* (1992) evaluated alternative sampling designs for assessing occupancy rates of boreal owls in nest boxes. They reported that a reasonable number of boxes could be used to adequately monitor numbers of cavity nesting species if occupancy rates of nest boxes were greater than 7%. They added, however, that demographic parameters estimated from nest box results may be a misleading indicator of habitat change, e.g., a greater occupancy rate would be expected with a decrease in habitat quality or availability of natural cavities. Moller (1994) described other potential difficulties in extrapolating demographic data gathered from birds using nest boxes to those using natural cavities, including differential effects of ectoparasites on reproductive success. In general, the use of nest boxes to assess population trend assumes that birds using nest boxes are representative of those using natural cavities. That is, changes in occupancy rates are assumed to mirror changes in the population as a whole. As Hayward *et al.* (1992) noted, occupancy rates could, in fact, be inversely correlated with local population numbers if loss of natural cavities causes an increase in use of nest boxes. Therefore, biologists should be extremely careful when interpreting results of a nest box monitoring program.

9.2.2. ADJUSTING FOR INCOMPLETE DETECTABILITY

Capture-recapture and distance sampling methods account for incomplete detectability of individuals or objects while providing unbiased abundance or density estimates. However, these techniques will only produce unbiased estimates when appropriately applied. They also are much more expensive than index methods. In the following we discuss the assumptions underling these techniques from the perspective of bird studies, and list and discuss avian studies that have applied these methods.

9.2.1.1. Capture–Recapture and Related Methods

Properly applying capture-recapture methods to estimating numbers of birds is challenging, mainly because of their mobility and behavior. See Nichols *et al.* (1981) for another review of possible applications of capture-recapture methods to avian populations.

Closed Population Models

The mobility of birds creates a dilemma for satisfying the closure assumption in many field applications. It is not uncommon for capture activities in studies of migratory species to occur during times when migrants are moving through an area of interest. Further, local transients or floaters may be moving through an area at any time. Therefore, length and timing of capture events must be cautiously planned and every attempt must be made to avoid violating the closure assumption.

The assumption of equal catchability in the Lincoln–Petersen model may be violated because of individual attributes (i.e., "heterogeneity") and behavior due to prior capture history (i.e., "trap response"). There are models available in the program CAPTURE (White *et al.*, 1982; Rexstad and Burnham, 1992) for handling heterogeneity, behavior, and time (i.e., catchability varies with time of capture) in studies with more than two capture events.

Attributes of individuals affecting their capture rates may include age, sex, breeding or social status, and core movement area relative to trap location (O'Brien *et al.*, 1985). Young birds may be more easily caught than older birds. Incubating females move less, and thus are less likely to be caught, than their mates. Also, floaters probably move around more than birds with territories. Finally, individuals with core movement areas that overlap a trapping site are more likely to encounter the trap than those whose territory only peripherally encompasses a site.

Trap happiness in birds could occur when traps are baited with food as an enticement (O'Brien *et al.*, 1985). Trap shyness may arise if capture sites are used over and over so that marked birds learn to avoid them (MacArthur and MacArthur, 1974; Karr, 1981). Both heterogeneity and trap response can be addressed by using different capture methods for each capture occasion such as mark–resight methods for certain two-sample studies. Catchabilities associated with different capture methods are dissimilar enough that heterogeneity is probably not a problem (Seber, 1982, p. 85). In addition, different methods likely preclude marked birds from forming a capture response.

Coming up with alternative capture methods that are viable becomes problematic for more than two trapping events. O'Brien *et al.* (1985) used a shooting harvest as a second method, although this is limited to game species. Perhaps voice recognition (Gilbert *et al.*, 1994; McGregor and Byle, 1992) could be used to "capture" individuals; even if feasible, though, this approach would probably be limited to singing or calling males so that an adequate number "captures" of the vocalizations may be obtained. There are nearly always practical constraints that severely limit the number of feasible capture methods available. We recommend some form of mark–resight approach with more than two occasions when feasible; White (1996) described four estimators for this approach.

Edge effect (White *et al.*, 1982; Fig. 9.4) is an important consideration in capture–recapture studies when home ranges of animals are not entirely contained within the selected plot or study area (i.e., a violation of the geographic closure assumption). A valid density estimate probably cannot

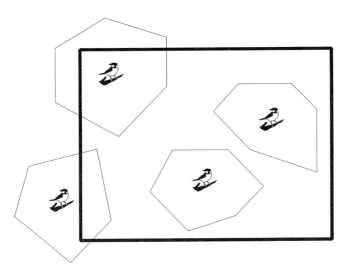

Figure 9.4 An example of edge effect where four birds may occur on a plot at any one time, but only two of them are assured of being within the boundaries. That is, territory size is large relative to plot size.

be obtained if size of the average home range is very large compared to size of the selected plot or study area (White *et al.*, 1982). Even minimum home range sizes of small songbirds [mean = 1.0 ha; adapted from Holling (1992)] should be considered large when compared to plot sizes (e.g., 16 ha) recommended as a standard for some studies (Ralph *et al.*, 1993). Even for larger plot sizes (e.g., 40 ha) such a situation would probably result in an overestimation of density because the number of individuals that have a chance of being caught is greater than the number whose home ranges are entirely contained within the plot.

There are a few options for addressing the problem of edge effect. A trapping web design (Anderson *et al.*, 1983) uses a combination of capture–recapture methodology and distance sampling to estimate densities, but would require many more trapping or netting stations than would be feasible to operate. White (1996) described a mark–resight estimator in program NOREMARK that adjusted for immigration and emigration in studies of radio-equipped animals. Although radiotelemetry technology has advanced enough so that even small birds [e.g., ≥7g; (Naef-Daenzer, 1993)] can be equipped with transmitters, the cost may be prohibitive for equipping enough birds to obtain reasonable abundance estimates. In a practical sense, closed population methods may be restricted to sedentary birds (O'Brien *et al.*, 1985), birds for which radiotelemetry is a feasible option, species that occur in discrete habitat units (e.g., marshes), or immobile objects like nests.

Rotella *et al.* (1995) employed a mark–resight approach using Chapman's (1951) modification of the Lincoln–Petersen estimator to calculate densities of female mallards on randomly selected strip transects. They wished to compare this density estimate with one obtained using an *ad hoc* visual survey method to assess its usefulness as a density index. Mallards were captured and equipped with radio transmitters and nasal markers. A visual survey from the ground was conducted that recorded all female mallards sighted. Radiotelemetry locations were used to confirm how many marked birds were present. Although the authors addressed a number of assumptions underlying valid use of Chapman's estimator, including equal detectability of marked and unmarked birds, they did not address the assumption of constant detectability of all birds throughout the strip transects. This assumption may have been violated given habitat heterogeneity (i.e., physical structure ranging from open water to vegetational cover). If so, their density estimate may have been biased, which would mean any comparisons between this estimate and the *ad hoc* estimate would be suspect. Given that the density estimate was reliable, the authors pointed out that theirs was only a 1-year study conducted on a single area, which meant that their results could not be extrapolated to other years or areas. This is a very important point. Observability of ducks may vary on the same site from year to year (Fig. 9.5), and among different sites in the same years or across years, with a corresponding effect on *ad hoc* estimates. Therefore, we caution against using a single year's data to extrapolate to other areas and time periods.

Some studies have evaluated the validity of a mark–resight approach to flocking birds in open habitats. Schroeder *et al.* (1992) compared the accuracy of Chapman's estimator to a three-sample estimator, selected using program CAPTURE, for estimating number of leks of greater and lesser prairie chickens. Results from complete ground counts were used as the

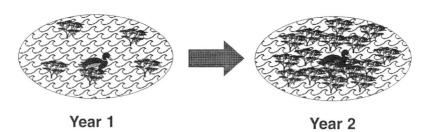

Year 1   **Year 2**

Figure 9.5   Density of vegetation can greatly affect visibility of waterfowl. Increases in water levels from increased precipitation can cover more area, and therefore flood more vegetation, on the same site from 1 year to the next.

standard of accuracy. They cautioned against applying the mark–resight approach in this situation because of the variability in lek numbers between the two surveys. Reversing the order of the survey results would have yielded different results. They agreed with White *et al.*'s (1982) suggestion that more effort should be placed in obtaining more initial observations than reobservations. O'Brien *et al.* (1985) compared abundance estimates of northern bobwhite generated via Chapman's estimator and models requiring more than two capture occasions in program CAPTURE. Quail were captured in baited traps during a 15- to 20-day period to obtain the capture data. The modified Lincoln–Petersen estimate was based on a harvested sample (second capture method) from banded and unbanded quail (grouped as first capture method). They concluded that a multiple capture–recapture model was not appropriate for quail because of biases introduced by unequal catchability. They discussed a number of factors affecting catchability of individuals. These studies point to the need for a critical assessment of assumptions for any enumeration method before it is incorporated into a full-scale monitoring program.

Pollock and Kendall (1987) discussed a mark–resight example using Chapman's modification for Henny and Anderson's (1979) survey data of osprey nests. Active nests were located and mapped during counts from the air and ground (or boat). Nests were "marked" during the first count and treated as resightings when recorded during the second count. Notation was redefined as follows: $\hat{N}_i$, estimator of abundance of occupied nest in the ith selected strip transect; n_{i1}, number of occupied nests seen and mapped by observer 1 in the ith selected strip transect; n_{i2}, number of occupied nests seen and mapped by observer 2 in the ith selected strip transect; and m_i, number of the same occupied nests seen and mapped by both observers in the ith selected strip transect. Notation used for a stratified sampling design would have an additional h subscript to denote stratum (e.g., $\hat{N}_{hi}$). Pollock and Kendall (1987) listed the assumptions that must be met to obtain valid abundance estimates. A particularly key assumption is that nests are equally detectable throughout the surveyed area. The counting method that is more efficient at locating nests should be used in the initial survey (White *et al.*, 1982).

The mark–resight approach with two different counting methods to estimate number of active nests is probably best suited for cliff-nesting species, particularly where cliffs are adjacent to a river or lake that may allow boat/raft counts. Feasibility of conducting ground counts will depend on habitat, terrain, accessibility, size of study area, and funding. Alternatively, perhaps two aerial counts could be used if flights are in opposite directions and two different observers are used (Example 9.2). The bias associated with applying the same counting method twice may be lessened

Example 9.2. Mark–Resight Study for Estimating Numbers of Active Nests (Breeding Pairs)

A 5-year study was initiated to monitor the breeding population of red-tailed hawks located in a 11,500-km^2 area in western Colorado. A mark–resight (modified Lincoln–Petersen) aerial survey was used to obtain estimates of active nests within strip transects chosen using stratified random sampling. A different observer was used on two separate flights down the midline of each selected strip and equal detectability of nests was assumed within each sampling occasion and stratum (see below). Flights were in opposite directions in hopes of lessening the effect of bias. Previous information indicated that about 70% of nests were visible from the air. Strips were 0.8 km wide (east–west) and 25 km long (north–south), which was based on an area large enough to contain 35 visible nests (i.e., 50 active nests with 70% visibility).

The target area was stratified based on proportion of deciduous component of forested habitats. Thirty-eight ($U_1 = 38$) strip transects had a larger deciduous component than the other 10 ($U_2 = 10$). Surveys were flown in late March (i.e., before leaf out) and during nest construction/egg laying (Preston and Beane, 1993) to maximize sighting probabilities of nests. Three strip transects were randomly chosen from each stratum ($u_1 = u_2 = 3$) and surveyed to provide the following pilot study data (calculations based on two-stage stratified sampling estimators in Appendix C, section C.2.2.2):

Stratum	$\hat{N}_{hi}$	$\hat{\text{Var}}(\hat{N}_{hi})$	$\hat{S}^2_{N_{hi}}$	$\overline{\hat{\text{Var}}(\hat{N}_{hi}/N_{hi})}$	$\hat{N}_h$	$\hat{\text{Var}}(\hat{N}_h)$
I	32.6, 55.1, 59.8	1.4, 15.2, 7.0	211.4	7.9	1868.5	94020.9
II	54.5, 47.0, 25.6	23.9, 9.8, 1.4	224.9	11.7	423.7	5364.7

Overall estimates for abundance and precision were: $\hat{N} = 1868.5 + 423.7 = 2292.2$, $\hat{\text{Var}}(\hat{N}) = 94,020.9 + 5364.7 = 99,385.6$, $\hat{\text{SE}}(\hat{N}) = 315.3$, and $\hat{\text{CV}}(\hat{N}) = 315.3/2292.2 = 0.138$. Sample sizes needed for an abundance estimate within 10% ($e = 0.1$) of the true abundance about 95% ($\alpha = 0.05$) of the time were 16 for stratum I and 5 for stratum II. Based on these sample sizes, total cost of the next survey was estimated to be $74,500, where $c_{11} = 1500, $c_{12} = 1000, $c_{21} = 2300, and $c_{22} = 1700.

---

if nests have different observabilities depending on flight direction. This is a distinct possibility for species that nest on cliffs, uneven terrain, or sometimes forested habitat. However, the assumption of equal detectability of nests still must be addressed. As with counts at nesting colonies, this technique will only provide an estimate of the breeding population; however, the nests located subsequently could be monitored to provide productivity estimates, or one could employ Manly *et al.*'s (1996) approach that models factors potentially affecting detectability so that counts can be adjusted. This technique uses two independent observers in the same flight.

Arnason *et al.* (1991) described an estimator of population size constructed from independent sightings of marked and unmarked animals in a closed population. Marked animals must be individually identifiable, but the number of them alive does not have to be known. This method does not require recaptures or removals of animals during a survey. The authors presented an example in which they estimated the number of immature bald eagles on a lake. Guidelines were developed for determining the number of sightings required to obtain acceptably precise abundance estimates.

Rosenberg *et al.* (1995) applied jackknife estimators to capture data from a small population characterized by low and heterogeneous capture probabilities. Reliable abundance estimates could be obtained by using low-order jackknife estimators, an *ad hoc* adjustment to variance estimators provided by the authors, and ≥ 12 capture occasions. We refer interested readers to their paper for further details.

Open Population Models

The Jolly–Seber model (Jolly, 1965; Seber, 1965) allows for births/permanent immigration and deaths/permanent emigration in the population of interest. This relaxation of the demographic component of the closure assumption comes at a price, however. That is, the extra model parameters needed to account for demographic openness result in abundance estimates with less precision than those based on closed models. In addition, the Jolly–Seber model assumes that capture probabilities are equal among individuals during any given capture occasion (Seber, 1982). Therefore, neither individual heterogeneity nor trap response is allowed. Unequal capture probabilities (i.e., from trapping response, capture heterogeneity, or both) can substantially bias abundance estimates (Gilbert, 1973). Further, the geographic closure assumption (e.g., no edge effect) must still be satisfied. Thus, model assumptions always should be tested for validity. Unfortunately, goodness-of-fit tests commonly used in testing for capture heterogeneity (e.g., Jolly, 1982; Pollock *et al.*, 1985; Brownie *et al.*, 1986) typically have low power, i.e., fail to detect heterogeneity when it is present. Therefore, we generally recommend against the use of open population models for population estimation unless capture probabilities are very high and resulting estimates reasonably precise.

Loery and Nichols (1985) used the Jolly–Seber model to estimate abundance of black-capped chickadees captured and recaptured in Potter traps (Lincoln and Baldwin, 1929, pp. 22–23) baited with seed during winter. Jolly's (1982) goodness-of-fit test was used to test for capture heterogeneity and temporary emigration. The authors concluded that the Jolly–Seber model adequately fit their data. They were able to detect a significant decline in chickadee populations during 1968–1983.

Lint *et al.* (1995) used Buckland's (1980) modification of the Jolly–Seber estimator on mark–recapture data to estimate abundance of male wild turkeys during a 9-year period. Assumptions of this estimator were deemed to be valid by the investigators. They used their abundance estimates as the standard against which accuracy of results from three index methods were compared. Number of harvested males and harvest effort were reported to adequately reveal population trends, although a specified level of precision was not stated (e.g., able to detect some percentage change in numbers). Their abundance estimates were not particularly precise; the average $\hat{CV}(\hat{N})$ over the 9 years was 31.0% (range = 12.2–71.8%), which is probably inadequate to detect any but a large change in numbers.

Franklin *et al.* (1990) used both the Jolly–Seber model on mark–recapture and mark–resight data and systematic counts of their study area to estimate densities of northern spotted owls during a 4-year period. Goodness-of-fit tests suggested by Pollock *et al.* (1985) and Brownie *et al.* (1986) were used to test model assumptions. High capture probabilities of owls resulted in unbiased and precise abundance estimates.

Baillie *et al.* (1986) described a methodology, called the Constant Effort Scheme (CES), for capturing birds with mist nets and conducting point counts at the same sites from year to year. Sites are chosen based on their apparent representativeness and ability to maximize captures. Each site is visited once during 12 consecutive 10-day periods beginning in spring (exact starting date depends on latitude of the study). The Jolly–Seber model is used to analyze capture data. In North America, the Monitoring Avian Productivity and Survivorship (MAPS) program has adopted the protocol of the Constant Effort Scheme (DeSante, 1992; Ralph *et al.*, 1993).

The CES/MAPS approach uses a nonrandom sampling scheme to select sites, thus making inferences to the sampled population, which is not explicitly defined, problematic. Another difficulty is the use of point counts and misnetting results to obtain abundance estimates. As we discussed earlier, methods that do not account for incomplete detectability of birds could easily lead to highly biased results. Validity of abundance estimates derived from modeling mist net data is in serious doubt if underlying assumptions are not met.

A key assumption underlying the Jolly–Seber model is equal catchability of individuals among netting/trapping occasions. Violation of this assumption can substantially affect validity of abundance estimates and, to a lesser extent, survival estimates (Lancia *et al.*, 1994). Baillie *et al.* (1986) noted a seasonal decline in captures of adult songbirds, a finding reported from other misnetting studies as well (Karr, 1981; Wooller, 1986). Baillie *et al.* (1986) considered a few potential causes of this decline and finally concluded that trap shyness of previously netted birds could have been an important contributing factor. Although trap shyness in songbirds has seldom been investigated (but see Buckland and Hereward, 1982), there is anecdotal evidence that it occurs (Baillie *et al.*, 1986). In addition, the violation of the geographic closure assumption is a distinct possibility during at least some of the sampling period. That is, local birds are very likely moving in and out of the netting area during the breeding season.

We cannot recommend the MAPS program for inferential monitoring of avian populations because of inherent problems associated with point counts and serious questions concerning equal catchability and geographic closure, and their effect on the validity of estimates obtained from the Jolly–Seber model. Moreover, there is not a defined sampling frame or a random selection process for choosing sites, therefore making valid inferences problematic. The usefulness of the MAPS program as an index monitoring tool should be evaluated against other index methods for cost effectiveness before it is implemented.

9.2.2.2. Distance Sampling

Distance sampling (Buckland *et al.*, 1993) refers to distance-based estimation methods derived from either perpendicular distances from a line [i.e., line transect method (Burnham *et al.*, 1980)] or radial distances from a point [i.e., point transect method (Reynolds *et al.*, 1980)] to an individual or object. Estimators based on distance sampling theory have a rigorous statistical foundation while accounting for incomplete and unequal detectability of individuals. If model assumptions are satisfied, these methods produce unbiased density estimates even if a low percentage of the population is detected (Buckland *et al.*, 1993).

Avian mobility presents special problems when one attempts to estimate bird densities using distance sampling methods. Therefore, close attention should be given to satisfying critical model assumptions. In order of greater to lesser importance, these assumptions are: all birds present on a transect line or point are detected; birds do not move prior to detection during a count; and linear distances (or angles, when appropriate) from a transect line or point to detected birds are accurately measured (Buckland *et al.*, 1993). Linear distance refers to the surface distance between the observer and the ground point directly beneath the detected bird.

Ground Counts

Detecting all individuals on a line or point could be difficult depending on the survey situation. Complex vegetational structure could prevent detection of birds on a line or point. Nonvocal birds in heavy understory could move off of a line or point without attracting the attention of the observer. Nonvocal individuals in the overstory could either move or remain and avoid notice (Fig. 9.6). Movement noises made by observers could trigger bird movement or concealment behavior (Pyke and Recher, 1985; Verner, 1985). Such movements may be evident in histograms of perpendicular distances such that the distance interval nearest the line has a lower frequency than intervals further from the line (Verner, 1985). If there are relatively few movements in response to the observer or the movements are random, then this assumption can be relaxed.

A commonly suggested remedy for observer-induced responses by birds during a count is to delay its initiation until birds have "settled down." DeSante (1986) and Knopf *et al.* (1988) waited 1 min before conducting counts at each point, whereas Bollinger *et al.* (1988) waited 4 min. However, evidence supporting this idea is anecdotal at best. Another possible option is to use distance categories constructed in such a way that the nearest one to the line is large enough to contain any observer-induced movements, but small enough so that detection rates are very high. Whether this is realistic will depend on the situation. A more extreme approach is to use only two distant categories (e.g., 0–30 and >30 m). However, the fit of the detection function cannot be adequately assessed with less than four or five distance categories. Buckland *et al.* (1993) further suggested using two sets of distance categories (e.g., 0–30, >30 m and 0–15, >15 m) and modeling the results separately; if both density estimates are reasonably close,

Figure 9.6 An example of a nonvocal bird located directly above an observer. Missing this individual would violate the assumption in distance sampling that all individuals on a line or point are detected.

then this may indicate that the detection function adequately fit these data, which would indicate reliable estimates.

Obtaining accurate distances to mobile individuals in heavy cover is problematic. A better approach would be to use five to seven distance categories, when feasible. The idea is that grouped distances are easier to estimate than exact distances. Buckland *et al.* (1993, p. 328) offered some suggestions on constructing these categories. Estimating distances by sight or ear increases measurement bias (i.e., response error), particularly when multiple observers are used (Scott *et al.*, 1981), and may require fairly extensive training (Kepler and Scott, 1981).

The choice of whether to use line transects or point transects (also called variable-circular plots) depends largely on their appropriateness under different topographical and vegetational conditions. All else being equal, line transects are more efficient because information is continually gathered along a line rather than only at discrete points (Buckland *et al.*, 1993). That is, travel time among points is unproductive because no information is gathered during this period. Additionally, travel among points is inefficient due to their random distribution across the target area, although avian studies using point transects sometimes have placed points along a line or set of lines and then incorrectly treated these points as independent (Buckland *et al.*, 1993, pp. 299–300). Finally, the point transect method usually requires about 25% more detections (minimum roughly 75–100) than line transects (minimum roughly 60–80) to obtain valid density estimates (Buckland *et al.*, 1993, p. 302).

Point transects were specifically developed for use in rugged terrain or densely vegetated habitats where transect lines were difficult to traverse (Reynolds *et al.*, 1980). Line transects also are difficult to apply in fragmented or heterogeneous habitats; count data would have to be separated by habitat because of the varying rates of detectability associated with specific habitats. This is not a problem with point transects because they "fit in" much easier into stratified habitats. Moreover, point transects offer a better opportunity to gather habitat data because they are smaller and more discrete than lines (Buckland *et al.*, 1993). Because of the various considerations involved in choosing between lines or point transects, decisions must be made on a study-by-study basis.

Guthery (1988) applied line transect methodology to northern bobwhite populations in rangelands in Texas. The author evaluated model assumptions by assessing their potential violations in this situation. He concluded that the line transect method appeared to be a reliable approach for estimating bobwhite densities in rangelands. The line transect method also was used by both Sherman *et al.* (1995) and Kelley (1996) to obtain density estimates for waterfowl in forested wetlands. Both studies concluded that line transects were a feasible method for obtaining density estimates for water-fowl inhabiting forested wetlands that are shallow enough to be wadeable.

Buckland *et al.* (1993, pp. 396–409) provided a detailed analysis and discussion of point transect data gathered for house wrens by Knopf (1986) and Sedgwick and Knopf (1987), and data collected for several species of songbirds by Knopf *et al.* (1988). We encourage readers to review these examples (also see Example 9.3).

Manci and Rusch (1988) broadcasted taped calls at each point transect to increase detectability of three species of rails. We recommend against this approach because birds have a tendency to move toward the observer in response to broadcasts (Johnson *et al.*, 1981; Marion *et al.*, 1981). An inflated estimate of density will result if these movements go undetected and birds are subsequently recorded.

Example 9.3. Point Transect Sampling for Estimating Songbird Densities

A park researcher was interested in monitoring population densities of ruby-crowned kinglets (rcki), yellow-rumped warblers (auwa), dark-eyed juncos (orju), and mountain chicka-dees (moch) in approximately 65,000 ha of undeveloped areas below timberline in Rocky Mountain National Park. Four surveys ($n = 4$) were planned, one every 3 years over a 12-year period. He calculated the $\hat{CV}(\hat{D})$ required to detect a 20% change ($r = 0.20$) in the population with $\alpha = 0.10$ and $\beta = 0.10$ [Eq. (6.7); program TRENDS]:

$$\hat{CV}(\hat{D}) \leq \sqrt{\frac{(0.2)^2(4)^3}{12(1.645 + 1.282)^2}} \leq 0.158.$$

The area was divided into five equal-sized (13,000 ha) strata within which an equal number of points were randomly selected. The strata were created without regard to variability or other attribute, but only to provide good distribution of selected points across the target area. Ten randomly selected points (u_0) were surveyed to obtain an estimate of the number of detections (n_0) for each species, which was used to compute sample sizes at the preset level of precision. The dispersion parameter (b) was set equal to 3. For example, the calculation for rcki was [Eq. (3.5)]:

$$u = \left(\frac{3}{[0.158]^2}\right)\left(\frac{10}{39}\right) = 31.$$

Species	n_0	u	$C(\$)$
rcki	39	31	23,250
auwa	28	43	32,250
orju	12	101	75,750
moch	12	101	75,750

Total cost was computed using an overhead cost (c_0) and a cost per unit c of \$750. Only \$30,000 was available in the budget, so data were only collected for kinglets in the full survey.

A way to avoid problems associated with avian mobility is to survey immobile objects like nests. This approach, however, is constrained by the visibility and density of nests. Nests are difficult to see in many situations. Exceptions to this include colonially nesting waterbirds, and some waterfowl and raptor species nesting in more open habitats. Yet, nest studies offer the added option of monitoring detected nests for obtaining estimates of nest success and productivity.

Aerial Counts

Aerial line transect surveys of birds are mostly limited to grassland or tundra species that are large enough to count, either singly or in groups, from an airplane or helicopter over relatively level terrain. Vegetation or rugged terrain may cause incomplete visibility so that not all birds on the flight line are seen. Even some aircraft designs may obstruct an observer's vision of the ground directly on the flight line.

Accuracy of estimated detection distances from an aircraft is adversely affected in uneven terrain. Distances should be taken at a constant height above the surface in order to be accurate. That is, the sighting distance from an aircraft to a point on the ground changes if the height of the aircraft changes. It is unlikely that a pilot can adjust rapidly enough when flying over rugged or uneven terrain. Therefore, one should try to control for terrain, high winds (i.e., turbulence), or other factors that may potentially affect an aircraft's height above the surface. Shupe *et al.* (1987) used both ground and aerial line transects to obtain density estimates of northern bobwhite on rangelands in Texas. They concluded that line transects counts from helicopters showed promise as a quick and efficient method for surveying northern bobwhites on rangelands.

Aerial line transect surveys of nests present added problems to those present in surveying individuals. That is, nests that are large and visible enough to be seen from the air almost always (except for nesting colonies) occur in densities too low to permit valid density estimates from a transect of reasonable length. Further, species with large, visible nests quite commonly nest in rugged terrain, which presents severe difficulties for accurate distance measurement from an aircraft.

Golden eagles exemplify some of the problems encountered when attempting to use aerial line transects to survey low-density species with large, visible nests. There are roughly 250 active nests of golden eagles during any given year in Colorado, and perhaps about 115 of these are located in the northwestern portion of the state [Craig (1981) as cited in Andrews and Righter (1992)]. Assume that the probability of seeing a nest is about 75% (adapted from Phillips *et al.*, 1984) and that an absolute minimum of 60 nests need to be seen for valid density estimates. Given this, we would have to survey an area large enough to contain 80 active nests (nearly one-third of the state total) and flat enough so that sighting distances could be accurately measured. If such an area existed in Colorado, which is questionable, the transect line(s) would likely be too long to realistically survey.

Survey Design

Buckland *et al.* (1993) devoted an entire chapter to the topic of survey design and field methods in distance sampling. We strongly urge readers to review this material. These authors' comments regarding some previous avian studies that used distance sampling methods should be of particular interest to ornithologists. Critical assumptions underlying distance sampling methods should be assessed in a pilot study before a full survey is conducted. A pilot study also will provide information needed to estimate the transect length or number of points required to obtain predetermined levels of precision of density estimates.

9.3. RECOMMENDATIONS

In this section, we offer some general recommendations for choosing among enumeration methods discussed in this chapter. We begin by making a few general comments and then present a dichotomous key for more detailed help.

9.3.1. GENERAL COMMENTS

To accurately detect a trend in numbers with a predetermined level of precision requires unbiased, or nearly unbiased, and precise estimates. Unfortunately, the cost and effort needed to obtain these data for most species often far exceed available funding. Thus, "cheap" methods such as indices will have to suffice for the majority of species. Relative indices will probably be useful only for detecting large differences in numbers (50% or more?), which may or may not be sufficient to meet monitoring objectives. In fact, because of the potential for biased estimates to confound trend results, index methods may fail to detect even fairly large trends.

9.3.2. DICHOTOMOUS KEY TO ENUMERATION METHODS

We have constructed the following dichotomous key primarily for use as a guide to methods for obtaining abundance/density estimates from either complete and incomplete counts. That is, this key may be used as an early step in the process to determine if either complete counts or methods that adjust for detectability are realistic for any species that require such detailed information. Further, this key should be used in conjunction with information presented earlier in this chapter (sections are in parentheses) and in previous chapters on sampling design, enumeration methods, and guidelines for planning surveys. In particular, assumptions required for valid abundance/density estimates should be assessed critically via a pilot study or some other means before these methods are incorporated into a full-scale monitoring program.

1. (a) A complete count of individuals, nests, or territories in all randomly chosen plots or surveyed areas is possible with available funding 2
   (b) A complete count is either impossible or too costly in all randomly chosen plots or surveyed areas...3
2. (a) All birds or nests can be counted within each plot ....
       Under very limited circumstances: roosts counts, colony counts, lek counts, intensive ground counts (9.1.2) and nest searches (e.g., cable-dragging)
   (b) All breeding territories, pairs and associated nests can be located within each plot .... Total mapping (9.1.2)
3. (a) Individuals can be caught and uniquely marked or nests can be located and mapped...4
   (b) Individuals cannot be caught, are too expensive to catch, or are not conducive to marking 10
4. (a) Element of interest is mobile 5
   (b) Element of interest is immobile (e.g., nest)...................................7
5. (a) Individuals are completely contained (i.e., no edge effect) either within entire area of interest if survey is over the entire sampling frame or within a given plot if surveys are performed at the plot level..6
   (b) Not as above 8
6. (a) Time period of study is short enough to treat population as closed in terms of births/immigration and deaths/emigration 7
   (b) Not as above 9
7. (a) There are at least 100 individuals contained within the survey area (e.g., plot) and their capture probabilities are at least 0.3 (preferably

0.5 or above for the range of population sizes encountered in many fish and wildlife studies); lower capture probabilities are acceptable with larger population sizes (program CAPTURE has a simulation routine to check if estimated capture probabilities are adequate) or ≥ 12 sampling occasions.. Capture–recapture/mark–resight models for 3 or more sampling occasions (9.2.2.1)

   (b) Not as above 10

8. (a) Individuals can be equipped with radio transmitters in a cost-efficient manner, and readily and accurately located via radiotelemetry......................... White (1996); program NOREMARK (9.2.2.1)

   (b) Not as above 10

9. (a) Population is geographically closed and there is no heterogeneity in capture probability or behavioral response to capture method ....Jolly–Seber open population model (9.2.2.1); abundance estimates probably imprecise but good for survival analysis

   (b) Population is subject to edge effect, there is heterogeneity of capture probabilities, and/or there is a trapping response 10

10. (a) Perpendicular distances (or distance categories) from either a line or a point to an individual, discrete group of individuals, or object can be recorded in a cost-efficient manner 11

   (b) Perpendicular distances are either unobtainable or too costly to obtain...14

11. (a) Every individual or object on a line or point can be located 12

   (b) Not all individuals or objects occurring on a line or point can be detected because of vegetational structure, undetected movements of individuals, or other reasons, and methods to adjust for incomplete detection are not feasible 14

12. (a) Individuals do not move in response to observer movements either along the transect line or among points or, if affected, individuals either return to original locations after a waiting period and/or remain within the same distance category......................................13

   (b) Not as above 14

13. (a) Adequate numbers of individuals, groups of individuals, or objects can be detected for reliable model selection in program DISTANCE and reasonably precise density estimates.................... Distance sampling methods; point/line transects for songbirds, line transects for visible birds or groups of birds (e.g., quail) in open habitats, etc. (9.2.2.2)

   (b) Not as above 14

14. (a) Only data on species occurrence required ....Presence–
absence methods (9.2.1.1)
   (b) Uncorrected counts of all individuals/objects detected on a
plot.. Relative index methods (9.2.1.2)

LITERATURE CITED

Anderson, D. R., Burnham, K. P., White, G. C., and Otis, D. L. (1983). Density estimation of small-mammal populations using a trapping web and distance sampling methods. *Ecology* **64:** 674–680.

Andrews, R., and Righter, R. (1992). *Colorado Birds: A Reference to Their Distribution and Habitat.* Denver Mus. Nat. Hist., Denver, CO.

Anthony, R. M., Anderson, W. H., Sedinger, J. S., and McDonald, L. L. (1995). Estimating populations of nesting brant using aerial videography. *Wildl. Soc. Bull.* **23:** 80–87.

Arnason, A. N., Schwarz, C. J., and Gerrard, J. M. (1991). Estimating closed population size and number of marked animals from sighting data. *J. Wildl. Mgmt.* **55:** 716–730.

Baillie, S. R., Green, R. E., Boddy, M., and Buckland, S. T. (1986). *An Evaluation of the Constant Effort Sites Scheme.* Rep. CES Rev. Grp., Norfolk, UK.

Balph, M. H., Balph, D. F., and Romesburg, H. C. (1979). Social status signaling in winter flocking birds: An examination of a current hypothesis. *Auk* **96:** 78–93.

Barker, R. J., Sauer, J. R., and Link, W. A. (1993). Optimal allocation of point-count sampling effort. *Auk* **110:** 752–758.

Bart, J., and Klosiewski, S. P. (1989). Use of presence-absence to measure changes in avian density. *J. Wildl. Mgmt.* **53:** 847–852.

Bart, J. and Schoultz, J. D. (1984). Reliability of singing bird surveys: Changes in observer efficiency with avian density. *Auk* **101:** 307–318.

Bart, J., Stehn, R. A., Herrick, J. A., Heaslip, N. A., Bookhout, T. A., and Stenzel, J. R. (1984). Survey methods for breeding yellow rails. *J. Wildl. Mgmt.* **48:** 1382–1386.

Bennett, A. J. (1989). Population size and distribution of Florida sandhill cranes in the Okefenokee Swamp, Georgia. *J. Field Ornithol.* **60:** 60–67.

Best, L. B. (1975). Interpretational errors in the "mapping method" as a census technique. *Auk* **92:** 452–460.

Best, L. B., and Fowler, R. (1981). Infrared emissivity and radiant surface temperatures of Canada and snow geese. *J. Wildl. Mgmt.* **45:** 1026–1029.

Best, L. B., and Petersen, K. L. (1982). Effects of stage of the breeding cycle on sage sparrow detectability. *Auk* **99:** 788–791.

Bibby, C. J., and Buckland, S. T. (1987). Bias of bird census results due to detectability varying with habitat. *Acta Oecol.-Oecol. Gen.* **8:** 103–112.

Bibby, C. J., Burgess, N. D., and Hill, D. A. (1992). *Bird Census Techniques.* Academic Press, London.

Blondel, J. (1975). L'analyse des peuplements d'oiseaux, element d'un diagnostic ecologique. 1. La methode des echantillonnages frequentiels progressifs (E. F. P.). *La Terre et la Vie* **29:** 533–589.

Blondel, J., Ferry, C., and Frochot, B. (1981). Point counts with unlimited distance. *Stud. Avian Biol.* **6:** 414–420.

Bollinger, E. K., Gavin, T. A., and McIntyre, D. C. (1988). Comparison of transects and circular-plots for estimating bobolink densities. *J. Wildl. Mgmt.* **52:** 777–786.

Bromaghin, J. F., and McDonald, L. L. (1993). A systematic-encounter-sampling design for nesting studies. *Auk* **110:** 646–651.

Brownie, C., Hines, J. E., and Nichols, J. D. (1986). Constant-parameter capture-recapture models. *Biometrics* **42:** 561–574.

Bub, H. (1991). *Bird Trapping and Bird Banding.* Cornell Univ. Press, Ithaca, NY. [Transl. F. Hamerstrom and K. Wuertz-Schaefer]

Buckland, S. T. (1980). A modified analysis of the Jolly–Seber capture-recapture model. *Biometrics* **36:** 419–435.

Buckland, S. T., Anderson, D. R., Burnham, K. P., and Laake, J. L. (1993). *Distance Sampling: Estimating Abundance of Biological Populations.* Chapman and Hall, New York.

Buckland, S. T., and Hereward, A. C. (1982). Trap-shyness of yellow wagtails *Motacilla flava flavissima* at a pre-migratory roost. *Ringing Migration* **4:** 15–23.

Burnham, K. P., Anderson, D. R., and Laake, J. L. (1980). Estimation of density from line transect sampling of biological populations. *Wildl. Monogr.* **72:** 1–202.

Bystrak, D. (1981). The North American Breeding Bird Survey. *Stud. Avian Biol.* **6:** 34–41.

Calder, W. A., III (1990). The scaling of sound output and territory size: Are they matched? *Ecology* **71:** 1810–1816.

Call, M. W. (1978). *Nesting Habitats and Surveying Techniques for Common Western Raptors.* U.S.D.I. Bur. Land Manage. Denver, CO. [Tech. Note T-N-316]

Call, M. W. (1981). *Terrestrial Wildlife Inventories: Some Methods and Concepts.* U.S.D.I. Bur. Land Manag., Denver, CO. [Tech. Rep. 349).

Cannon, R. W., and Knopf, F. L. (1981). Lek numbers as a trend index to prairie grouse populations. *J. Wildl. Mgmt.* **45:** 776–778.

Chapman, D. G. (1951). Some properties of the hypergeometric distribution with applications to zoological censuses. *Univ. Calif. Publ. Stat.* **1:** 131–160.

Cowardin, L. M., and Blohm, R. J. (1992). Breeding population inventories and measures of recruitment. In *Ecology and Management of Breeding Waterfowl* (B. D. J. Batt, A. D. Afton, M. G. Anderson, C. D. Ankney, D. H. Johnson, J. A. Kadlec, and G. L. Krapu, Eds.), pp. 423–445. Univ. Minn. Press, Minneapolis, MN.

Craig, G. R. (1981). *Raptor Investigations.* Job Progress Report, Colo. Div. Wildl. Res. Rep., Denver, CO.

Cyr, A. (1981). Limitation and variability in hearing ability in censusing birds. *Stud. Avian Biol.* **6:** 327–333.

Dawson, D. K. (1981). Sampling in rugged terrain. *Stud. Avian Biol.* **6:** 311–315.

DeSante, D. F. (1986). A field test of the variable circular-plot censusing method in a Sierran subalpine forest habitat. *Condor* **88:** 129–142.

DeSante, D. F. (1992). Monitoring avian productivity and survivorship (MAPS): A sharp, rather than blunt, tool for monitoring and assessing bird populations. In *Wildlife 2001: Populations* (D. R. McCullough and R. H. Barrett, Eds.), pp. 511–521. Elsevier Applied Science, Essex, UK.

Diehl, B. (1981). Bird populations consist of individuals differing in many respects. *Stud. Avian Biol.* **6:** 225–229.

Dodd, M. G., and Murphy, T. M. (1995). Accuracy and precision of techniques for counting great blue heron nests. *J. Wildl. Mgmt.* **59:** 667–673.

Dolton, D. D. (1993). The call-count survey: Historic development and current procedures. In *Ecology and Management of the Mourning Dove* (T. S. Baskett, M. W. Sayre, R. E. Tomlinson, and R. E. Mirarchi, Eds.), pp. 233–252. Stackpole Books, Harrisburg, PA.

Droege, S. (1990). The North American breeding bird survey. In *Survey Designs and Statistical*

Methods for the Estimation of Avian Population Trends (J. R. Sauer and S. Droege, Eds.), pp. 1–4. U.S.D.I. Fish Wildl. Serv., Washington, DC. [Biol. Rep. 90(1)]

Emlen, J. T. (1971). Population densities of birds derived from transect counts. *Auk* **88:** 323–342.

Emlen, J. T. (1977). Estimating breeding season bird densities from transect counts. *Auk* **94:** 455–468.

Emlen, J. T., and DeJong, M. J. (1992). Counting birds: The problem of variable hearing abilities. *J. Field Ornithol.* **63:** 26–31.

Eng, R. L. (1986a). Waterfowl. In *Inventory and Monitoring of Wildlife Habitat* (A. Y. Cooperrider, R. J. Boyd, and H. R. Stuart, Eds.), pp. 371–386. U.S.D.I. Bur. Land Manage. Serv. Ctr., Denver, CO.

Eng, R. L. (1986b). Upland game birds. In *Inventory and Monitoring of Wildlife Habitat* (A. Y. Cooperrider, R. J. Boyd, and H. R. Stuart, Eds.), pp. 407–428. U.S.D.I. Bur. Land Manage. Serv. Ctr., Denver, CO.

Ferry, C., and Frochot, B. (1970). L'airfaune nidofreatrice d'une foret de chenes pedoncules en bourgogne etude de deux successions ecologiques. *La terre et la Vie* **24:** 153–250.

Franklin, A. B., Ward, J. P., Gutierrez, R. J., and Gould, G. I. (1990). Density of northern spotted owls in northwest California. *J. Wildl. Mgmt.* **54:** 1–10.

Fuller, M. R., and Mosher, J. A. (1981). Methods of detecting and counting raptors: A review. *Stud. Avian Biol.* **6:** 235–246.

Fuller, M. R., and Mosher, J. A. (1987). Raptor survey techniques. In *Raptor Management Techniques Manual* (B. A. Geron Pendleton, B. A. Millsap, and K. W. Kline, Eds.), pp. 37–65. Inst. Wildl. Res., Nat. Wildl. Fed., Washington, DC.

Garner, D. L., Underwood, H. B., and Porter, W. F. (1995). Use of modern infrared thermography for wildlife population surveys. *Environ. Mgmt.* **19:** 233–238.

Gates, J. M. (1966). Crowing counts as indices to cock pheasant populations in Wisconsin. *J. Wildl. Mgmt.* **30:** 735–744.

Gibbs, J. P., Woodward, S., Hunter, M. L., and Hutchinson, A. E. (1988). Comparison of techniques for censusing great blue heron nests. *J. Field Ornithol.* **59:** 130–134.

Gilbert, G., McGregor, P. K., and Tyler, G. (1994). Vocal individuality as a census tool: Practical considerations illustrated by a study of two rare species. *J. Field Ornithol.* **65:** 335–348.

Gilbert, R. O. (1973). Approximations of the bias in the Jolly–Seber capture-recapture model. *Biometrics* **29:** 501–526.

Glahn, J. F. (1974). Study of breeding rails with recorded calls in northcentral Colorado. *Wilson Bull.* **86:** 206–214.

Greeley, F., Labisky, R. F., and Mann, S. H. (1962). *Distribution and Abundance of Pheasants in Illinois.* Ill. Nat. Hist. Surv. Biol. Notes No. 147, Urbana, IL.

Griese, H. J., Ryder, R. A., and Braun, C. E. (1980). Spatial and temporal distribution of rails in Colorado. *Wilson Bull.* **92:** 96–102.

Guthery, F. S. (1988). Line transect sampling of bobwhite density on rangeland: Evaluation and recommendations. *Wildl. Soc. Bull.* **16:** 193–203.

Haramis, G. M., Goldsberry, J. R., McAuley, D. G., and Derleth, E. L. (1985). An aerial photographic census of Chesapeake Bay and North Carolina canvasbacks. *J. Wildl. Mgmt.* **49:** 449–454.

Hayward, G. D., Steinhorst, R. K., and Hayward, P. H. (1992). Monitoring boreal owl populations with nest boxes: Sample size and cost. *J. Wildl. Mgmt.* **56:** 777–785.

Henny, C. J., and Anderson, D. W. (1979). Osprey distribution, abundance, and status in western North America: III. The Baja California and Gulf of California population. *Bull. S. Calif. Acad. Sci.* **78:** 89–106.

Higgins, K. F., Kirsch, L. M., and Ball, I. J., Jr. (1969). A cable-chain device for locating duck nests. *J. Wildl. Mgmt.* **33:** 1009–1011.

Hogstad, O. (1984). The reliability of the mapping and standard check methods in making censuses of willow warbler *Phylloscopus trochilus* populations during the breeding season. *Fauna Norv. Ser. C Cinclus* **7:** 1–6.

Holling, C. S. (1992). Cross-scale morphology, geometry, and dynamics of ecosystems. *Ecol. Monogr.* **62:** 447–502.

Holmes, R. T., and Sturgess, F. W. (1975). Bird community dynamics and energetics in a northern hardwood ecosystem. *J. Anim. Ecol.* **44:** 175–200.

Jarvinen, O., and Vaisanen, R. A. (1975). Estimating relative densities of breeding birds by the line transect method. *Oikos* **26:** 316–322.

Jarvinen, O., and Vaisanen, R. A. (1976). Estimating relative densities of breeding birds by the line transect method. IV. Geographical constancy of the proportion of main belt observations. *Ornis Fenn.* **53:** 87–91.

Jenni, L., Leuenberger, M., and Rampazzi, F. (1996). Capture efficiency of mist nets with comments on their role in the assessment of passerine habitat use. *J. Field Ornithol.* **67:** 263–274.

Johnson, D. H. (1995). Point counts of birds: What are we estimating? In *Monitoring Bird Populations by Point Counts* (C. J. Ralph, J. R. Sauer, and S. Droege, Eds.), pp. 117–123. U.S.D.A. For. Serv., Albany, CA. [Gen. Tech. Rep. PSW-GTR-149]

Johnson, R. R., Brown, B. T., Haight, L. T., and Simpson, J. M. (1981). Playback recordings as a special avian censusing technique. *Stud. Avian Biol.* **6:** 68–75.

Jolly, G. M. (1965). Explicit estimates from capture-recapture data with both death and immigration—Stochastic Model. *Biometrika* **52:** 225–247.

Jolly, G. M. (1982). Mark-recapture models with parameters constant in time. *Biometrics* **38:** 301–321.

Karr, J. R. (1981). Surveying birds with mist nets. *Stud. Avian Biol.* **6:** 62–67.

Keister, G. P., Jr. (1981). *An Assessment of Bald Eagle Communal Roosting in Northwestern Washington.* Wash. Dep. Game, Olympia, WA.

Kelley, J. R., Jr. (1996). Line-transect sampling for estimating wood duck density in forested wetlands. *Wildl. Soc. Bull.* **24:** 32–36.

Kendall, W. L., Peterjohn, B. G., and Sauer, J. R. (1996). First-time observer effects in the North American Breeding Bird Survey. *Auk* **113:** 823–829.

Kennedy, P. L., and Stahlecker, D. W. (1993). Responsiveness of nesting Northern Goshawks to taped broadcasts of 3 conspecific calls. *J. Wildl. Mgmt.* **57:** 249–257.

Kepler, C. B., and Scott, J. M. (1981). Reducing bird count variability by training observers. *Stud. Avian Biol.* **6:** 366–371.

Kerbes, R. H. (1975). *The Nesting Population of Lesser Snow Geese in the Eastern Arctic: A Photographic Inventory of June 1973.* [Can. Wildl. Serv. Rep. Ser. 35]

Kimmel, J. T., and Yahner, R. H. (1990). Response of northern goshawks to taped conspecific and great horned owl calls. *J. Raptor Res.* **24:** 107–112.

Knopf, F. L. (1986). Changing landscapes and the cosmopolitism of the eastern Colorado avifauna. *Wildl. Soc. Bull.* **14:** 132–142.

Knopf, F. L., Sedgwick, J. A., and Cannon, R. W. (1988). Guild structure of a riparian avifauna relative to seasonal cattle grazing. *J. Wildl. Mgmt.* **52:** 280–290.

Kochert, M. N. (1986). Raptors. In *Inventory and Monitoring of Wildlife Habitat* (A. Y. Cooperrider, R. J. Boyd, and H. R. Stuart, Eds.), pp. 313–349. U.S.D.I. Bur. Land Manage. Serv. Ctr., Denver, CO.

Labisky, R. F. (1957). Relation of hay harvesting to duck nesting under a refuge-permittee system. *J. Wildl. Mgmt.* **21:** 194–200.

Lancia, R. A., Nichols, J. D., and Pollock, K. H. (1994). Estimating the number of animals in wildlife populations. In *Research and Management Techniques for Wildlife and Habitats* (T. A. Bookhout, Ed.), fifth ed., pp. 215–253. The Wildlife Society, Bethesda, MD.

Lincoln, F. C., and Baldwin, S. P. (1929). *Manual for Bird Banders.* U.S.D.A., Washington, DC. [Misc. Publ. No. 58]

Lint, J. R., Leopold, B. D., and Hurst, G. A. (1995). Comparison of abundance indexes and population estimates for wild turkey gobblers. *Wildl. Soc. Bull.* **23:** 164–168.

Loery, G., and Nichols, J. D. (1985). Dynamics of a black-capped chickadee population, 1958–1983. *Ecology* **66:** 1195–1203.

MacArthur, R. H., and MacArthur, A. T. (1974). On the use of mist nets for population studies. *Proc. Natl. Acad. Sci. USA* **71:** 3230–3233.

Manci, K. M., and Rusch, D. H. (1988). Indices to distribution and abundance of some inconspicuous waterbirds on Horicon Marsh. *J. Field Ornithol.* **59:** 67–75.

Manly, B. F. J., McDonald, L. L., and Garner, G. W. (1996). Maximum likelihood estimation for the double-count method with independent observers. *J. Agric. Biol. Environ. Stat.* **1:** 170–189.

Manuwal, D. A., and Carey, A. B. (1991). Methods for measuring populations of small, diurnal forest birds. U.S.D.A. For. Serv., Portland, OR. [Gen. Tech. Rep. PNW-GTR-278]

Marion, W. R., O'Meara, T. E., and Maehr, D. S. (1981). Use of playback recordings in sampling elusive or secretive birds. *Stud. Avian Biol.* **6:** 81–85.

Martin, S. A., and Knopf, F. L. (1981). Aerial survey of greater prairie chicken leks. *Wildl. Soc. Bull.* **9:** 219–221.

Martin, T. E., and Geupel, G. R. (1993). Nest-monitoring plots: Methods for locating nests and monitoring success. *J. Field Ornithol.* **64:** 507–519.

McGregor, P. K., and Byle, P. (1992). Individually distinctive bittern booms: Potential as a census tool. *Bioacoustics* **4:** 93–109.

Merikallio, E. (1958). Finnish birds; their distribution and numbers. *Fauna Fenn.* **5:** 1–181.

Moller, A. P. (1994). Facts and artefacts in nest-box studies: Implications for studies of birds of prey. *J. Raptor Res.* **28:** 143–148.

Morozov, N. S. (1994). Reliability of the mapping method for censusing blue tits *Parus caeruleus. Ornis Fenn.* **71:** 102–108.

Morrell, T. E., Yahner, R. H., and Harkness, W. L. (1991). Factors affecting detection of great horned owls by using broadcast vocalizations. *Wildl. Soc. Bull.* **19:** 481–488.

Mosher, J. A., Fuller, M. R., and Kopeny, M. (1990). Surveying woodland raptors by broadcast of conspecific vocalizations. *J. Field Ornithol.* **61:** 453–461.

Naef-Daenzer, B. (1993). A new transmitter for small animals and enhanced methods of home-range analysis. *J. Wildl. Mgmt.* **57:** 680–689.

Nichols, J. D., Noon, B. R., Stokes, S. L., and Hines, J. E. (1981). Remarks on the use of mark-recapture methodology in estimating avian population size. *Stud. Avian Biol.* **6:** 121–126.

O'Brien, T. G., Pollock, K. H., Davidson, W. R., and Kellogg, F. E. (1985). A comparison of capture-recapture with capture-removal for quail populations. *J. Wildl. Mgmt.* **49:** 1062–1066.

Palmer, D. A., and Rawinski, J. J., (1986). A technique for locating boreal owls in the fall in the Rocky Mountains. *Colo. Field Ornithol. J.* **20**(2): 38–41.

Pardieck, K., and Waide, R. B. (1992). Mesh size as a factor in avian community studies using mist nets. *J. Field Ornithol.* **63:** 250–255.

Parr, D. E., and Scott, M. D. (1978). Analysis of roosting counts as an index to wood duck population size. *Wilson Bull.* **90:** 423–437.

Pendleton, G. W. (1995). Effects of sampling strategy, detection probability, and independence of counts on the use of point counts. In *Monitoring Bird Populations by Point Counts* (C. J. Ralph, J. R. Sauer, and S. Droege, Eds.), U.S.D.A. For. Serv., Albany, CA. [Gen. Tech. Rep. PSW-GTR-149]

Petraborg, W. H., Wellein, E. G., and Gunvalson, V. E. (1953). Roadside drumming counts, a spring census method for ruffed grouse. *J. Wildl. Mgmt.* **17:** 292–295.

Petty, S. J., Shaw, G., and Anderson, D. I. K. (1994). Value of nest boxes for population studies and conservation of owls in coniferous forests in Britain. *J. Raptor Res.* **28:** 134–142.

Phillips, R. L., McEneaney, T. P., and Beske, A. E. (1984). Population densities of breeding golden eagles in Wyoming. *Wildl. Soc. Bull.* **12:** 269–273.

Pollock, K. H., Hines, J. E., and Nichols, J. D. (1985). Goodness-of-fit tests for open capture-recapture models. *Biometrics* **41:** 399–410.

Pollock, K. H., and Kendall, W. L. (1987). Visibility bias in aerial surveys: A review of estimation procedures. *J. Wildl. Mgmt.* **51:** 501–510.

Preston, C. R., and Beane, R. D. (1993). Red-tailed hawk (*Buteo jamaicensis*). In *The birds of North America* (A. Poole, and F. Gill, Eds.), No. 52, pp. 1–24. Acad. Nat. Sci., Philadelphia, PA/Am. Ornithol. Union, Washington, DC.

Pyke, G. H., and Recher, H. F. (1985). Estimated forest bird densities by variable distance point counts. *Aust. Wildl. Res.* **12:** 307–319.

Ralph, C. J., Geupel, G. R., Pyle, P., Martin, T. E., and DeSante, D. F. (1993). Handbook of field methods for monitoring landbirds. U.S.D.A. For. Serv., Albany, CA. [Gen. Tech. Rep. PSW-GTR-144]

Ralph, C. J., Sauer, J. R., and Droege, S. (1995). Monitoring bird populations by point counts. U.S.D.A. For. Serv., Albany, CA. [Gen. Tech. Rep. PSW-GTR-149]

Ramsey, F. L., and Scott, J. M. (1981). Tests of hearing ability. *Stud. Avian Biol.* **6:**341–345.

Rappole, J. H., McShea, W. J., and Vega-Rivera, J. (1993). Evaluation of two survey methods in upland avian breeding communities. *J. Field Ornithol.* **64:** 55–70.

Remsen, J. V., Jr., and Good, D. A. (1996). Misuse of data from mist-net captures to assess relative abundance in bird populations. *Auk* **113:** 381–398.

Rexstad, E. A., and Burnham, K. P. (1992). *User's Guide for Interactive Program CAPTURE.* Colo. Coop. Fish Wildl. Res. Unit, Colorado State Univ., Fort Collins.

Reynolds, R. T., Scott, J. M., and Nussbaum, R. A. (1980). A variable circular-plot method for estimating bird numbers. *Condor* **82:** 309–313.

Richards, D. G. (1981). Environmental acoustics and censuses of singing birds. *Stud. Avian Biol.* **6:** 297–300.

Robbins, C. S. (1970). Recommendations for an international standard for a mapping method for bird census work. *Aud. Field Notes* **24:** 723–726.

Robbins, C. S. (1981a). Effect of time of day on bird activity. *Stud. Avian Biol.* **6:** 275–286.

Robbins, C. S. (1981b). Bird activity levels related to weather. *Stud. Avian Biol.* **6:** 301–310.

Robinson, S. K. (1992). Population dynamics of breeding neotropical migrants in a fragmented Illinois landscape. In *Ecology and Conservation of Neotropical Migrant Landbirds* (J. M. Hagan, III, and D. W. Johnston, Eds.), pp. 408–418. Smithsonian Institution Press, Washington, DC.

Rodgers, J. A., Jr., Linda, S. B., and Nesbitt, S. A. (1995). Comparing aerial estimates with ground counts of nests in wood stork colonies. *J. Wildl. Mgmt.* **59:** 656–666.

Rosenberg, D. K., Overton, W. S., and Anthony, R. G. (1995). Estimation of animal abundance when capture probabilities are low and heterogeneous. *J. Wildl. Mgmt.* **59:** 252–261.

Rotella, J. J., Devries, J. H., and Howerter, D. W. (1995). Evaluation of methods for estimating density of breeding female mallards. *J. Field Ornithol.* **66:** 391–399.

Rotella, J. J., and Ratti, J. T. (1986). Test of a critical density index assumption: A case study with gray partridge. *J. Wildl. Mgmt.* **50:** 532–539.

Sauer, J. R., and Droege, S. (Eds.) (1990). Survey designs and statistical methods for the estimation of avian population trends. U.S.D.I. Fish Wildl. Serv., Washington, DC. [Biol. Rep. 90(1)]

Sauer, J. R., Peterjohn, B. G., and Link, W. A. (1994). Observer differences in the North American Breeding Bird Survey. *Auk* **111:** 50–62.

Sayre, M. W., Atkinson, R. D., Baskett, T. S., and Haas, G. H. (1978). Reappraising factors affecting mourning dove perch cooing. *J. Wildl. Mgmt.* **42:** 884–889.

Schroeder, M. A., Giesen, K. M., and Braun, C. E. (1992). Use of helicopters for estimating numbers of greater and lesser prairie-chicken leks in eastern Colorado. *Wildl. Soc. Bull.* **20:** 106–113.

Schueck, L. S., and Marzluff, J. M. (1995). Influence of weather on conclusions about effects of human activities on raptors. *J. Wildl. Mgmt.* **59:** 674–682.

Scott, J. M., and Ramsey, F. L. (1981). Effects of abundant species on the ability of observers to make accurate counts of birds. *Auk* **98:** 610–612.

Scott, J. M., Ramsey, F. L., and Kepler, C. B. (1981). Distance estimation as a variable in estimating bird numbers from vocalizations. *Stud. Avian Biol.* **6:** 334–340.

Seber, G. A. F. (1965). A note on the multiple-recapture census. *Biometrika* **52:** 249–259.

Seber, G. A. F. (1982). *The Estimation of Animal Abundance and Related Parameters,* second ed. Griffin, New York.

Sedgwick, J. A., and Knopf, F. L. (1987). Breeding bird response to cattle grazing of a cottonwood bottomland. *J. Wildl. Mgmt.* **51:** 230–237.

Sherman, D. E., Kaminski, R. M., and Leopold, B. D. (1995). Winter line-transect surveys of wood ducks and mallards in Mississippi greentree reservoirs. *Wildl. Soc. Bull.* **23:** 155–163.

Shupe, T. E., Guthery, F. S., and Beasom, S. L. (1987). Use of helicopters to survey northern bobwhite populations on rangeland. *Wildl. Soc. Bull.* **15:** 458–462.

Skirven, A. A. (1981). Effect of time of day and time of season on the number of observations and density estimates of breeding birds. *Stud. Avian Biol.* **6:** 271–274.

Smith, D. R., Reinecke, K. J., Conroy, M. J., Brown, M. W., and Nassar, J. R. (1995). Factors affecting visibility rate of waterfowl surveys in the Mississippi alluvial valley. *J. Wildl. Mgmt.* **59:** 515–527.

Speich, S. M. (1986). Colonial waterbirds. In *Inventory and Monitoring of Wildlife Habitat* (A. Y. Cooperrider, R. J. Boyd, and H. R. Stuart, Eds.), U.S.D.I. Bur. Land Manage. Serv. Ctr., Denver, CO.

Spinner, G. P. (1946). Improved methods for estimating numbers of waterfowl. *J. Wildl. Mgmt.* **10:** 365–365.

Strong, L. L., Gilmer, D. S., and Brass, J. A. (1991). Inventory of wintering geese with a multispectral scanner. *J. Wildl. Mgmt.* **55:** 250–259.

Sweeney, T. M., and Fraser, J. D. (1986). Vulture roost dynamics and monitoring techniques in southwest Virginia. *Wildl. Soc. Bull.* **14:** 49–54.

Sykes, P. W., Jr. (1979). Status of the Everglade Kite in Florida—1968–1978. *Wilson Bull.* **91:** 495–511.

Temple, S. A., and Temple, B. L. (1976). Avian population trends in central New York State, 1935–1972. *J. Field Ornithol.* **47:** 238–257.

Thomas, G. E. (1993). Estimating annual total heron population counts. *Appl. Stat.* **42:** 473–486.

Tomialojc, L. (1980). The combined version of the mapping method. In *Bird Census Work and Nature Conservation* (H. Oelke, Ed.), pp. 92–106. Univ. Gottengen, Bundesrepublik, Germany.

Verner, J. (1985). Assessment of counting techniques. In *Current Ornithology* (R. F. Johnston, Ed.), Vol. 2, pp. 247–302. Plenum, New York.

Verner, J., and Milne, K. A. (1990). Analyst and observer variability in density estimates from spot mapping. *Condor* **92:** 313–325.

von Haartman, L. (1956). Territory in the pied flycatcher (*Muscicapa hypoleuca*). *Ibis* **98:** 460–475.

Waide, R. B., and Narins, P. M. (1988). Tropical forest bird counts and the effect of sound attenuation. *Auk* **105:** 296–302.

White, G. C. (1996). NOREMARK: Population estimation from mark-resighting surveys. *Wildl. Soc. Bull.* **24:** 50–52.

White, G. C., Anderson, D. R., Burnham, K. P., and Otis, D. L. (1982). Capture-recapture and removal methods for sampling closed populations. Los Alamos Natl. Lab., Los Alamos, NM. [LA-8787-NERP]

White, G.C., and Bennetts, R. E. (1996). Analysis of frequency count data using the negative binomial distribution. *Ecology* **77:** 2549–2557.

Wiens, J. A. (1969). An approach to the study of ecological relationships among grassland birds. *Ornithol. Monogr.* **8:** 1–93.

Wiens, J. A. (1989). *The Ecology of Bird Communities, Vol. 2, Foundations and Patterns.* Cambridge Univ. Press, Cambridge, UK.

Williams, A. B. (1936). The composition and dynamics of a breech-maple climax community. *Ecol. Monogr.* **6:** 317–408.

Wilson, D. M., and Bart, J. (1985). Reliability of singing bird surveys: Effects of song phenology during the breeding season. *Condor* **87:** 69–73.

Wooller, R. D. (1986). Declining rates of capture of birds in mist-nets. *Corella* **10:** 63–64.

Zicus, M. C., and Hennes, S. K. (1987). Use of nest boxes to monitor cavity-nesting waterfowl populations. *Wildl. Soc. Bull.* **15:** 525–532.

Chapter 10

Mammals

10.1. Complete Counts
    10.1.1. Entire Sampling Frame
    10.1.2. Portion of the Sampling
           Frame
10.2. Incomplete Counts
    10.2.1. Index Methods
    10.2.2. Adjusting for Incomplete
           Detectability

10.3. Recommendations
    10.3.1. General Comments
    10.3.2. Dichotomous Key to
           Enumeration Methods
Literature Cited

This chapter provides an overview of some potential methods to survey mammals. It is organized by methods as opposed to size of the mammal we are considering. We expect readers to browse through the material to help them decide on an approach for the mammal species in which they are interested. At the end of the chapter, we have provided a dichotomous key to enumeration methods as a general aid to biologists for choosing an appropriate technique. See Wilson *et al.* (1996) for a comprehensive review of enumeration methods for mammals. Information regarding ecology and distribution of mammals within Colorado may be obtained from Armstrong (1972), Fitzgerald *et al.* (1994), and Warren (1942).

10.1. COMPLETE COUNTS

10.1.1. Entire Sampling Frame

A complete count across an entire area is rarely possible under current technology. However, technologically innovative methods have been evaluated for counting mammals, such as remote sensing techniques for ungulates (Wyatt et al., 1980) and ultraviolet photography (Lavigne and Oritsland, 1974). Infrared thermal imaging has been applied to small to medium-sized mammals by Boonstra et al. (1994) and to large animals (e.g., deer) by Garner et al. (1995).

10.1.2. PORTION OF THE SAMPLING FRAME

Some mammals, particularly large mammals, lend themselves to complete enumeration on the plot. The most common example of this approach is the aerial counting of large herbivores. Typically, sampling units are quadrats or strips, although irregularly shaped sampling units have been used for moose (Gasaway et al., 1986). Examples of species that were counted by observers from the air include mule deer (Kufeld et al., 1980; Bartmann et al., 1986), white-tailed deer (Beasom et al., 1986; DeYoung et al., 1989) moose (Gasaway et al., 1986); and pronghorn (Pojar et al., 1995). Helicopters are typically used for mule deer and moose, whereas fixed-wing aircraft often are used for animals in more open habitats like pronghorn (Example 10.1).

Numerous procedures have been developed to correct for visibility bias in complete coverage of a quadrat. For elk, Samuel et al. (1987) developed a sightability model based on logistic regression. This model then is used to correct the count from a sample of quadrats to obtain an unbiased estimate of population size (Steinhorst and Samuel, 1989).

Example 10.1. Complete Counts on Randomly Selected Quadrats

A population study of pronghorn was conducted within a 1295-km^2 area in eastern Colorado. Population estimates were obtained from helicopter counts within randomly chosen 2.59-km^2 quadrats as described by Pojar et al. (1995). Counts within quadrats were treated as if they were complete. Quadrats were placed into 1 of 2 strata depending on proportion of agricultural land contained within them. One-hundred-fifty ($U_1 = 150$) quadrats had $\geq 25\%$ agricultural land and 350 ($U_2 = 350$) had less. Five quadrats were randomly selected and surveyed from each stratum to provide preliminary estimates within 10% ($e = 0.1$) of the true number of pronghorn about 90% of the time ($\alpha = 0.10$). This approach corresponds to a one-stage stratified random sample (formulas listed in Appendix C, section C.2.1.2: stratified random sample). Results of the preliminary study are given below.

Stratum	N_{hi}	$\overline{N}_h$	$\hat{S}^2_{N_{hi}}$	$\hat{N}_h$	$\hat{\mathrm{Var}}(\hat{N}_h)$	u_h
I	9, 3, 6, 7, 0	5.0	12.5	750	54,375	52
II	4, 2, 0, 8, 3	3.4	8.8	1190	212,520	81

Overall estimates for abundance and precision were: $\hat{N} = 750 + 1190 = 1940$; $\hat{Var}(\hat{N}) = 54{,}375 + 212{,}520 = 266{,}895$; $\hat{SE}(\hat{N}) = 516.6$; and $\hat{CV}(\hat{N}) = 516.6/1940 = 0.266$. Calculation of sample sizes (u_1 and u_2 in table above) revealed that 133 quadrats would need to be sampled to meet relative error requirements under optimum allocation. Total cost of the full survey was computed using $c_1 = \$500$, $c_2 = \$750$, and $c_0 = \$10{,}000$, and was calculated to be $\$96{,}750$. Results from a full survey were: $\hat{N} = 2015$; $\hat{Var}(\hat{N}) = 124{,}345$; $\hat{SE}(\hat{N}) = 352.6$; and $\hat{CV}(\hat{N}) = 0.175$. For a 5-year study ($n = 5$), with $\alpha = \beta = 0.05$, the biologist would be able to detect a 19.5% change in population numbers [Eq. (6.8); program TRENDS]:

$$r \geq \sqrt{\frac{156}{5^3}}(0.175) = 0.195.$$

10.2. INCOMPLETE COUNTS

Because complete counts across the entire sampling frame or delineated study area are rarely possible, we must rely on incomplete counts to satisfy our monitoring needs. These incomplete counts could be in the form of an incomplete count over the entire frame, complete counts within a portion of plots or study area, or incomplete counts over a portion of plots or study area. This last scenario is common in population studies. Incomplete counts can further be partitioned by whether the counting method had an adjustment for incomplete detectability of individuals. Techniques that do not account, or do not adequately account, for incomplete and/or unequal detectability are generally referred to as index methods, which we will discuss first.

10.2.1. INDEX METHODS

Index methods are an attractive option for assessing abundance or density because they are relatively cheap and easy to use. These techniques may be divided into presence–absence methods, which are usually used only to assess distribution and to construct species lists, and relative index methods, which assume that a constant fraction of a population is counted among areas and/or across time. Relative index methods require some type of standardization in effort and counting conditions in order to be comparable. We list and describe some of the more common index approaches used for assessing mammal populations under current technology. Index methods may prove to be the only feasible option, both logistically and financially, for the majority of species if a large number of species must be monitored.

10.2.1.1. Presence–Absence

A common procedure to obtain a list of small mammal species occurring in an area is to sample with drift fences and pitfall traps (Friend, 1984; Bury and Corn, 1987). Bury and Corn (1987) concluded that 10 days of trapping was only adequate to detect common species, whereas about 60 days were required to compile a relatively complete species list (>85% of the species captured compared to 180 days).

Other approaches can be used to determine if a species is present, i.e., tracks in snow or ground, hair sampling tubes (Scotts and Craig, 1988), scat identification, and territory marking. Motion- and heat-sensitive cameras also can be used to record rare species that might come to a bait or other lure. For example, T. D. I. Beck (personal communication) has obtained photographs of mountain lions attracted to bait stations designed for black bears. Because many mammals are secretive and/or nocturnal, evidence of their presence often is used instead of actually seeing them. Tracks, territory marking, and defecations often are used.

10.2.1.2. Indices of Relative Abundance or Density

A common method for quantifying abundance is through a count of individuals, physical sign (e.g., tracks), vocalizations and so forth. It is hoped that changes in these counts over space or time accurately represent actual changes in population numbers. However, unless this assumption is properly tested, we really do not know if relative indices are valid for use in monitoring populations. These methods are usually much cheaper and easier to conduct than methods that account for differences in detectability, hence their popularity.

Visual Cues

Tracks Animals generally leave tracks, so numerous surveys have been conducted based on track counts. Typically, we have no method to correct track counts to the actual population, i.e., we have no calibration of the length of a set of tracks or the number of tracks made per animal. If we assume that the relationship between the index (track count) and the population size remains constant over time and a range of population sizes, then the index provides a reasonable method to examine trends in population size through time or across several areas.

Various methods have been used to observe tracks. Surveys from aircraft (Becker, 1991) can be used to count the number of tracks in fresh snow crossing the flight path, a line-intercept sampling technique. Boyce (1989, pp. 19–21) used counts of elk tracks in snow as an index of the number of elk coming to the Jackson Hole Elk Refuge. Merrill et al. (1994) used track counts with mule deer. Mooty *et al.* (1984) used tracks in dirt roads to evaluate white-tailed deer abundance, as did Van Dyke *et al.* (1986) for mountain lions. Tracking also has been used for small mammals (Marten, 1972).

Linhart and Knowlton (1975) developed lines of scent stations, where a scent lure was used to attract predators, mainly coyotes, to a circle of earth that had been cleared so that animals visiting the station would leave discernable tracks. They used the proportion of stations showing visitations by the species of interest as an index (SSI) of population abundance. Typically, scent stations are placed on transects in groups of 10, and transects are separated so that individual animals cannot visit more than one transect (Hon, 1979; Roughton and Sweeny, 1982; Conner *et al.*, 1983). However, Linhart and Knowlton (1975) used transects with 50 stations for coyotes. Scent station surveys have been used to estimate relative abundance for bobcats (Johnson and Pelton, 1981; Diefenbach *et al.*, 1994), red and gray foxes (Wood, 1959), cottontail rabbit (Drew *et al.*, 1988), raccoons and opossums (Conner *et al.*, 1983), and black bears (Lindzey and Thompson, 1977).

In one of the few tests of the relationship between SSI and population size, Diefenbach *et al.* (1994) noted a positive relationship ($r^2 = 0.45$, $P = 0.007$) between population size and SSI. However, predictions of population size using individual scent-station surveys had poor precision. Analysis of statistical power indicated that four replicate scent-station surveys had an 80% probability of detecting only large ($\geq 25\%$) changes in populations of high density (0.5 bobcats/km^2). They recommend that (1) multiple scent-station surveys be conducted each year to monitor changes in bobcat population; (2) SSI values be calculated as proportions and transformed to reduce heteroscedasticity; (3) each stratum in a sampling design should contain as many stations as possible to minimize the problem of discrete data (No. of visits) analyzed as a continuous variable (proportion of stations visited); (4) scent stations should be placed as far apart as logistically feasible to minimize multiple visits by individual bobcats; and (5) results of the power analysis should be used as a minimum guideline for estimating sample-size requirements.

Roughton and Sweeny (1982) recommended that transects of 10 scent stations be operated for one night to increase sample sizes, enhance sampling distribution, and minimize weather interference. Previous work with scent stations to estimate relative coyote abundance had used 50 stations per transect monitored for four nights. Cluster-sampling analysis indicated that the information obtained from each 50-station transect observed for four nights could be obtained from 1.5 to 3.8 10-station transects observed for one night. They also recommended the use of Fisher's randomization test to test for differences in visitation rates. The major advantage of this randomization test is that no distributional assumptions are required about the data.

Pellet Groups   Abundance of ungulates has been monitored by their fecal droppings (Neff, 1968; Batcheler, 1975; White and Eberhardt, 1980; Freddy and Bowden, 1983; Rowland et al., 1984), as have snowshoe hares, jack rabbits, and cottontail rabbits. This approach requires that the defecation rate of the animals remains the same across animal densities, a questionable assumption when you consider that food quality and quantity may change as a function of density. Certainly, food intake will change with season (Rogers, 1987), so that comparisons across time should be within the same season.

Attempts to measure defecation rate (e.g., Smith, 1964; Neff et al., 1965; Crete and Bedard, 1975; Joyal and Richard, 1986; Rogers, 1987) have demonstrated high variation among individuals and across seasons. Further, pen studies give very different estimates than free-ranging animals. As a result, this index is suspect because changes in the age structure of the population or changes in food quality/quantity would be interpreted as changes in density.

Spotlight Surveys and Roadside Counts   Spotlight surveys have been used to estimate relative abundance of deer (McCullough, 1982; Fafarman and DeYoung, 1986; Cypher, 1991), cottontails (Lord, 1959), jack rabbits (Smith and Nydegger, 1985), feral pigs (Choquenot et al., 1990), and raccoons. The most serious bias of this index is that visibility changes with vegetation density (Whipple et al., 1994). As a result, changes in the index due to vegetation influences may appear as changes in animal density.

Nighttime surveys with lights can be converted from an index to a population estimate with line transect methodology (section 10.2.2.2). An assumption that must be considered in using spotlights to conduct line transect surveys is that the lines are randomly placed. Thus, roadside surveys would likely produce biased estimates of population density because of habitat differences associated with roads. Daytime roadside counts have been used for numerous species, e.g., cottontail rabbits (Kline, 1965), deer, moose, and bears (Fraser, 1979).

Auditory Cues

Howling surveys have been used to estimate relative abundance of wolves (Harrington and Mech, 1982; Fuller and Sampson, 1988), and coyotes (Wenger and Cringan, 1978; Okoniewski and Chambers, 1984). Both of these species often respond vocally to a siren or human howling. Because of logistical and statistical constraints, Fuller and Sampson (1988) do not recommend the technique as practical for surveying large (e.g., state- or province-wide) areas, but do suggest that simulated howling is useful for locating packs in smaller areas. High variation among animals (Wenger and Cringan, 1978) makes the use of this technique suspect for estimating relative abundance of canids.

10.2.2. ADJUSTING FOR INCOMPLETE DETECTABILITY

When relevant assumptions are met, methods that account for incomplete detectability produce unbiased population estimates. They are typically more intensive and costly to apply than index methods. Assumptions underlying these methods always should be critically evaluated for validity before these methods are incorporated into a monitoring program.

10.2.2.1. Capture–Recapture and Related Methods

Capture–recapture and mark–resight surveys provide one of the more reliable approaches to estimating population size for mammals. We will start our review of these methods with the common trapping grid used for small and medium-sized mammals, and then expand the discussion to resightings of marked or otherwise individually identifiable animals.

Trapping Grids

Use of a grid of traps to capture, mark, and recapture mammals is a long-standing technique. Otis *et al.* (1978), White *et al.* (1982), and Skalski and Robson (1992) provide lengthy sections on design of capture–recapture surveys. Basically, high capture probabilities (>0.3) with a large trapping grid (100–400 traps) and with 5–10 nights of trapping are required to obtain adequate population estimates. However, the design can be modified depending on the nature of the animal being trapped, as discussed in the above sources.

Numerous references are available on increasing trap effectiveness with different baits, prebaiting (Chitty and Kempson, 1949; Zejda and Holisova, 1971), trap type (e.g., snap traps, live traps, pitfall traps, and funnel traps) (Petersen, 1980; Williams and Braun, 1983; Bury and Corn, 1987). For estimating the population, the grid is assumed to be demographically and geographically closed, so that closed-population estimators can be used

(Otis *et al.*, 1978; White *et al.*, 1982). The issue of demographic closure is not difficult to meet because the trapping interval can be short enough to preclude much, if any, mortality, and immatures recruited to the population can be recognized and excluded from the sample.

Geographic closure, however, is a more difficult problem. Animals at the edge of the grid, with only a portion of their home range included in the grid, cause difficulty with defining the exact grid size. The population estimate from capture–recapture estimators for the grid includes some undetermined number of animals that only partially inhabit the grid. If the population estimate, $\hat{N}$, is used to compute a naive estimate of density for the grid, $\hat{D}$ as $\hat{N}/A$, a biased estimate of $\hat{D}$ results because the effective area of the grid is $>A$, causing an "edge effect." A nested grid approach developed by Otis *et al.* (1978) and White *et al.* (1982) sometimes works, where $\hat{A}$ is estimated from assuming a fixed boundary strip of width W around nested subgrids within the larger grid. Program CAPTURE computes $\hat{D}$ and $\hat{W}$ for trapping grid data. Typically, grids on the order of 15 × 15 are needed to adequately estimate D. However, Wilson and Anderson (1985a) reported that computer simulations using the trapping web estimator still provided a biased estimate of population density because the edge effect often was not corrected in their simulations. Population estimates were greater than the actual population size because $\hat{A}$ often was not estimated larger than A. They suggested that, although theoretically sound, the nested subgrid approach requires more data than are likely to be collected in practice.

Another approach to correcting the area of the grid that has been used for estimating density is assessment lines (Kaufman *et al.*, 1971; Smith *et al.*, 1971; Swift and Steinhorst, 1976; O'Farrell *et al.*, 1977). The idea is to trap and mark animals on a grid for some period, then evaluate the movement away from the grid with lines of traps pointed from the grid to surrounding habitat. We consider the nested grid approach to be superior to the assessment line approach because the nested grid approach is based on a more rigorous statistical modeling procedure.

An alternative to the nested grid approach is a trapping web, first proposed by Anderson *et al.* (1983). The trapping web consists of lines of traps emanating at equal angular increments from a center point, like spokes from a wheel. On each line, traps are place at equal distances, creating rings of traps. Anderson *et al.* (1983) developed an estimator of density based on point transect theory using program TRANSECT. Simulation of the procedure by Wilson and Anderson (1985b) suggested that unbiased estimates of density were obtained. However, the goodness-of-fit test of the models in TRANSECT often rejected, leaving the user to question the validity of the inferences. Part of the lack-of-fit problem is associated with the lack of a rigorous point transect estimator in TRANSECT, a problem now fixed with program DISTANCE.

A field test of the method using known populations of ground-dwelling darkling beetles (Tenebrionidae: *Eleodes* spp.) was performed in a shrub–steppe ecosystem in southwestern Wyoming (Parmenter *et al.*, 1989). The main problem noted was the lack of fit to the model to the outermost rings, thought to be because of the enclosure boundaries. However, Anderson *et al.* (1983) also discussed an edge effect of the grid, noting that traps "on the margin of the grid ... have fewer neighboring traps with which to compete for available animals." Link and Barker (1994) have proposed a geometric analysis that avoids the need to estimate a point on a density function, and incorporates the data from the outermost rings of traps, explaining the large numbers of captures in these rings rather than truncating them from the analysis. This method needs to be simulated using procedures similar to Wilson and Anderson (1985b) to evaluate its statistical properties more fully.

Trapping grids traditionally have been used with small mammals, but some examples exist with medium-sized mammals, such as snowshoe hares (Keith *et al.*, 1968), cottontail rabbits (Geis, 1955; Edwards and Eberhardt, 1967; Chapman and Trethewey, 1972), gray squirrels (Flyger, 1959), striped skunks (Greenwood *et al.*, 1985), raccoons and Virginia opossums (Hallett *et al.*, 1991), and swift foxes (J. P. Fitzgerald, personal communication). Usually with medium-sized mammals, no attempt is made to correct the population estimate for the grid area and hence estimate density. However, use of radio collars with medium-sized mammals provides a method to correct the population estimates for area, because whether marked animals are on or off the study grid can be determined by radio location (e.g., White, 1996; program NOREMARK).

Trapping at Roosts

Bats offer unique survey problems among mammals because of their ability to fly and their small size (in North America). Thomas and LaVal (1988) reviewed survey methods that have been applied to estimating bat populations. We will discuss an untried capture–recapture approach for estimating numbers of bats roosting in large concentrations in caves, abandoned mines, and similar locations. Applying such an approach to species roosting in smaller numbers and in numerous dispersed locations, like under the bark of trees, would be essentially impossible under present technology.

Once a roost has been located, some type of trapping device to capture bats can be set up along the usual entrance/exit to the roost. The trap or traps should be placed randomly somewhere within this perceived route of travel and close enough to the entrance to maximize the chance of trapping individuals (Fig. 10.1). The probability of placing a trap in a particular spot can be weighted by expected numbers of individuals thought to travel through this spot. This idea can be expanded to roost sites that have multiple entrances, where probability of trap placement could be based on perceived numbers of bats using the entrance/exit. We are assuming that the entrance/exit is too large to cover with one trap. Even if the opening is small enough, bats may be reluctant to leave the roost after the initial capture if trapping is subsequently attempted shortly thereafter (K. Navo, personal communication).

The difficulty with this approach is that bats often will avoid traps after the initial capture (Stevenson and Tuttle, 1981; Thomas and LaVal, 1988). Also, trapping operations that are temporally spaced too closely together will probably increase the likelihood of trap avoidance, whereas trapping spaced too widely apart risks violating the closure assumption. Further, heterogeneity in trapping behavior violates the assumption of equal catchability in the Lincoln–Petersen model (this problem is avoided with a mark–resight approach—see below); hence, a minimum of at least three, if not four, capture occasions must be conducted to adequately fit a capture–recapture model given that numbers of individuals and capture probabilities are adequate. Also, using the same type of trapping technique each time will cause the estimator to be biased (but see mark-resight surveys below).

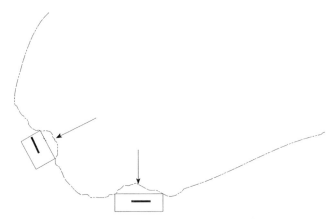

Figure 10.1 A hypothetical cave (dashed outline) containing a bat roost. The arrows show the two openings, and primary routes of travel, that the bats use to enter and exit the cave. The two bold lines represent traps that have been placed randomly within the areas of likely capture (rectangles). Random trap placement can be weighted so that the randomly selected point is more likely to occur within the primary routes of travel.

Moreover, if individuals travel between/among roosts in a locality then the closure assumption would be violated as well. Radio-tagging and then following a sample of bats from a roost may help assess the validity of the geographic closure assumption.

When unbiased or nearly unbiased abundance estimates are needed, our recommendation is to try this general approach at one roost and evaluate its validity before committing much time and money to it. Information from individuals during a pilot study at the very least will provide an index to numbers, which is all that other bat survey methods usually provide anyway.

Resighting Surveys of Marked Animals

For the capture–recapture estimators discussed in the previous section, unmarked animals are assumed to be marked when captured. In this section we consider methods where an initial sample of animals is marked, but then subsequent surveys do not actually capture and mark animals, but only estimate the proportion of marked animals in the population. The method depends on being able to mark a large number of animals, and on having a population that is relatively easy to observe. The most reliable application of this approach depends on using radios for marking animals, so that the exact number of marks available to be observed will be known. To correct for individual heterogeneity of animal sightings, marks should be individually identifiable.

One mark–resight approach involves two or more observers independently recording locations of animal groups during aerial surveys. Records then are matched to determine the number of groups seen by both observers, and only by one of the observers. From these data, an estimate of the number of groups not seen by either observer can be computed based on the Lincoln–Petersen method (Magnusson *et al.*, 1978).

For large mammals, aerial surveys typically are used, e.g., mule deer (Bartmann *et al.*, 1987), moose (Bowden and Kufeld, 1995), mountain sheep (Neal *et al.*, 1993; J. George, personal communication; Example 10.2), and bears (Miller *et al.*, 1997). Ground-based surveys also could be used if the observers were careful to sample the entire population, e.g., coyotes (Hein and Andelt, 1995). This approach also could be applied with night observations, e.g., a sample of swift foxes marked with radios that are individually identifiable with a spotlight. Taking the approach even further, Minta and Mangel (1989) estimated the population size of badgers by tracking them to their dens after a snow storm. They then checked the entrance of the den with a radio receiver to determine if the occupant was marked. Another innovative application has developed with motion/heat sensing cameras to photograph animals. Marks must be visible on the photograph to determine

Example 10.2. Mark–Resight Study of Bighorn Sheep

George *et al.* (1996) conducted a mark–resight study of bighorn sheep in the Kenosha and Tarryall Mountains, Colorado. The entire study area was small enough to be surveyed; hence, this was a one-stage design in which the entire study area was essentially treated as a single, large sampling unit. Sheep were captured and equipped with radio collars that were visibly distinguishable from the air. Resighting data were collected during two helicopter flights in the Kenosha Mountains and during three helicopter flights in the Tarryall Mountains. We will restrict this example to their data collected in the Kenosha Mountains. Program NOREMARK (White, 1996) computed similar abundance and confidence interval estimates based on the joint hypergeometric maximum likelihood estimator and Bowden's estimator. The estimates based on Bowden's estimator were $\hat{N} = 96$ with a 90% maximum likelihood profile CI of 80–116.

the status of the individual, and preferably marked animals should be individually identifiable. This approach has been used by Mace *et al.* (1994) to estimate population size for grizzy bears, and by T. D. I. Beck (personal communication) to estimate black bear population size.

The mark-resight approach has great promise for use with volunteers, particularly in refuge situations where the animals are particularly visible. With Bowden's estimator, volunteers only need to tally the number of unmarked animals observed, and the number of observations of each of the marked animals observed. The same animal may be seen more than one time by an observer; this does not present a problem because of the way this estimator is formulated.

As an untried approach, we suggest that the mark–resight method could be useful in estimating bat populations if a means of individually identifying individuals can be used. That is, bats could be initially captured at a roost using a trap or net, and marked so that they can be visually distinguishable during surveys on a subsequent night. This approach may not be realistic with present technology, but may be possible in the future.

Variations of the same kind of statistical methodology are provided by estimating the capture probability of a sample of radio-marked animals known to be on the study area. Hallett *et al.* (1991) used regular trapping methods for raccoons and Virginia opossums with a known number of radio-collared animals on the study area. The capture probability then is estimated for each occasion as the number of the radio-collared animals captured divided by the number known to be on the study area. A similar procedure was used by Greenwood *et al.* (1985). The estimators in program NOREMARK (White, 1996) would work well for this type of survey.

An even more extreme extension of this type of estimator is provided by Conner and Labisky (1985). They estimated raccoon population size by capturing and radioisotope-tagging 48 raccoons. Samples of scats then were collected, and the proportion of these scats that came from the 48 radioisotope-marked raccoons was used to estimate population size. If a unique radioisotope marker could be applied to each of the marked animals, then Bowden's estimator in program NOREMARK would work well to estimate the population size. Other examples of the use of radioisotope tagging are for bears (Pelton and Marcum, 1977), bobcats (Pelton, 1979), river otters (Shirley *et al.*, 1988), and coyotes (Crabtree *et al.*, 1989). Skalski (1991) has developed an alternative estimator appropriate for this type of marked sign survey.

"Natural" Marks

Sometimes, animals are individually identifiable naturally because of unique markings. Examples include African elephants (i.e., ear shapes) (Moss, 1988) and black rhinocerous (i.e., horn shapes and ear tears) (Emslie, 1993). If all the animals can be individually identified when they are first seen, then a capture–recapture model can be used, i.e., program CAPTURE. If some of the animals are not individually identifiable, then a reference list of marked animals is kept, and the mark–resight methodology of program NOREMARK is appropriate.

An extension of the natural mark approach is to use a DNA marker. Animals are sampled by some method to obtain a tissue sample. DNA from this tissue then is sequenced to provide a unique mark for the individual. This approach is being used by British Columbia to estimate grizzly bear populations. Bears are attracted to bait where barbed wire is used to collect a hair sample. The drawback of using DNA as the marking method is the cost of sequencing each sample. However, the major benefit is that animals do not actually have to be captured and marked, so there is no loss of marks, and the logistics of marking are greatly simplified.

10.2.2.2. Distance Sampling

Line transect estimators (Buckland *et al.*, 1993) will only work properly for mammals that are visible and can be seen before they move away from the observer. Most mammals surveyed with line transect methodology are done so with the aid of aircraft, i.e., helicopters with mule deer (White *et al.*, 1989) and red deer (Trenkel *et al.*, 1997), and fixed-wing aircraft with pronghorn (Johnson *et al.*, 1991). However, some additional innovations are possible. As suggested by Burnham and Anderson (1984), perpendicular distance data always should be taken with strip transect counts to provide better efficiency and to test validity of the assumptions of the strip transect. Night time counts with spot lights are one example of the potential application of line transects. Swift and kit foxes could be surveyed at night from 4-wheel drive vehicles traveling on back country roads. A key assumption of this approach is that the animals do not move as a result of the presence of the observer. Distances can be paced or accurately measured to the location where the animal was first sighted.

Line transects have been little used on medium-sized to small mammals, although Healy and Welsh (1992) used this technique on gray squirrels. In general, this size of mammal is not that visible, and so the probability of sighting the animal on the line is not 1, as required by this method to give unbiased estimates. However, even if the sighting probability at zero distance is not quite 1, the method still should be useful as an index for comparing surveys across time and space as long as the lowered sighting probability on the line is consistent across surveys. Another example of a walking line transect survey is provided by Southwell (1994) for macropods. Line transects also can be used to estimate the abundance of animal sign, such as antler rubs by deer (Miller *et al.*, 1987) or animal remains (i.e., carcass surveys; e.g., applied to birds by Philibert *et al.*, 1993), although this still is an index because the relationship between the abundance of sign or remains and the true abundance is unknown.

10.2.2.3. Line-Intercept Methods

Becker (1991) developed a method to estimate wolverine and lynx populations in Alaska using tracks in snow. Becker's method provides a population estimate, so it is better than an index based on tracks. His method requires that good snow conditions be present during the course of the survey and that all animal tracks intersected during the sampling process are observed. Good snow conditions are defined to be fresh snow of sufficient depth so that presnowstorm and postsnowstorm animal tracks can be distinguished. Two general sampling designs were presented: the first assumed that animal tracks can be observed and followed both to the animal's location at the end of the snowstorm and to its present location; the second assumed that the number of different animals encountered along a set of transects can be determined and that it is possible to get movement data from a random sample of radio-collared animals. To sample tracks, transects are selected and surveyed, either by ground travel or slow-moving aircraft.

Critical assumptions for this method are that all animals move during the course of the survey, all animal tracks of the species of interest are readily recognizable, all animal tracks are continuous, animal movements are independent of the sampling process, pre- and post-snowstorm tracks can be distinguished, all animal tracks that cross-sampled transects are observed, the study area is rectangular in shape, and all the transects are oriented perpendicular to a specified reference axis (x axis) (Becker, 1991). Van Sickle and Lindzey (1991) suggested a similar procedure for estimating cougar population size. Ballard *et al.* (1995) used Becker's method to estimate densities of wolves in northwest Alaska.

Becker's approach is probably only suitable in areas that are open enough so that tracks can be followed and the animal located by air. If there is too much forest cover, then tracks cannot be spotted or, if spotted, probably cannot be followed from one end to the other; accurately identifying tracks from the air could be problematic in areas that contain species leaving similar-looking tracks. Ground surveys via snowmobile will likely be precluded in areas of rugged terrain. Moreover, ground truthing a track will almost certainly cause a response from the tracked animal and hence change the track length, which would violate a key assumption. Thus, model assumptions should be given careful thought to ensure that this method is suitable for a particular species within a certain habitat or habitats.

10.3. RECOMMENDATIONS

In this section we offer some general recommendations for choosing among enumeration techniques discussed in this chapter. We begin by making a few general comments and then present a dichotomous key for more detailed help.

10.3.1. GENERAL COMMENTS

To accurately detect a trend in numbers with a predetermined level of precision requires unbiased, or nearly unbiased, and precise estimates. Unfortunately, the cost and effort needed to obtain these data for most species often far exceeds available funding. Thus, "cheap" methods such as indices may have to suffice for the majority of species.

10.3.2. DICHOTOMOUS KEY TO ENUMERATION METHODS

We have constructed the following dichotomous key primarily for use as a general guide to methods for obtaining abundance estimates from either complete or incomplete counts. That is, this key may be used as an early step in the process to determine if either complete counts or methods that adjust for detectability are realistic for any species that require such detailed information. Further, this key should be used in conjunction with information presented earlier in this chapter (sections are in parentheses) and in previous chapters on sampling design, enumeration methods, and guidelines for planning surveys. In particular, assumptions required for valid abundance estimates should be critically assessed via a pilot study or some other means before these methods are incorporated into a full-scale monitoring program.

1. (a) A complete count of individuals or objects in all randomly chosen plots or surveyed areas is possible with available funding ....
....Quadrat counts (10.1.2)
   (b) A complete count is either impossible or too costly in all randomly chosen plots or surveyed areas..2
2. (a) Individuals can be caught and uniquely marked...........................3
   (b) Individuals cannot be caught, are too expensive to catch, or are not conducive to marking 9
3. (a) Individuals are completely contained (i.e., no edge effect) within either the entire area of interest if the survey is over the entire sampling frame or within a given plot if surveys are performed at the plot level 4
   (b) Not as above 7
4. (a) Time period of study is short enough to treat population as closed in terms of births/immigration and deaths/emigration 5
   (b) Not as above 9
5. (a) Individuals can be recaptured or resighted................................6
   (b) Not as above 9
6. (a) There are at least 100 individuals contained within the survey area (e.g., plot) and their capture/sighting probabilities are at least 0.3 (preferably 0.5 or above for the range of population sizes encountered in nearly all fish and wildlife studies); lower capture probabilities are acceptable with larger population sizes (program CAPTURE has a simulation routine to check if estimated capture probabilities are adequate)
       or ≥ 12 sampling occasions..Capture–
       recapture or mark–resight (9.2.2.1); Rosenberg *et al.* (1995)
   (b) Not as above 9
7. (a) Trapping can be done using a trapping web design with adequate numbers of traps...Trapping webs (9.2.2.1)
   (b) Not as above 8
8. (a) Individuals can be equipped with radio transmitters in a cost-efficient manner, and readily and accurately located via

radiotelemetry .... Correction for immigration/
emigration in mark–resight; program NOREMARK (10.2.2.1)
  (b) Not as above 9
9. (a) Perpendicular distances (or distance categories) from either a line
       or point to an individual, discrete group of individuals, or object
       can be recorded in a cost-efficient manner 10
  (b) Perpendicular distances or distance categories are either
       unobtainable or too costly to obtain 12
10. (a) Every individual or object on a line or point can be located and is
        not affected by observer presence 11
  (b) Not all individuals or objects occurring on a line or point can
       be detected because of vegetational structure, undetected move-
       ments of individuals, or other reasons, and no methods for ad-
       justing for incomplete detectability on a line or point are feasible
      12
11. (a) Adequate numbers of individuals, groups of individuals, or objects
        can be detected for reliable model selection in program
        DISTANCE and reasonably precise density estimates...................
        ....Distance sampling methods (10.2.2.2)
  (b) Not as above 12
12. (a) Individuals leave identifiable tracks that can be followed from one
        end to the other without causing animal movement or can be
        estimated from movement patterns from a random sample of radio-
        collared animals 13
  (b) Tracks not identifiable because of substrate or size of track, or
       tracks cannot be followed without causing further movement by
       individual 14
13. (a) Fresh tracks are distinguishable from older tracks either because
        of weather effects on substrate (e.g., snow) or older tracks are
        marked by an observer...
        .... Becker's (1991) line-intercept estimator (10.2.2.3)
  (b) Not as above 14
14. (a) Only data on species occurrence required ....
        .... Presence–absence methods (10.2.1.1)
  (b) Uncorrected counts of all individuals/objects detected on a plot
       .... Relative index methods (10.2.1.2)

LITERATURE CITED

Anderson, D. R., Burnham, K. P., White, G. C., and Otis, D. L. (1983). Density estimation of small-mammal populations using a trapping web and distance sampling methods. *Ecology* **64:** 674–680.

Armstrong, D. M. (1972). *Distribution of Mammals in Colorado.* Mus. Nat. Hist. Monogr. 3, Univ. of Kansas, Lawrence, KS.

Ballard, W. B., McNay, M. E., Gardner, C. L., and Reed, D. J. (1995). Use of line-intercept track sampling for estimating wolf densities. In *Ecology and Conservation of Wolves in a Changing World* (L. N. Carbyn, S. H. Fritts, and D. R. Seip, Eds.), pp. 469–480. Can. Circum. Inst., Edmonton, AB, Canada. [Occas. Publ. No. 642].

Bartmann, R. M., Carpenter, L. H., Garrott, R. A., and Bowden, D. C. (1986). Accuracy of helicopter counts of mule deer in pinyon-juniper woodland. *Wildl. Soc. Bull.* **14:** 356–363.

Bartmann, R. M., White, G. C., Carpenter, L. H., and Garrott, R. A. (1987). Aerial mark-recapture estimates of confined mule deer in pinyon-juniper woodland. *J. Wildl. Mgmt.* **51:** 41–46.

Batcheler, C. L. (1975). Development of a distance method for deer census from pellet groups. *J. Wildl. Mgmt.* **39:** 641–652.

Beasom, S. L., Leon, F. G., III, and Synatzske, D. R. (1986). Accuracy and precision of counting white-tailed deer with helicopters at different sampling intensities. *Wildl. Soc. Bull.* **14:** 364–368.

Becker, E. F. (1991). A terrestrial furbearer estimator based on probability sampling. *J. Wildl. Mgmt.* **55:** 730–737.

Boonstra, R., Krebs, C. J., Boutin, S., and Eadie, J. M. (1994). Finding mammals using infrared thermal imaging. *J. Mammal.* **75:** 1063–1068.

Bowden, D. C., and Kufeld, R. C. (1995). Generalized mark-resight population size estimation applied to Colorado moose. *J. Wildl. Mgmt.* **59:** 840–851.

Boyce, M. S. (1989). *The Jackson Hole Elk Herd.* Cambridge Univ. Press, Cambridge, UK.

Buckland, S. T., Anderson, D. R., Burnham, K. P., and Laake, J. L. (1993). *Distance Sampling: Estimating Abundance of Biological Populations.* Chapman and Hall, London.

Burnham, K. P., and Anderson, D. R. (1984). The need for distance data in transect counts. *J. Wildl. Mgmt.* **48:** 1248–1254.

Bury, R. B., and Corn, P. S. (1987). Evaluation of pitfall trapping in northwestern forests: Trap arrays with drift fences. *J. Wildl. Mgmt.* **51:** 112–119.

Chapman, J. A., and Trethewey, D. E. C. (1972). Factors affecting trap responses of introduced eastern cottontail rabbits. *J. Wildl. Mgmt.* **36:** 1221–1226.

Chitty, D., and Kempson, D. A. (1949). Prebaiting small mammals and a new design of live trap. *Ecology* **30:** 536–542.

Choquenot, D., Kay, B., and Lukins, B. (1990). An evaluation of warfarin for the control of feral pigs. *J. Wildl. Mgmt.* **54:** 353–359.

Conner, M. C., and Labisky, R. F. (1985). Evaluation of radioisotope tagging for estimating abundance of raccoon populations. *J. Wildl. Mgmt.* **49:** 326–332.

Conner, M. C., Labisky, R. F., and Progulske, D. R., Jr. (1983). Scent-station indices as measures of population abundance for bobcats, raccoons, gray foxes, and opossums. *Wildl. Soc. Bull.* **11:** 146–152.

Crabtree, R. L., Burton, F. G., Garland, T. R., Cataldo, D. A., and Rickard, W. H. (1989). Slow-release radioisotope implants as individual markers for carnivores. *J. Wildl. Mgmt.* **53:** 949–954.

Crete, M., and Bedard, J. (1975). Daily browse consumption by moose in the Gaspe Peninsula, Quebec. *J. Wildl. Mgmt.* **39:** 368–373.

Cypher, B. L. (1991). A technique to improve spotlight observations of deer. *Wildl. Soc. Bull.* **19:** 391–393.

DeYoung, C. A., Guthery, F. S., Beasom, S. L., Coughlin, S. P., and Heffelfinger, J. R. (1989). Improving estimates of white-tailed deer abundance from helicopter surveys. *Wildl. Soc. Bull.* **17:** 275–279.

Diefenbach, D., Conroy, M. J., Warren, R. J., James, W. E., Baker, L. A., and Hon, T. (1994). A test of the scent-station survey technique for bobcats. *J. Wildl. Mgmt.* **58:** 10–17.

Drew, G. S., Fagre, D. B., and Martin, D. J. (1988). Scent-station surveys for cottontail rabbit populations. *Wildl. Soc. Bull.* **16:** 396–398.

Edwards, W. R., and Eberhardt, L. L. (1967). Estimating cottontail abundance from live-trapping data. *J. Wildl. Mgmt.* **31:** 87–96.

Emslie, R. H. (1993). *Rhino Version 1.2/1.21 Reference Manual.* Ecoscot Consultancy Services, Pilanesberg, South Africa.

Fafarman, K. R., and DeYoung, C. A. (1986). Evaluation of spotlight counts of deer in south Texas. *Wildl. Soc. Bull.* **14:** 180–185.

Fitzgerald, J. P., Meaney, C. A., and Armstrong, D. M. (1994). *Mammals of Colorado.* Denver Mus. Nat. Hist. and Univ. Press of Colorado, Niwot, CO.

Flyger, V. F. (1959). A comparison of methods for estimating squirrel populations. *J. Wildl. Mgmt.* **23:** 220–223.

Fraser, D. (1979). Sightings of moose, deer, and bears on roads in northern Ontario. *Wildl. Soc. Bull.* **7:** 181–184.

Freddy, D. J., and Bowden, D. C. (1983). Sampling mule deer pellet-group densities in juniper-pinyon woodland. *J. Wildl. Mgmt.* **47:** 476–485.

Friend, G. T. (1984). Relative efficiency of two pitfall-drift fence systems for sampling small vertebrates. *Aust. Zool.* **21:** 423–433.

Fuller, T. K., and Sampson, B. A. (1988). Evaluation of a simulated howling survey for wolves. *J. Wildl. Mgmt.* **52:** 60–63.

Garner, D. L., Underwood, H. B., and Porter, W. F. (1995). Use of modern infrared thermography for wildlife population surveys. *Environ. Mgmt.* **19:** 233–238.

Gasaway, W. C., DuBois, S. D., Reed, J. D., and Harbo, S. J. (1986). *Estimating Moose Population Parameters from Aerial Surveys.* [Biological Papers of the University of Alaska No. 22].

Geis, A. D. (1955). Trap response of the cottontail rabbit and its effect on censusing. *J. Wildl. Mgmt.* **19:** 466–472.

George, J. L., Miller, M. W., White, G. C., and Vayhinger, J. (1996). Comparison of mark-resight population size estimators for bighorn sheep in alpine and timbered habitats. [Unpublished manuscript]

Greenwood, R. J., Sargeant, A. B., and Johnson, D. H. (1985). Evaluation of mark-recapture for estimating striped skunk abundance. *J. Wildl. Mgmt.* **49:** 332–340.

Hallett, J. G., O'Connell, M. A., Sanders, G. D., and Seidensticker, J. (1991). Comparison of population estimators for medium-sized mammals. *J. Wildl. Mgmt.* **55:** 81–93.

Harrington, F. H., and Mech, L. D. (1982). An analysis of howling response parameters useful for wolf pack censusing. *J. Wildl. Mgmt.* **46:** 686–693.

Healy, W. M., and Welsh, C. J. E. (1992). Evaluating line transects to monitor gray squirrel populations. *Wildl. Soc. Bull.* **20:** 83–90.

Hein, E. W., and Andelt, W. F. (1995). Estimating coyote density from mark-resight surveys. *J. Wildl. Mgmt.* **59:** 164–169.

Hon, T. (1979). Relative abundance of bobcats in Georgia: Survey techniques and preliminary results. In *Bobcat Research Conference Proceedings* (L. G. Blum and P. C. Escherich, Eds.), pp. 104–106. [Natl. Wildl. Fed., Tech. Rep. Ser. 6]

Johnson, B. K., Lindzey, F. G., and Guenzel, R. J. (1991). Use of aerial line transect surveys to estimate pronghorn populations in Wyoming. *Wildl. Soc. Bull.* **19:** 315–321.

Johnson, K. G., and Pelton, M. R. (1981). A survey of procedures to determine relative abundance of furbearers in the southeastern United States. *Proc. Annual Conf. Southeast Assoc. Fish Wildl. Agencies* **35:** 261–272.

Joyal, R., and Richard, J. (1986). Winter defecation output and bedding frequency of wild, free-ranging moose. *J. Wildl. Mgmt.* **50:** 734–736.

Kaufman, D. W., Smith, G. C., Jones, R. M., Gentry, J. G., and Smith, M. H. (1971). Use of assessment lines to estimate density of small mammals. *Acta Theriol.* **16:** 122–147.

Keith, L. B., Meslow, E. C., and Rongstad, O. J. (1968). Techniques for snowshoe hare population studies. *J. Wildl. Mgmt.* **32:** 801–812.

Kline, P. D. (1965). Factors influencing roadside counts of cottontails. *J. Wildl. Mgmt.* **29:** 665–671.

Kufeld, R. C., Olterman, J. H., and Bowden, D. C. (1980). A helicopter quadrat census for mule deer on Uncompahgre Plateau, Colorado. *J. Wildl. Mgmt.* **44:** 632–639.

Lavigne, D. M., and Oritsland, N. A. (1974). Ultraviolet photography: A new application for remote sensing of mammals. *J. Fish. Res. Bd. Can.* **52:** 939–951.

Lindzey, F. G., and Thompson, S. K. (1977). Scent station index of black bear abundance. *J. Wildl. Mgmt.* **41:** 151–153.

Linhart, S. B., and Knowlton, F. F. (1975). Determining the relative abundance of coyotes by scent station lines. *Wildl. Soc. Bull.* **3:** 119–124.

Link, W. A., and Barker, R. J. (1994). Density estimation using the trapping web design: A geometric analysis. *Biometrics* **50:** 733–745.

Lord, R. D., Jr. (1959). Comparison of early morning and spotlight roadside censuses for cottontails. *J. Wildl. Mgmt.* **23:** 458–460.

Mace, R. D., Minta, S. C., Manley, T. L., and Aune, K. E. (1994). Estimating grizzly bear population size using camera sightings. *Wildl. Soc. Bull.* **22:** 74–83.

Magnusson, W. E., Caughley, G. J., and Grigg, G. C. (1978). A double-survey estimate of population size from incomplete counts. *J. Wildl. Mgmt.* **42:** 174–176.

Marten, G. G. (1972). Censusing mouse populations by means of tracking. *Ecology* **53:** 859–867.

McCullough, D. R. (1982). Evaluation of night spotlighting as a deer study technique. *J. Wildl. Mgmt.* **46:** 963–973.

Merrill, E. H., Hemker, T. P., Woodruff, K. P., and Kuck, L. (1994). Impacts of mining facilities on fall migration of mule deer. *Wildl. Soc. Bull.* **22:** 68–73.

Miller, K. V., Kammermeyer, K. E., Marchinton, R. L., and Moser, E. B. (1987). Population and habitat influences on antler rubbing by white-tailed deer. *J. Wildl. Mgmt.* **51:** 62–66.

Miller, S. D., White, G. C., Sellers, R. A., Reynolds, H. V., Schoen, J. W., Titus, K., Barnes, V. G., Jr., Smith, R. B., Nelson, R. R., Ballard, W. B., and Schwartz, C. C. (1997). Brown and black bear density estimation in Alaska using radiotelemetry and replicated mark-resight techniques. *Wildl. Monogr.* **133.**

Minta, S., and Mangel, M. (1989). A simple population estimate based on simulation for capture-recapture and capture-resight data. *Ecology* **70:** 1738–1751.

Mooty, J. J., Karns, P. D., and Heisey, D. M. (1984). The relationship between white-tailed deer track counts and pellet-group surveys. *J. Wildl. Mgmt.* **48:** 275–279.

Moss, C. (1988). *Elephant Memories: Thirteen Years in the Life of an Elephant Family.* Morrow, New York.

Neal, A. K., White, G. C., Gill, R. B., Reed, D. F., and Olterman, J. H. (1993). Evaluation of mark-resight model assumptions for estimating mountain sheep numbers. *J. Wildl. Mgmt.* **57:** 436–450.

Neff, D. J. (1968). The pellet-group count technique for big game trend, census, and distribution: A review. *J. Wildl. Mgmt.* **32:** 597–614.

Neff, D. J., Wallmo, O. C., and Morrison, D. C. (1965). A determination of defecation rate for elk. *J. Wildl. Mgmt.* **29:** 406–407.

O'Farrell, M. J., Kaufman, D. W., and Lundahl, D. W. (1977). Use of live-trapping with assessment line method for density estimation. *J. Mammal.* **58:** 575–582.

Okoniewski, J. C., and Chambers, R. E. (1984). Coyote vocal response to an electronic siren and human howling. *J. Wildl. Mgmt.* **48:** 217–222.

Otis, D. L., Burnham, K. P., White, G. C., and Anderson, D. R. (1978). Statistical inference from capture data on closed animal populations. *Wildl. Monogr.* **62.**

Parmenter, R. P., MacMahon, J. A., and Anderson, D. R. (1989). Animal density estimation using a trapping web design: Field validation experiments. *Ecology* **70:** 169–179.

Pelton, M. R. (1979). Potential use of radioisotopes for determining densities of bobcats. *Bobcat Res. Conf. Natl. Wildl. Fed. Sci. Tech. Ser.* **6:** 97–100.

Pelton, M. R., and Marcum, L. C. (1977). Potential use of radioisotopes for determining densities of black bears and other carnivores. In *Proceedings, 1975 Predator Symposium* (R. L. Phillips and C. Jonkel, Eds.), pp. 221–236. Montana For. Conserv. Exp. Station, Univ. Montana, Missoula.

Petersen, M. K. (1980). A comparison of small mammal populations sampled by pitfall and live-traps in Durango, Mexico. *Southwest Nat.* **25:** 122–124.

Philibert, H., Wobeser, G., and Clark, R. G. (1993). Counting dead birds: Examination of methods. *J. Wildl. Dis.* **29:** 284–289.

Pojar, T. M., Bowden, D. C., and Gill, R. B. (1995). Aerial counting experiments to estimate pronghorn density and herd structure. *J. Wildl. Mgmt.* **59:** 117–128.

Rogers, L. L. (1987). Seasonal changes in defecation rates of free-ranging white-tailed deer. *J. Wildl. Mgmt.* **51:** 330–333.

Rosenberg, D. K., Overton, W. S., and Anthony, R. G. (1995). Estimation of animal abundance when capture probabilities are low and heterogeneous. *J. Wildl. Mgmt.* **59:** 252–261.

Roughton, R. D., and Sweeny, M. W. (1982). Refinements in scent-station methodology for assessing trends in carnivore populations. *J. Wildl. Mgmt.* **46:** 217–229.

Rowland, M. M., White, G. C., and Karlen, E. M. (1984). Use of pellet-group plots to measure trends in deer and elk populations. *Wildl. Soc. Bull.* **12:** 147–155.

Samuel, M. D., Garton, E. O., Schlegel, M., and Carson, R. G. (1987). Visibility bias during aerial surveys of elk in northcentral Idaho. *J. Wildl. Mgmt.* **51:** 622–630.

Scotts, D. J., and Craig, S. A. (1988). Improved hair-sampling tube for the detection of rare mammals. *Aust. Wildl. Res.* **15:** 469–472.

Shirley, M. G., Linscombe, R. G., Kinler, N. W., Knaus, R. M., and Wright, V. L. (1988). Population estimates of river otters in a Louisiana coastal marshland. *J. Wildl. Mgmt.* **52:** 512–515.

Skalski, J. R. (1991). Using sign counts to quantify animal abundance. *J. Wildl. Mgmt.* **55:** 705–715.

Skalski, J. R., and Robson, D. S. (1992). *Techniques for Wildlife Investigations: Design and Analysis of Capture Data.* Academic Press, San Diego.

Smith, A. D. (1964). Defecation rates of mule deer. *J. Wildl. Mgmt.* **28:** 435–444.

Smith, G. W., and Nydegger, N. C. (1985). A spotlight, line-transect method for surveying jack rabbits. *J. Wildl. Mgmt.* **49:** 699–702.

Smith, M. H., Blessing, R., Chelton, J. G., Gentry, J. B., Golly, F. B., and McGinnis, J. T. (1971). Determining density for small mammal populations using a grid and assessment lines. *Acta Theriol.* **16:** 105–125.

Southwell, C. (1994). Evaluation of walked line transect counts for estimating macropod density. *J. Wildl. Mgmt.* **58:** 348–356.

Steinhorst, R. K., and Samuel, M. D. (1989). Sightability adjustment methods for aerial surveys of wildlife populations. *Biometrics* **45:** 415–426.

Stevenson, D. E., and Tuttle, M. D. (1981). Survivorship in the endangered gray bat (*Myotis grisescens*). *J. Mammal.* **62:** 244–257.

Swift, D. M., and Steinhorst, R. K. (1976). A technique for estimating small mammal population densities using a grid and assessment lines. *Acta Theriol.* **21:** 471–480.

Thomas, D. W., and LaVal, R. K. (1988). Survey and census methods. In *Ecological and Behavioral Methods for the Study of Bats* (T. H. Kunz, Ed.), pp. 77–89. Smithsonian Institution Press, Washington, DC.

Trenkel, V. M., Buckland, S. T., and Elston, D. A. (1997). Evaluation of aerial line transect methodology for estimating red deer (*Cervus elaphus*) abundance in Scotland. *J. Environ. Mgmt.* **50:** 39–50.

Van Dyke, F. G., Brocke, R. H., and Shaw, H. G. (1986). Use of road track counts as indices of mountain lion presence. *J. Wildl. Mgmt.* **50:** 102–109.

Van Sickle, W. D., and Lindzey, F. G. (1991). Evaluation of a cougar population estimator based on probability sampling. *J. Wildl. Mgmt.* **55:** 738–743.

Warren, E. R. (1942). *The Mammals of Colorado: Their Habits and Distribution,* second ed. Univ. of Oklahoma Press, Norman, OK.

Wenger, C. R., and Cringan, A. T. (1978). Siren-elicited coyote vocalizations: An evaluation of a census technique. *Wildl. Soc. Bull.* **6:** 73–76.

Whipple, J. D., Rollins, D., and Schacht, W. H. (1994). A field simulation for assessing accuracy of spotlight deer surveys. *Wildl. Soc. Bull.* **22:** 667–673.

White, G. C. (1996). Program NOREMARK: Population estimation from mark-resight surveys. *Wildl. Soc. Bull.* **24:** 50–52.

White, G. C., Anderson, D. R., Burnham, K. P., and Otis, D. L. (1982). *Capture-Recapture and Removal Methods for Sampling Closed Populations.* Los Alamos Natl. Lab., Los Alamos, NM. [LA-8787-NERP]

White, G. C., Bartmann, R. M., Carpenter, L. H., and Garrott, R. A. (1989). Evaluation of aerial line transects for estimating mule deer densities. *J. Wildl. Mgmt.* **53:** 625–635.

White, G. C., and Eberhardt, L. E. (1980). Statistical analysis of deer and elk pellet-group data. *J. Wildl. Mgmt.* **44:** 121–131.

Williams, D. F., and Braun, S. E. (1983). Comparison of pitfall and conventional traps for sampling small mammal populations. *J. Wildl. Mgmt.* **47:** 841–845.

Wilson, D. E., Cole, F. R., Nichols, J. D., Rudran, R., and Foster, M. S. (Eds.) (1996). *Measuring and Monitoring Biological Diversity—Standard Methods for Mammals.* Smithsonian Institution Press, Washington, DC.

Wilson, K. R., and Anderson, D. R. (1985a). Evaluation of a nested grid approach for estimating density. *J. Wildl. Mgmt.* **49:** 675–678.

Wilson, K. R., and Anderson, D. R. (1985b). Evaluation of a density estimator based on a trapping web and distance sampling theory. *Ecology* **66:** 1185–1194.

Wood, J. E. (1959). Relative estimates of fox population levels. *J. Wildl. Mgmt.* **23:** 53–63.

Wyatt, C. L., Trivedi, M., and Anderson, D. R. (1980). Statistical evaluation of remotely sensed thermal data for deer census. *J. Wildl. Mgmt.* **44:** 397–402.

Zejda, J., and Holisova, V. (1971). Quadrat size and the prebaiting effect in trapping small mammals. *Ann. Zool. Fenn.* **8:** 14–16.

Appendix A

Glossary of Terms

Abundance total number of individuals, objects, or items of interest within some defined area and time period. Also known as absolute abundance.

Alpha value (α) (1) type I error rate, i.e., the probability of rejecting a null hypothesis when it is actually "true"; (2) value used to set the level of confidence (i.e., confidence coefficient, $1 - \alpha$) in a confidence interval.

Alternative hypothesis in statistical testing, the option representing some difference or change; this option is only accepted with a prespecified high degree of certainty (alpha or type I error rate) that a difference exists, beyond random chance alone, based on rejection of the null hypothesis.

Beta value (β) Type II error rate, i.e., the probability of failing to reject a null hypothesis when it is actually false. The statistical power of a test is $1 - \beta$.

Bias a persistent statistical error associated with parameter estimates whose source is not random chance. Mathematically, bias is the difference between the expected value of a pa-rameter estimate and the true value of the parameter.

Census a complete count of individuals, objects, or items within a specified area and time period. A census generally refers to a complete count of elements within a sampled population and/or target population; however, this term also may be applied at the unit level to represent a complete count of elements within a sampling unit, such as a "plot census." "Census" is frequently misused as a synonym for "survey" (cf. survey).

Closed population a fixed group of individuals within a defined area and time period, i.e., there are no births, deaths, immigration, and emigration.

Coefficient of variation (CV) ratio of a standard error of a parameter estimate to the parameter estimate. The coefficient of variation is used in computing sample sizes and as a measure of relative precision when comparing degree of variation between or among sets of data.

Confidence interval (CI) an interval around a parameter estimate that provides a measure of confidence regarding how close a sample-based estimate is to the true parameter. The usual two-sided, symmetrical confidence interval around the parameter estimate is generated by adding and subtracting the quantity computed from the product of the standard error and the t value or z value corresponding to the prespecified $(1 - \alpha)\%$ confidence level (α is frequently set at 0.05). For example, a 95% confidence interval will contain, on average, the true parameter of interest 95 of 100 times if 100 such intervals were calculated in a like manner. That is, confidence refers to the procedure of obtaining an interval rather than the interval itself. There is not a 95% probability that the true parameter occurs in the interval; either a parameter is in the interval or it is not.

Correlative study see observational study.

Coverage (area) a description of how well a random sample of plots is spatially arranged across a sampling frame. A sample exhibiting good coverage is one whose sampling units are geographically spread out so that they capture the underlying variability of the elements. Good coverage is important when elements are spatially clustered because a given simple random sample of plots may not include plots containing elements. Coverage is largely irrelevant if elements are randomly distributed, which is unlikely for most biological populations.

Coverage (of CI) a measure of how closely a particular type of confidence interval approaches the nominal level of coverage. For instance, one would expect a 95% confidence interval based on a standard normal to con-

tain, on average, the true parameter of interest in 95 out of 100 confidence intervals constructed in a like manner under similar conditions. A given type of confidence interval will exhibit poor coverage if the assumed sampling distribution (i.e., the one used to construct the interval) differs greatly from the true sampling distribution.

Coverage error a source of bias from a portion of the target population either being excluded from the sampling frame (e.g., gaps among sampling units) or being counted more than once (e.g., overlapping sampling units)

Degrees of freedom (*df*) a quantity used to define the shape of a t distribution. In general, it is the sample size minus the number of parameters in a model.

Density total number of individuals or objects of interest per unit area (also known as absolute density).

Detectability probability of correctly noting the presence of an element within some specified area and time period.

Edge effect an underestimation of abundance or density caused when the perceived area of detection (e.g., trapping grid) is smaller than the effective area of detection. This commonly occurs in studies of mobile individuals that may move back and forth across the boundary of a sampling unit or study area.

Element an individual, object, or item of interest that is directly measured, counted, or recorded.

Enumeration variation variance associated with incomplete counts or captures of individuals within a sampling unit (i.e., within-unit variance) or study area.

Estimate a numerical value calculated from sample data collected from a sampled population and used to represent the parameter of interest.

Estimator a mathematical formula used to calculate an estimate. An unbiased estimator will produce unbiased estimates when appropriate assumptions are satisfied.

Experiment strictly speaking, a study in which elements or experimental units from the same population are randomly assigned to treatment and nontreatment (or control) groups. The treatment should be the only factor, other than random error, affecting the parameter of interest. The purpose of an experiment is to test for a cause and effect relationship between the treatment and any resultant changes in the parameter of interest. An experiment attempts to address the "why?" question (e.g., Why is a population declining?). An experiment is rarely possible under field conditions because of the absence of a true "control." Also called a true experiment or controlled experiment (cf. observational study, quasi-experiment).

Experimental design the structure and layout of an experiment. The fundamental components of an experimental design are randomization, replication, and control. Randomization refers to the random assignment of elements or experimental units to treatment and nontreatment groups. Replication is the use of multiple groups, containing different elements or experimental units, for each

treatment/nontreatment category. A "control" refers to a nontreatment group that only differs from the treatment group(s) by the specific treatment, i.e., the experimenter has accounted for and removed effects of any potential factors that could confound results.

Field experiment see quasi-experiment.

Index a relative measure used as an indicator of the true state of nature.

Index monitoring an assessment protocol that collects data that usually represent at best a rough guess at population trends (and at worst may lead to an incorrect conclusion).

Index of relative abundance index of a proportional measure of the number of individuals within an area.

Index of relative density index of a proportional measure of the number of individuals per unit area.

Inferential monitoring an assessment protocol that uses unbiased or nearly unbiased estimators of spatial distribution and abundance/density that can be validly expanded to the entire area of interest for testing for trends.

Monitoring a repeated assessment of some quantity, attribute, or task for the purpose of detecting a change in average status within a defined area over time.

Nonrandom sampling a subjective choice of samples and associated sampling units. Nonrandom samples are not associated with a known or knowable sampling distribution; hence, statements of probability or certainty cannot be attached to the resulting estimates and valid inferences about the sampled and/or target populations cannot be made.

Nonresponse error   the component of nonsampling error that occurs if all elements included in a sample are not included in count results. This is a common occurrence in plot counts of animals. Incomplete detectability produces nonresponse error (cf. detectability).

Nonsampling error   bias affecting sample estimates whose source is outside of the sample selection process.

Null hypothesis   in statistical hypothesis testing, the option representing no difference or change. This is the hypothesis actually tested (cf. alternative hypothesis).

Observational study   a study in which the population of interest is sampled for the purpose of inference; any treatment grouping is usually done by the individuals themselves or by where they occur rather than having a treatment applied to them (but see quasi-experiment). A number of confounding factors could potentially affect the magnitude and direction of changes in the parameter of interest over time. In monitoring, for instance, an investigator can only properly report that a trend in population estimates was detected, and not the cause of the trend. One example is to conclude that a decline in population numbers in an area was due to a habitat change when in fact it could have been mainly due to some other, unknown factor such as an outbreak of a deadly disease (cf. experiment).

One-tailed test   a statistical test of a null hypothesis with a one-sided alternative hypothesis (e.g., x is greater than y).

Open population   a group of individuals whose number and composition are not fixed within a defined area and time period, i.e., there could be births, immigration, deaths, and/or emigration.

Parameter   an unknown numerical quantity or constant associated with some measure of a target population.

Permanent plot   a sampling unit that is continually surveyed during the course of a study, i.e., is not replaced by newly selected sampling units over time.

Plot   a sampling unit of some defined area or volume.

Population trend   an important average change in magnitude and direction of some population parameter within a specified area across multiple time intervals.

Power $(1 - \beta)$ the probability of detecting a statistically significant difference in a test of the null hypothesis given that this difference is present.

Precision   the degree of spread in estimates generated from repeated samples.

Probability distribution   the probability structure generated from all possible values of some random variable.

Process variation   a component of variation that arises from environmental and demographic processes affecting animal numbers and spatial distribution (i.e., spatial and temporal variation), which is distinct from variation from random selection of sampling units or counts of individuals.

Quadrat   a regularly shaped, four-sided plot.

Quasi-experiment a specific type of observational study in which some treatment or manipulation is applied to individuals or, more often, to habitats containing individuals, not under direct control of the investigator. That is, individuals are not randomly assigned to treatment groups, but rather treatments are randomly assigned to preexisting groups of individuals that may be influenced by a number of outside influences affecting the parameter of interest (i.e., those outside the control, and perhaps knowledge, of the investigator). Moreover, all potentially relevant influences cannot be entirely controlled, or even identified, by the investigator. As such, there is a potential for ecological and biological factors other than treatments to affect or confound results. Although a strict cause and effect relationship cannot be addressed by a quasi-experiment as in a true experiment, the strength of inference related to some cause is still much stronger than in a simple observational study in which there is no manipulation or treatment. That is, quasi-experiments address the question, "Why is a population declining?", but with less rigor (i.e., greater potential for incorrect conclusions) than a true experiment. However, this is usually the best that can be done in field situations. Quasi-experiments are frequently referred to as field experiments (cf. observational study, experiment).

Random sample a collection of sampling units chosen based on some known chance of selection. Random selection allows some probability or degree of certainty to be attached to resulting sample estimates in order to assess their usefulness.

Relative error (*e*) a measure of precision used in computing sample sizes, often represented as a percentage. Mathematically, it is the absolute difference between a parameter estimate and a parameter (e.g., $\hat{N} - N$) divided by the parameter (e.g., N). For example, one would set $e = 0.10$ to compute a sample size that, on average, would result in a parameter estimate within $\pm 10\%$ of the true parameter.

Response error the component of nonsampling error that comes about from mismeasuring or misrecording information associated with an element that has been detected during a survey.

Sample a group of sampling units selected during a survey.

Sampled population all elements associated with sampling units listed or mapped within the sampling frame.

Sampling design protocol for obtaining parameter estimates for a sampled population. The purpose of a sampling design is to make inferences to the sampled population, usually in conjunction with an observational study. In monitoring, a sampling design lays the groundwork for obtaining a set of parameter estimates for addressing the question, "Is there a trend?" A sampling design also may be used in an experimental-type approach for obtaining parameter estimates within treatment groups (cf. experimental design).

Sampling distribution the probability distribution of a sample estimate based on probability of occurrence of estimates generated by all possible samples of a given size.

Sampling frame a complete list or mapping of sampling units.

Sampling plan strategy for selecting plots and sampling estimators; a component of a sampling design.

Sampling unit a unique set usually of one or more elements, although in area sampling a sampling unit (e.g., plot of ground) may not contain any elements.

Sampling variation a measure of the degree of spread whose source is solely from random chance associated with the selection procedure (i.e., among-unit variation) and/or counting protocol (i.e., enumeration variation).

Sampling with partial replacement in area sampling, a design that uses a mixture of randomly selected permanent and temporary plots for surveying a population over time. This approach is preferred if the initial set of permanent plots either is consciously treated differently than other plots, which would yield biased estimates, or becomes radially different over time compared to unsampled plots. In the latter case, the estimator still would be unbiased, but a given sample estimate of trend would likely differ greatly from the true trend.

Selection bias persistent error arising from a nonrandom choice of sampling units, e.g., nonrandom sampling.

Spatial distribution geographical range of locations or areas occupied by a species.

Spatial variation a geographically based source of variation associated with

collecting survey data from more than one study area or sampling frame.

Standard deviation square root of variance of individual items in a probability distribution. In this case, "distribution" refers to either the true or population distribution, such as the distribution of all plot abundances, N_i (called the population standard deviation), or the distribution within a single sample, such as the distribution of items within a single plot sample (called the sample standard deviation or just "standard deviation"; cf. standard error).

Standard error square root of variance; the standard deviation of a sampling distribution of sample estimates. The "population standard error" describes this measure for a sampling distribution of all possible sample estimates. An estimator of this quantity, called the sample standard error (or just "standard error"), may be obtained from a single sample and, for infinite populations, is equal to the sample standard deviation divided by the square root of sample size. The standard error is especially useful for computing a confidence interval for a parameter estimate (cf. standard deviation).

Survey an incomplete count of individuals, objects, or items within a specific area and time period (cf. census).

Target population all elements representing the species of interest within some defined area and time period.

Temporal variation a numerical spread in parameter estimates due to collecting data from different time periods.

Temporary plot a sampling unit that is only surveyed during part of a study, i.e., it is replaced by a newly selected sampling unit for at least 1 time period. The extreme case is to choose a new sample of plots for each time period.

Trend an important change in average status of some quantity, attribute, or task within a defined time period (see population trend).

Two-tailed test a statistical test of a null hypothesis with a two-sided alternative hypothesis (e.g., x is not equal to y).

Type I error rejection of a null hypothesis when it is actually true.

Type II error failing to reject a null hypothesis when it is actually false.

Variance a measure of precision; average of squared differences between a set of values and the mean of the distribution of those values.

Appendix B

Glossary of Notation

This appendix contains notation we have used throughout this book. There is overlap in use of some of the following notation and therefore the reader should determine the appropriate definition based on the context in which the notation is used.

B.1. STANDARD NOTATION

A	area contained within a sampling frame or defined study unit
a	total area of sampling units within a sample (i.e., total area surveyed)
B	effect of initial capture on recapture probability under capture–recapture model M_{tb}
b	dispersion parameter in distance sampling
C	(1) proportionality constant
	(2) confidence (confidence interval)
	(3) total cost of conducting a survey or census
C_l	term contained in formula for the log transformation of a confidence interval for capture estimators
c	(1) cost of conducting counts on a sampling unit under a one-stage sampling design
	(2) recapture probability
c_0	overhead cost
c_1	cost of collecting data among sampling units in a two-stage sampling design
c_2	cost of collecting data within units in a two-stage sampling design
CI	confidence interval

CV   coefficient of variation
$\hat{CV}(\hat{D})$ estimator of the coefficient of variation of $\hat{D}$
$\hat{CV}(\hat{N})$ estimator of the coefficient of variation of $\hat{N}$
D density (a parameter)
$\hat{D}$ estimator of density
d_k horizontal distance from one end of the kth track to the other (line-intercept sampling)
df degrees of freedom
$E(\hat{N})$ expected value of $\hat{N}$
e relative error, such as $(\hat{N} - N)/N$; often given as a percentage
$f(x)$ probability density function of x
$f(y)$ probability density function of perpendicular distances (line transects) or detection distances (point transects) in distance sampling
$f(0)$ value of $f(y)$ at zero distance from a line transect (i.e., value on the line itself) in distance sampling
$g(y)$ detection function; probability of detection for an animal or object at distance y from a line or point transect in distance sampling
$g(0)$ probability that an animal or object exactly on a line or point is detected in distance sampling
g_k size of group associated with the kth track in line-intercept sampling
H number of strata
h identifier variable for individual strata; e.g., the number of animals or objects of interest in the hth stratum would be denoted N_h
$h(0)$ slope of the probability density function of detection distances at zero distance from a point transect (i.e., value at the point itself) in distance sampling
i general identifier variable; often used to denote individual sampling units; e.g., the number of animals or objects of interest in the ith unit is represented by N_i
j general identifier variable; often used to denote individual subunits such as N_{ij}, which is the number of animals in the jth subunit of the ith unit
k (1) clumping parameter in the negative binomial distribution
                      (2) identifier variable for tracks in line-intercept sampling
k_s sampling unit interval for systematic sampling (i.e., 1-in-k_s)
k_s^* sampling unit interval in repeated systematic sampling (also the total number of systematic samples available)

L	length of entire line transect surveyed in distance sampling
$\mathcal{L}$	symbol for the likelihood function
L_0	length of line transect surveyed in a pilot study in distance sampling
L_L	lower limit of a confidence interval
l_i	length of ith line transect segment in distance sampling
ln	natural logarithm
M_0	capture–recapture model that assumes all animals have the same initial capture probability on each occasion and marked animals have the same capture probability as those not yet marked
M_t	capture–recapture model that allows each animal to have the same capture probability within a given capture occasion, although this probability may be different among occasions (i.e., capture probabilities may change over *time*)
M_b	capture–recapture model that allows for capture probabilities to change due to *behavioral* responses to marking within each capture occasion, although initial capture probabilities are assumed to be the same on all occasions
M_h	capture–recapture model that allows each animal to have its own capture probability (i.e., *heterogeneity*), although this probability does not change after capture or among occasions
M_{bh}	capture–recapture model that allows each animal to have its own capture probability that is constant across occasions, but that changes from the initial capture
M_{th}	capture–recapture model that allows each animal to have its own capture probability, which can change over time
M_{tb}	capture–recapture model that allows for capture probabilities to change after initial capture within a given occasion and over time
M_{tbh}	capture–recapture model for which no estimator currently exists; combines the characteristics of models M_t, M_b, and M_h
MNA	minimum number of animals known alive
m	number of repetitions in a ranked set sample
m_2	number of marked animals caught during second capture occasion
m_j	number of sampling units containing animals or objects of interest in the jth network in adaptive cluster sampling
N	number of animals or objects of interest (i.e., abundance) in a sampling frame or defined area (a parameter)

$\hat{N}$ — estimator of abundance within a defined area for a given time period

$\overline{N}$ — sample mean number of animals or objects of interest from complete counts on selected units

$\overline{N}_{\text{true}}$ — mean number of animals or objects of interest over all sampling units (i.e., population mean)

$\hat{\overline{N}}$ — estimator of $\overline{N}$, i.e., based on incomplete counts on selected units

$\overline{N}*$ — sample mean number of animals or objects of interest from complete counts on subsampled units in double sampling

$\hat{\overline{N}}*$ — estimator of $\overline{N}*$, i.e., based on incomplete counts of subsampled units, in double sampling

$\hat{N}_j$ — estimator of number of animals or objects of interest in the jth systematic sample

$\overline{N}_j$ — sample mean number of animals or objects of interest in the jth systematic sample

$N_{j[i]}$ — number of animals or objects of interest within the jth ranked plot selected in the ith repetition in ranked set sampling

n — (1) number of detections during the regular survey in distance sampling
(2) number of trials in the probability density function for the binomial distribution
(3) number of observations or time periods
(4) number of rows or columns (simple Latin square sample +1)

n_s — number of distinct samples selected (e.g., systematic samples)

n_0 — number of detections during pilot survey in distance sampling

n_1 — number of animals captured during initial capture occasion in capture–recapture

n_2 — number of animals captured during second capture occasion in capture–recapture

P — P value in statistical hypothesis testing, i.e., the probability of observing, under the null hypothesis, a value as extreme or more extreme than the computed value

p — (1) probability that some event occurs as defined by the context of its usage, such as the probability of capture or the probability of detection
(2) proportion of sampling units containing the animals or objects of interest (i.e., presence–absence survey)

$\hat{p}$ — estimator of p

$p(s)$ — probability of selection for a given set of sampling units

q	number of transects surveyed in each systematic sample in line-intercept sampling
R	variance inflation constant
r	rate constant, such as the rate of decline of a population
S	Mann–Kendall statistic
S_j	jth systematic sample (line-intercept sampling)
$\hat{S}^2_{N_i}$	sample variance for a one-stage sampling design; the N_i subscript denotes a complete count on the ith selected plot
$\hat{S}^2_{\hat{N}_i}$	sample variance for a two-stage sampling design; the $\hat{N}_i$ subscript denotes an incomplete count on the ith selected plot
$\hat{S}_{N_i}$	sample standard deviation for a one-stage sampling design
$\hat{S}_{\hat{N}_i}$	sample standard deviation for a two-stage sampling design
SD	standard deviation
SE	standard error
$\hat{SE}(\hat{D})$	estimator of the standard error of $\hat{D}$
$\hat{SE}(\hat{N})$	estimator of the standard error of $\hat{N}$
T_j	deviation from the mean for the jth capture occasion
t	(1) t value
	(2) time
t^*_b	bootstrap t value
U	number of sampling units in a sampling frame
U_L	upper limit of a confidence interval
u	number of sampling units in a sample
V	number of subunits within each sampling unit in two-stage cluster sampling
V_T	number of subunits within a sampling frame
v	number of subunits randomly selected from each selected unit in two-stage cluster sampling
v_T	number of subunits included in a sample
$\hat{Var}(\hat{D})$	estimator of variance of $\hat{D}$
$Var(N)$	variation in number of animals or objects of interest among all sampling units (i.e., among-unit variance)
$Var(\hat{N})$	variance of incomplete counts among all sampling units (i.e., enumeration variance)
$\hat{Var}(\hat{N})$	estimator of variance of total number of animals or objects both among and within sampling units
$\hat{Var}(\hat{N}_i/N_i)$	estimator of enumeration variance within the ith selected unit
$\overline{\hat{Var}(\hat{N}_i/N_i)}$	sample mean of the estimator of enumeration variance of the i selected units
W	width of a study area or sampling unit in line-intercept sampling

w	(1) strip width of a line transect survey
	(2) weighting variable
w_i	average of the counts within the network containing the ith initially selected unit
X	predictor variable in regression
X_{ij}	matrix of capture histories
x	(1) number of successes in n trials in the probability density function for the binomial distribution
	(2) the horizontal axis (i.e., abscissa)
Y	number of sampling units containing animals or objects of interest
$\hat{Y}$	estimator of Y
Y_i	notation for whether an animal or object of interest is present ("1") or absent ("0") on the ith selected unit
y	the vertical axis (i.e., ordinate)
z	(1) z value
	(2) z test

B.2. GREEK NOTATION

α	probability of a type I error
β	probability of a type II error
β_0	y intercept in regression
β_1	slope of the line in regression
ε_i	residual error in regression
μ	(1) population mean of a normal distribution
	(2) mean capture probability across all occasions for initial captures
Π	multiplication operator
π	pi, approximately equal to 3.14159265
Σ	summation operator
$\hat{\sigma}_i^2$	estimator of sampling variance for time i
$\hat{\sigma}_{time}^2$	estimator of temporal variance
$\hat{\sigma}_{total}^2$	estimator of total variance
τ_{jhi}	total count of animals or objects of interest in stratum j for the network containing the ith initially selected unit in stratum h (stratified adaptive cluster sample)
χ^2	chi squared
ψ_i	network that contains unit i in adaptive cluster sampling

Appendix C

Sampling Estimators

This appendix contains a list of selected one-stage estimators for calculating estimates of the proportion of sampling units occupied (p) as well as one-stage and two-stage estimators[1] for number of individuals (N) for some species of interest in a defined area and time period. We refer interested readers to Cochran (1977), Gilbert (1987), and Sukhatme et al. (1984) for three-stage estimators.

This appendix follows the same notation presented throughout this book and in Appendix B; hence, notation used in formulas has been modified from the cited sources. We list only variance estimators because standard error can be computed from the square root of the variance. Estimators for a single systematic sample are the same as for a simple random sample. Also, data on proportions are assumed to be collected from equal-sized sampling units. Finally, sample size formulas are presented in terms of parameters; estimators based on previous data or knowledge should be substituted for the appropriate parameters.

---

[1] Our broad definitions of one- and two-stage samples follow those of Hankin (1984) and Skalski (1994), which were adapted for fish and wildlife applications. Conversely, statistical sampling texts strictly define a one-stage sample as a random selection of plots that have complete counts conducted on them, and a two-stage sample strictly as a two-stage cluster sample.

C.1. PROPORTIONS—ONE-STAGE ESTIMATORS

C.1.1. SIMPLE RANDOM SAMPLE

Estimators for the proportion of sampling units occupied by some species of interest are

$$\hat{Y} = \sum_{i=1}^{u} Y_i,$$

$$\hat{p} = \frac{\hat{Y}}{u}$$

and

$$\hat{Var}(\hat{p}) = \left(1 - \frac{u}{U}\right)\frac{\hat{p}(1 - \hat{p})}{u - 1},$$

where Y_i takes the value of either 1 (present) or 0 (absent) for the ith selected unit, $\hat{Y}$ is the number of surveyed units that have at least 1 individual of interest (an estimator of the total number of occupied units), u is number of units in the sample, and U is the total number of units in the sampling frame (Cochran, 1977, pp. 50–52). The formula (Cochran, 1977, p. 76) for computing sample size is

$$u = \frac{\dfrac{t_{1-\alpha/2,u-1}^2(1 - p)}{e^2 p}}{1 + \dfrac{1}{U}\left(\dfrac{t_{1-\alpha/2,u-1}^2(1 - p)}{e^2 p} - 1\right)},$$

where e is relative error and t is the critical value associated with a 2-tailed (i.e., $1 - \alpha/2$) Student's t distribution based on $u - 1$ degrees of freedom. The t value must be solved iteratively (we do not know the value of u for the degrees of freedom) by starting with the value for $z_{1-\alpha/2}$ and repeatedly solving for u until a stable value is reached. This usually takes about three or four iterations. After we have estimated the sample size needed to satisfy what percentage (e) of the true number we wish to be within at a specified α level, we can compute the projected total costs C of a given survey using $C = c_0 + uc$, where c_0 is overhead cost and c is the cost for collecting data on the plots.

C.1.2. STRATIFIED RANDOM SAMPLE

Formulas (Scheaffer *et al.*, 1990, p. 117) for calculating estimates of proportions under a stratified random sampling design require an added "*h*" subscript to denote which stratum the estimator is associated with (e.g., Y_{hi} is the presence–absence value of the *i*th sampled unit in stratum *h*), and *H* denotes number of strata

$$\hat{Y}_h = \sum_{i=1}^{u_h} Y_{hi},$$

$$\hat{p}_h = \frac{\hat{Y}_h}{u_h},$$

$$\hat{p} = \frac{\sum_{h=1}^{H} U_h \hat{p}_h}{U}$$

and

$$\hat{\text{Var}}(\hat{p}) = \frac{1}{U^2} \sum_{h=1}^{H} U_h^2 \left(1 - \frac{u_h}{U_h}\right) \left[\frac{\hat{p}(1 - \hat{p})}{u_h - 1}\right].$$

To calculate the presumed optimum sample size necessary to be within some percentage of the true proportion about $(1 - \alpha)\%$ of the time, we would use

$$u = \frac{\left(\sum_{h=1}^{H} \frac{U_h}{U} \sqrt{p_h(1 - p_h)}\right)^2}{\left(\frac{ep}{t_{1-\alpha/2,u-1}}\right)^2 + \frac{1}{U} \sum_{h=1}^{H} \frac{U_h}{U} [p_h(1 - p_h)]},$$

where *e* and $t_{1-\alpha/2,u-1}$ are defined as before (adapted from Cochran, 1977, p. 110), and $t_{1-\alpha/2,u-1}$ must be solved iteratively. The formula for the sample size needed for each stratum at a minimum variance and a fixed cost is

$$u_h \doteq u \left(\frac{U_h \sqrt{\frac{p_h(1 - p_h)}{c_h}}}{\sum_{h=1}^{H} U_h \sqrt{\frac{p_h(1 - p_h)}{c_h}}}\right),$$

where c_h is the cost of collecting data in the *h*th stratum (Cochran, 1977, p. 109). Total survey costs may be computed from $C = c_0 + \sum_{h=1}^{H} u_h c_h$ (Cochran, 1977, p. 96).

C.2. ABUNDANCE ESTIMATORS—EQUAL-SIZED SAMPLING UNITS

This section lists sampling estimators most commonly used with one- and two-stage designs in which sampling units either are equal sized or are treated as equal sized. A density estimator is easily converted from abundance using the relationship, $\hat{D} = \hat{N}/A$, where A is the area contained within the sampling frame or defined boundary. Its variance estimator can be obtained from $\hat{\mathrm{Var}}(\hat{D}) = \hat{\mathrm{Var}}(\hat{N})/A^2$. Therefore, we only list abundance estimators. The one-stage estimators correspond to the situation in which a sample of units is chosen and a complete count is conducted on each of these units. We do not include estimators for the other one-stage scenario (as we defined it), namely, conducting an incomplete count over the entire study area (e.g., aerial mark–resight study). Note that notation defined in section C.1, such as u, U, e, and $t_{1-\alpha/2,u-1}$, will be used without further definition.

C.2.1. ONE-STAGE ESTIMATORS

C.2.1.1. Simple Random Sample

Formulas for calculating abundance estimates for a simple random are

$$\overline{N} = \frac{\sum\limits_{i=1}^{u} N_i}{u},$$

$$\hat{N} = U \times \overline{N},$$

$$\hat{\mathrm{Var}}(\hat{N}) = U^2 \left[\left(1 - \frac{u}{U} \right) \frac{\hat{S}_{N_i}^2}{u} \right]$$

and

$$\hat{S}_{N_i}^2 = \frac{\sum\limits_{i=1}^{u} (N_i - \overline{N})^2}{u - 1}.$$

We define N_i as the total count of animals on the ith selected plot, $\overline{N}$ as the average of the N_i, $\hat{N}$ as the estimator of the total number of animals in the sampling frame, $\hat{\mathrm{Var}}(\hat{N})$ as the estimator of variance of $\hat{N}$, and $\hat{S}_{N_i}^2$

as the estimator of the sample variance of the N_i. The following equation (Thompson, 1992, p. 32) is used to calculate sample size (u),

$$u = \frac{1}{\left(\dfrac{eN}{t_{1-\alpha/2,u-1}US_{N_i}}\right)^2 + \dfrac{1}{U}}.$$

The cost function is the same as presented in section C.1.1.

C.2.1.2. Stratified Random Sample

Estimators (Cochran, 1977, pp. 90–93) for a stratified random sample, with either a simple random sample or single systematic sample within each stratum, are

$$\overline{N}_h = \frac{\sum\limits_{i=1}^{u_h} N_{hi}}{u_h},$$

$$\hat{N}_h = U_h \times \overline{N}_h,$$

$$\text{Var}(\hat{N}_h) = U_h^2 \left[\left(1 - \frac{u_h}{U_h}\right) \frac{\hat{S}_{N_{hi}}^2}{u_h} \right]$$

and

$$\hat{S}_{N_{hi}}^2 = \frac{\sum\limits_{i=1}^{u_h} (N_{hi} - \overline{N}_h)^2}{u_h - 1}.$$

To obtain estimates of overall abundance and variance, we simply sum estimators across strata (H), i.e., $\hat{N} = \sum_{h=1}^{H} \hat{N}_h$ and $\text{Var}(\hat{N}) = \sum_{h=1}^{H} \text{Var}(\hat{N}_h)$. As stated previously, an additional h subscript is required to denote particular strata. Estimators under a stratified, repeated systematic sampling design are obtained by adding an h subscript to estimators listed in section C.2.1.3. The formula (Cochran, 1977, p. 106) for a presumed optimum sample size for a fixed u is

$$u = \frac{\left(\sum\limits_{h=1}^{H} U_h S_{N_{hi}}\right)^2}{\left(\dfrac{eN}{t_{1-\alpha/2,u-1}}\right)^2 + \sum\limits_{h=1}^{H} U_h S_{N_{hi}}^2}.$$

Three ways of allocating the estimated sample size among strata include proportional, Neyman, and optimum allocations. Proportional allocation simply assigns the number of selected units based on relative size of each stratum, i.e., $u_h = uU_h/U$. The formula for Neyman allocation (after Cochran, 1977, p. 98), which accounts for variation among strata, is

$$u_h = u \left(\frac{U_h S_{N_{hi}}}{\sum\limits_{h=1}^{H} U_h S_{N_{hi}}} \right).$$

The formula (after Cochran, 1977, p. 98) for optimum allocation, which balances stratum variation and sampling costs, is

$$u_h = u \left(\frac{\dfrac{U_h S_{N_{hi}}}{\sqrt{c_h}}}{\sum\limits_{h=1}^{H} \dfrac{U_h S_{N_{hi}}}{\sqrt{c_h}}} \right).$$

The cost function used to compute total costs is the same one as in section C.1.2.

C.2.1.3. Repeated Systematic Samples with Random Starts

Estimators for repeated systematic sampling are similar to those for simple random sampling except they are based on multiple samples, each with a random starting point, rather than multiple observations within only one sample; hence, notation must be modified (Scheaffer *et al.*, 1990, pp. 221–222) to

$$\overline{N}* = \frac{\sum\limits_{j=1}^{n_s} \overline{N}_j}{n_s},$$

$$\hat{N} = U \times \overline{N}*$$

and

$$\text{Var}(\hat{N}) = U^2 \left(1 - \frac{u}{U}\right) \frac{\sum\limits_{j=1}^{n_s} (\overline{N}_j - \overline{N}*)^2}{n_s(n_s - 1)},$$

where $\overline{N}_j$ is the arithmetic mean of the jth systematic sample, n_s is the number of systematic samples selected, and $\overline{N}^*$ is the overall mean across all systematic samples chosen. The sampling interval for a repeated systematic sample (k_s^*) is obtained by multiplying the number of systematic samples selected by the sampling interval for a single systematic sample, i.e., $k_s^* = n_s k_s$. The formula (Levy and Lemeshow, 1991, p. 91) for calculating the number of systematic samples required is

$$n_s = \frac{\dfrac{U}{u}}{\left(\dfrac{e\overline{N}^*}{t_{1-\alpha/2,u-1}}\right)^2 \left(\dfrac{\dfrac{U}{u}-1}{\sum\limits_{j=1}^{n_s'} (\overline{N}_j - \overline{N}^*)^2/(n_s'-1)}\right) + 1},$$

where n_s' is the number of systematic samples used in the pilot study. The cost equation would be $C = c_0 + c_i$, where c_i is the cost of the ith systematic sample.

C.2.1.4. Ranked Set Sampling

Ranked set sampling was explained in detail in Chapter 2. The mean abundance estimator (Stokes, 1986, p. 585) is

$$\overline{N} = \frac{\sum\limits_{i=1}^{m} \sum\limits_{j=1}^{n_s} N_{[j]i}}{mn_s},$$

where $N_{[j]i}$ is the count within the jth ranked plot selected in the ith repetition. Abundance is calculated as before, i.e., $\hat{N} = U \times \overline{N}$. The formula for sample variance (Stokes, 1986, p. 587) is

$$\hat{S}_{N_{[j]i}}^2 = \frac{\sum\limits_{i=1}^{m} \sum\limits_{j=1}^{n_s} (N_{[j]i} - \overline{N})^2}{mn_s - 1}.$$

The sample variance then is inserted into the usual variance estimator, $\hat{\text{Var}}(\hat{N})$, as presented in section C.2.1.1. There is no formula for calculating sample size under this design.

C.2.1.5. Adaptive Cluster Sampling

Adaptive cluster sampling was described in Chapter 2. We present estimators for simple random and stratified random designs. We refer interested readers to Thompson (1992) and Thompson and Seber (1996) for other designs. There are no sample size formulas for these designs.

Simple Adaptive Cluster Sample

The formula (Thompson, 1992, p. 271) for w_i is

$$w_i = \frac{\sum\limits_{j \in \psi_i} N_j}{m_i},$$

where ψ_i is the network that contains unit i, m_i is the number of units in that network, and N_j is the count of the jth unit in that network. The formula (Thompson, 1992, p. 271) for $\overline{N}$ is

$$\overline{N} = \frac{\sum\limits_{i=1}^{u} w_i}{u},$$

where u is the number of initially chosen sampling units, and $\hat{N} = U \times \overline{N}$. The variance estimator for $\overline{N}$ (Thompson, 1992, p. 271) is

$$\hat{\mathrm{Var}}(\hat{N}) = \left(1 - \frac{u}{U}\right) \frac{\sum\limits_{i=1}^{u} (w_i - \overline{N})^2}{u(u-1)}$$

and $\hat{\mathrm{Var}}(\hat{N}) = U^2 \times \hat{\mathrm{Var}}(\overline{N})$.

Stratified Adaptive Cluster Sample

The formula (Thompson, 1992, p. 306) for w_{hi} is

$$w_{hi} = \frac{\left(\dfrac{u_h}{U_h}\right) \sum\limits_{j=1}^{H} \tau_{jhi}}{\sum\limits_{j=1}^{H} \dfrac{u_j}{U_j} (m_{jhi})},$$

where u_h is the number of initially selected plots in stratum h, U_h is the total number of plots in stratum h, τ_{jhi} is the total count within stratum j for the network containing the ith initially selected plot in stratum h (strata are additionally referenced with a j in case 1 or more plots with animals within the ith network are inside a different stratum than the hth stratum containing the initially selected plot), u_j is number of initially selected plots in the jth stratum, U_j is the total number of plots in stratum j, and m_{jhi} is the number of plots that contribute to the τ_{jhi}. The estimator of total number of animals within each stratum (Thompson, 1992, p. 307) is

$$\hat{N}_h = \frac{U_h}{u_h} \left(\sum_{i=1}^{u_h} w_{hi} \right)$$

and the estimated overall abundance is simply the sum of the $\hat{N}_h$ as in other stratified designs. The variance estimator for $\hat{N}_h$ is defined in terms of w_{hi} (Thompson, 1992, p. 307)

$$\hat{\text{Var}}(\hat{N}_h) = U_h^2 \left(1 - \frac{u_h}{U_h} \right) \frac{\hat{S}^2_{w_{hi}}}{u_h},$$

where

$$\hat{S}^2_{w_{hi}} = \frac{\sum_{i=1}^{u_h} (w_{hi} - \overline{w}_h)^2}{u_h - 1}$$

and $\overline{w}_h$ is the sum of the w_{hi} divided by u_h. Again, overall estimates of abundance and associated variance are obtained by summing over stratum estimates.

C.2.2. Two-Stage Estimators

In this section, we list selected estimators that are used in designs that have both among-unit and within-unit (i.e., enumeration) levels of variance. There are two different scenarios in which within-unit variances arise. The first is the standard scenario presented in statistical texts in which chosen sampling units are partitioned into subunits and then a random sample of subunits is selected. Complete counts must be conducted on selected subunits for estimators to be unbiased. The second scenario is where methods that account for incomplete detectability of individuals are used on selected units (e.g., Hankin, 1984; Skalski, 1994). An example of the latter is to use capture–recapture estimators to obtain abundance and variance estimates on selected units.

C.2.2.1. Complete Counts within Randomly Selected Subunits of Chosen Units

Conducting complete counts on a sample of subunits within selected units often is called a two-stage cluster sample. There are few instances in studies of fish and wildlife populations where two-stage cluster sampling applies, but we will present estimators for simple random sampling and stratified random sampling as references. A more extensive list of estimators is available in Sukhatme *et al.* (1984).

Simple Random Sample at Both Stages

Abundance and variance estimators (Cochran, 1977, pp. 276–278) for designs that first select a simple random sample of units and then a simple random sample of subunits within those units are

$$\overline{N}_i = \frac{\sum_{j=1}^{v} N_{ij}}{v},$$

$$\hat{N}_i = V \times \overline{N}_i,$$

$$\hat{\overline{N}} = \frac{\sum_{i=1}^{u} \hat{N}_i}{u},$$

$$\hat{N} = U \times \hat{\overline{N}},$$

$$\widehat{\mathrm{Var}}(\hat{N}) = U^2 V^2 \left[\left(1 - \frac{u}{U}\right) \frac{\hat{S}_{\hat{N}_i}^2}{u} + \frac{u}{U} \left(1 - \frac{v}{V}\right) \frac{\hat{S}_{N_{ij}}^2}{uv} \right],$$

$$\hat{S}_{\hat{N}_i}^2 = \frac{\sum_{i=1}^{u} (\hat{N}_i - \hat{\overline{N}})^2}{u - 1},$$

and

$$\hat{S}_{N_{ij}}^2 = \frac{\sum_{i=1}^{u} \sum_{j=1}^{v} (N_{ij} - \overline{N}_i)^2}{u(v - 1)},$$

where N_{ij} is the estimator for the number of animals on the jth subunit within the ith selected unit, $\overline{N}_i$ is the estimator for the average count of animals among censused subunits in the ith selected unit, $\hat{N}_i$ is the estimator of the number of animals within the ith selected unit, $\hat{\overline{N}}$ is the sample mean of incomplete plot counts, v is the number of subunits sampled from each selected unit, and V is the number of subunits within every sampling unit.

To compute sample size we use the formula (adapted from Levy and Lemeshow, 1991, p. 225),

$$u = \frac{\dfrac{U^2 S_{N_i}^2}{N^2} + \dfrac{1}{v}\left(1 - \dfrac{v}{V}\right)\dfrac{V^2 S_{N_{ij}}^2}{N_i^2}}{\dfrac{e^2}{t_{1-\alpha/2,u-1}^2} + \dfrac{U S_{N_i}^2}{N^2}},$$

where the optimum number of subunits (Cochran, 1977, p. 289) is

$$v = \frac{S_{N_{ij}}\sqrt{c_1/c_2}}{\sqrt{(S_{N_i}^2 - S_{N_{ij}}^2)/V}}.$$

The associated cost function (Cochran, 1977, p. 289) is $C = c_1 u + c_2 uv$, where c_1 is the cost associated with sampling at the unit level and c_2 is the cost at the subunit level.

Stratified Random Sample of Units and Simple Random Sample of Subunits

The estimators for a stratified random sample of units would be the same as above, with the addition of an h subscript, except for the sample size equations

$$\left(\frac{eN}{z_{1-\alpha/2}}\right)^2 = \sum_{h=1}^{H} \frac{1}{u_h}\left[U_h V_h^2(U_h - u_h)S_{N_{hi}}^2 + \frac{U_h^2 V_h(V_h - v_h)S_{N_{hij}}^2}{v_h}\right]$$

and the cost equation (Cochran, 1977, p. 289) $C = \sum_{h=1}^{H} c_{1h}u_h + \sum_{h=1}^{H} c_{2h}u_h v_h$, where these terms are defined in relation to strata. Estimates of overall abundance and variance are obtained by summing over the estimates from each stratum as described before.

C.2.2.2. Incomplete Counts over Entire Area of Selected Units

The common situation in animal population studies is to have a two-stage design in which methods that correct for incomplete detectability are used on selected units (e.g., capture–recapture methods conducted within each chosen unit). We list estimators for designs that use simple random sampling and stratified random sampling for selecting units. Estimators for relevant adaptive cluster designs were presented in Thompson and Seber (1996). Within-unit estimates are obtained from abundance ($\hat{N}_i$) and variance ($\hat{V}ar(\hat{N}_i/N_i)$) estimators associated with the particular method used to survey each unit. Estimators for average abundance across units (i.e.,

$\overline{\hat{N}} = (\Sigma_{i=1}^{u} \hat{N}_i) \div u)$, and overall abundance (i.e., $\hat{N} = U \times \overline{\hat{N}}$) remain as before.

Simple Random Sample of Units

Variance estimators (Skalski, 1994, p. 194) under a design with a simple random sample of units at the first stage are

$$\hat{\text{Var}}(\hat{N}) = U^2 \left[\left(1 - \frac{u}{U} \right) \frac{\hat{S}_{\hat{N}_i}^2}{u} + \frac{\overline{\hat{\text{Var}}(\hat{N}_i/N_i)}}{U} \right],$$

$$\hat{S}_{\hat{N}_i}^2 = \frac{\sum\limits_{i=1}^{u} (\hat{N}_i - \overline{\hat{N}})^2}{u - 1}$$

and

$$\overline{\hat{\text{Var}}(\hat{N}_i/N_i)} = \frac{\sum\limits_{i=1}^{u} \hat{\text{Var}}(\hat{N}_i/N_i)}{u}.$$

We use the following formula (Skalski, 1994, p. 199) to compute sample size,

$$u = \frac{U(S_{\hat{N}_i}^2 + \overline{\text{Var}(\hat{N}_i/N_i)})}{U \left(\dfrac{e\overline{N}}{t_{1-\alpha/2, u-1}} \right)^2 + S_{\hat{N}_i}^2}.$$

Note that $S_{\hat{N}_i}^2$ is estimated by $\hat{S}_{\hat{N}_i}^2 = [\Sigma_{i=1}^{u} (\hat{N}_i - \overline{\hat{N}})^2/(u - 1)] - \hat{\text{Var}}(\hat{N}_i/N_i)$. The cost function would be modified from the one listed under simple random sampling for a two-stage cluster sample to $C = c_1 u + c_2 u$, where c_2 now is defined as the cost associated with obtained abundance estimates within each selected unit (i.e., there is no subunit cost). Skalski (1985) presented cost functions for conducting capture–recapture surveys.

Stratified Random Sample of Units

As with other stratified designs, we simply add an h subscript to terms and estimators presented under simple random sampling except for the sample size and cost equations, which require a bit more modification. As always, overall abundance and variance estimates are obtained by simply adding up the individual stratum estimates. The variance estimator for each stratum is

$$\hat{\text{Var}}(\hat{N}_h) = U_h^2 \left[\left(1 - \frac{u_h}{U_h} \right) \frac{\hat{S}^2_{\hat{N}_{hi}}}{u_h} + \frac{\overline{\hat{\text{Var}}(\hat{N}_{hi}/N_{hi})}}{U_h} \right],$$

where

$$\hat{S}^2_{\hat{N}_{hi}} = \frac{\sum_{i=1}^{u_h} (\hat{N}_{hi} - \overline{\hat{N}}_h)^2}{u_h - 1}$$

and

$$\overline{\hat{\text{Var}}(\hat{N}_{hi}/N_{hi})} = \frac{\sum_{i=1}^{u_h} \hat{\text{Var}}(\hat{N}_{hi}/N_{hi})}{u_h}.$$

The sample size formula (Skalski, 1994, p. 200) is

$$\left(\frac{eN}{z_{1-\alpha/2}} \right)^2 = \sum_{h=1}^{H} \left[\frac{U_h^2 \left(1 - \frac{u_h}{U_h} \right) S^2_{N_{hi}} + \overline{\text{Var}(\hat{N}_{hi}/N_{hi})}}{u_h} \right],$$

where $S^2_{N_{hi}}$ is estimated by $\hat{S}^2_{\hat{N}_{hi}} = [\sum_{i=1}^{u_h} (\hat{N}_{hi} - \overline{\hat{N}}_h)^2/(u_h - 1)] - \hat{\text{Var}}(\hat{N}_{hi}/N_{hi})$. The sample size equation may be solved by using an optimization procedure, which is available in many spreadsheet programs. The cost equation is modified from the one presented for a stratified design under a two-stage cluster sample to

$$C = \sum_{h=1}^{H} c_{1h}u_h + \sum_{h=1}^{H} c_{2h}u_h,$$

where c_{2h} is the cost of obtaining estimates within selected units in the hth stratum.

C.3. ABUNDANCE ESTIMATORS—UNEQUAL-SIZED SAMPLING UNITS

Estimators based on equal-sized units can be imprecise when applied to situations in which sampling units vary greatly in size and animal numbers are strongly correlated with unit size. In such situations, it is better to use estimators that account for varying sizes of units, i.e., sampling with probability proportional to size. Two approaches to defining and delineating unequal-sized sampling units are to: (1) partition units into some number of equal-sized subunits and base probabilities of selection on relative numbers of subunits; or (2) use the area contained within a given unit to determine its probability of selection relative to areas of other sampling units, i.e., using a Horvitz–Thompson estimator (Horvitz and Thompson, 1952). The second option is usually more applicable to fish and wildlife population studies because partitioning units into subunits implies that complete counts are feasible within subunits, which is not often true. However, we offer estimators for both situations. A more complete review of different estimators for the first situation is given by Sukhatme *et al.* (1984). Sample size formulas for two-stage cluster samples with units of unequal sizes are presented in Singh and Chaudhary (1986).

C.3.1. ONE-STAGE DESIGNS

One-stage designs discussed in this section assume that an initial sample of units is randomly chosen and then complete counts are conducted within each selected unit. We list estimators for both simple random and stratified random sampling.

C.3.1.1. Partitioning Sampling Units into Subunits

The probability of selecting a sampling unit when each has been partitioned into equal-sized subunits is v_i/V_T, where v_i is the number of subunits within the ith sampling unit, and V_T is the total number of subunits in the sampling frame. Associated abundance and variance estimators (Scheaffer, 1990, p. 270) under simple random sampling are

$$\hat{N} = \frac{V_T}{u} \sum_{i=1}^{u} \frac{N_i}{v_i}$$

and

$$\hat{Var}(\hat{N}) = \frac{V_T^2}{u} \left(\frac{\sum_{i=1}^{u} (\overline{N}_i - \overline{N})^2}{u - 1} \right),$$

where $\overline{N}_i = N_i/v_i$ and $\overline{N} = \sum_{i=1}^{u} \overline{N}_i/u$. Estimators for stratified random sampling would have an additional h subscript and stratum estimators would be added together to obtain overall estimates.

C.3.1.2. Sampling Units Not Partitioned

One may assign probabilities of selection based on the area of a given sampling unit relative to all other units. Abundance and variance estimators based on this approach (Horvitz and Thompson, 1952, pp. 669–670) for simple random sampling are

$$\hat{N} = \sum_{i=1}^{u} \frac{N_i}{p_i}$$

and

$$\text{V}\hat{\text{a}}\text{r}(\hat{N}) = \sum_{i=1}^{u} \left(\frac{1 - p_i}{p_i^2} \right) N_i^2 + \sum_{i=1}^{u} \sum_{j \neq i} \left(\frac{p_{ij} - p_i p_j}{p_i p_j} \right) \frac{N_i N_j}{p_{ij}},$$

where p_i is the probability (based on size) that the ith unit will be included in the sample, p_j is the probability that the jth unit will be included in the sample, and p_{ij} is the probability that both the ith and jth units are included in the sample. Estimators for stratified designs directly follow as described before.

C.3.2. TWO-STAGE DESIGNS

Two-stage designs based on unequal-sized units assume an incomplete count within each selected unit. This may be accomplished from either a complete count obtained from a random sample of equal-sized subunits or an incomplete count (i.e., corrected for incomplete detectability) within a portion of selected units.

C.3.2.1. A Random Sample of Subunits from Chosen Units

Appropriate abundance and variance estimators (Levy and Lemeshow, 1991, pp. 246–247) are

$$\hat{N} = \left(\frac{U}{u} \right) \sum_{i=1}^{u} \frac{V_i}{v_i} \sum_{j=1}^{v_i} N_{ij}$$

and

$$\text{V\^ar}(\hat{N}) = \left(\frac{U^2\overline{V}}{u\overline{v}}\right)\left(1 - \frac{v_T}{V_T}\right)\left[\frac{\sum\limits_{i=1}^{u} (\hat{N}_i - \hat{\overline{N}})^2}{u - 1}\right],$$

where $\overline{v}$ is the average number of subunits sampled from each unit (i.e., $\overline{v} = \sum_{i=1}^{u} v_i/u$), $\overline{V}$ is the average number of subunits per unit over the entire sampling frame (i.e., $\overline{V} = \sum_{i=1}^{U} v_i/U$), and v_T is the total number of subunits included in the sample. All other terms are defined as before. Stratified estimators follow as described previously.

C.3.2.2. Incomplete Counts over Part of Selected Units

Skalski (1994, p. 196) presented the following estimators for two-stage designs in which incomplete counts were conducted in selected units that are not partitioned into subunits,

$$\hat{N} = \sum_{i=1}^{u} \frac{\hat{N}_i}{p_i}$$

and

$$\text{V\^ar}(\hat{N}) = \sum_{i=1}^{u} \left(\frac{1 - p_i}{p_i^2}\right)\hat{N}_i^2 + 2\sum_{i=1}^{u}\sum_{j>1}^{u}\left(\frac{p_{ij} - p_i p_j}{p_i p_j}\right)\frac{\hat{N}_i \hat{N}_i}{p_{ij}} + \sum_{i=1}^{u}\frac{\text{V\^ar}(\hat{N}_i)}{p_i}.$$

Estimators for a stratified random sampling design would have additional h subscripts and would be summed across strata to obtain overall abundance and variance estimates.

LITERATURE CITED

Cochran, W. G. (1977). *Sampling Techniques,* third ed. Wiley, New York.
Gilbert, R. O. (1987). *Statistical Methods for Environmental Pollution Monitoring.* Van Nostrand Reinhold, New York.
Hankin, D. G. (1984). Multistage sampling designs in fisheries research: Applications in small streams. *Can. J. Fish. Aquat. Sci.* **41:** 1575–1591.
Horvitz, D. G., and Thompson, D. J. (1952). A generalization of sampling without replacement from a finite universe. *J. Am. Stat. Assoc.* **47:** 663–685.
Levy, P. S., and Lemeshow, S. (1991). *Sampling of Populations: Methods and Applications.* Wiley, New York.

Scheaffer, R. L., Mendenhall, W., and Ott, L. (1990). *Elementary Survey Sampling,* fourth ed. PWS-Kent, Boston.

Singh, D., and Chaudhary, F. S. (1986). *Theory and Analysis of Sample Survey Designs,* Wiley Eastern, New Delhi, India.

Skalski, J. R. (1985). Construction of cost functions for tag-recapture research. *Wildl. Soc. Bull.* **13:** 273–283.

Skalski, J. R. (1994). Estimating wildlife populations based on incomplete area surveys. *Wildl. Soc. Bull.* **22:** 192–203.

Stokes, S. L. (1986). Ranked set sampling. In *Encyclopedia of Statistical Sciences* (S. Kotz and N. L. Johnson, Eds.), Vol. 7, pp. 585–588. Wiley, New York.

Sukhatme, P. V., Sukhatme, B. V., Sukhatme, S., and Asok, C. (1984). *Sampling Theory of Surveys with Applications,* third ed. Iowa State Univ. Press, Ames, IA.

Thompson, S. K. (1992). *Sampling.* Wiley, New York.

Thompson, S. K., and Seber, G. A. F. (1996). *Adaptive Sampling.* Wiley, New York.

Appendix D

Common and Scientific Names
of Cited Vertebrates

Group	Common name	Scientific name
Fishes[1]	Burbot	*Lota lota*
	Crappie	
	Black	*Pomoxis nigromaculatus*
	White	*Pomoxis annularis*
	Dace, longnose	*Rhinichthys cataractae*
	Salmon	
	Chinook	*Oncorhynchus tshawytscha*
	Coho	*Oncorhynchus kisutch*
	Shad, gizzard	*Dorosoma cepedianum*
	Trout	
	Brown	*Salmo trutta*
	Colorado River cutthroat	*Oncorhynchus clarki*
	Cutthroat	*Oncorhynchus clarki*
	Rainbow	*Oncorhynchus mykiss*
Amphibians and reptiles[2]	Frog, northern leopard	*Rana pipiens*
	Hellbender	*Cryptobranchus alleghaniensis*
	Lizard, side-blotched	*Uta stansburiana*
	Racer	*Coluber constrictor*
	Rattlesnake, western	*Crotalus viridis*
	Salamander, Red Hills	*Phaeognathus hubrichti*
	Snake	
	Northern copperbelly water	*Nerodia erythrogaster*
	Western terrestrial garter	*Thamnophis elegans*
	Toad	
	Boreal	*Bufo boreas*
	Common	*Bufo bufo*
	Woodhouse's	*Bufo woodhousei*
	Tortoise	
	Desert	*Gopherus agassizii*
	Gopher	*Gopherus polyphemus*
	Turtle, painted	*Chrysemys picta*

(continues)

Group	Common name	Scientific name
Birds[3]	Bobwhite, northern	*Colinus virginianus*
	Brant, black	*Branta bernicla*
	Bunting, lark	*Calamospiza melanocorys*
	Chickadee	
	Black-capped	*Parus atricapillus*
	Mountain	*Parus gambeli*
	Crane	
	Sandhill	*Grus canadensis*
	Whooping	*Grus americana*
	Dove, mourning	*Zenaida macroura*
	Duck, wood	*Aix sponsa*
	Eagle	
	Bald	*Haliaeetus leucocephalus*
	Golden	*Aquila chrysaetos*
	Goshawk, northern	*Accipter gentilis*
	Grouse	
	Ruffed	*Bonasa umbellus*
	Sage	*Centrocercus urophasianus*
	Hawk, red-tailed	*Buteo jamaicensis*
	Heron, great blue	*Ardea herodias*
	Junco, dark-eyed	*Junco hyemalis*
	Kinglet, ruby-crowned	*Regulus calendula*
	Lark, horned	*Eremophila alpestris*
	Mallard	*Anas platyrhynchos*
	Meadowlark, western	*Sturnella neglecta*
	Osprey	*Pandion haliaetus*
	Owl	
	Boreal	*Aegolius funereus*
	Great horned	*Bubo virginianus*
	Mexican spotted	*Strix occidentalis*
	Northern spotted	*Strix occidentalis*
	Patridge, gray	*Perdix perdix*
	Pheasant, ring-necked	*Phasianus colchicus*
	Prairie-chicken	
	Greater	*Tympanuchus cupido*
	Lesser	*Tympanuchus pallidicinctus*
	Rail, yellow	*Coturnicops noveboracensis*
	Sparrow	
	Brewer's	*Spizella breweri*
	Chipping	*Spizella passerina*
	Turkey, wild	*Meleagris gallopavo*
	Vulture, turkey	*Cathartes aura*
	Warbler, yellow-rumped	*Dendroica coronata*
	Wren, house	*Troglodytes aedon*

(continues)

Group	Common name	Scientific name
Mammals[4]	Badger	*Taxidea taxus*
	Bear	
	Black	*Ursus americanus*
	Grizzly	*Ursus arctos*
	Bobcat	*Lynx rufus*
	Chipmunk, least	*Tamius minimus*
	Cottontail, eastern	*Sylvilagus floridanus*
	Coyote	*Canis latrans*
	Deer	*Odocoileus* spp.
	Mule	*Odocoileus hemionus*
	Red	*Cervus elaphus*
	White-tailed	*Odocoileus virginianus*
	Elephant, African	*Loxodonta africanus*
	Elk	*Cervus elaphus*
	Ferret, black-footed	*Mustela nigripes*
	Fox	
	Gray	*Urocyon cinereoargenteus*
	Kit	*Vulpes velox*
	Red	*Vulpes vulpes*
	Swift	*Vulpes velox*
	Hare, snowshoe	*Lepus americanus*
	Jackrabbit	*Lepus* spp.
	Kangaroo rat, Ord's	*Dipodomys ordii*
	Lynx	*Lynx lynx*
	Moose	*Alces alces*
	Mountain lion	*Felis concolor*
	Mouse	
	Deer	*Peromyscus maniculatus*
	Great Basin pocket	*Perognathus parvus*
	Harvest	*Reithrodontomys* spp.
	Northern grasshopper	*Onychomys leucogaster*
	Silky pocket	*Perognathus flavus*
	Opossum, Virginia	*Didelphis virginiana*
	Otter, river	*Lutra canadensis*
	Pig, feral	*Sus scrofa*
	Prairie dog, white-tailed	*Cynomys leucurus*
	Pronghorn	*Antilocapra americana*
	Raccoon	*Procyon lotor*
	Rhinoceros, black	*Diceros bicornis*
	Sheep, bighorn	*Ovis canadensis*
	Skunk, striped	*Mephitis mephitis*
	Squirrel	
	gray	*Sciurus carolinensis*
	thirteen-lined ground	*Spermophilus tridecimlineatus*

(*continues*)

Group	Common name	Scientific name
Vole		
	Meadow	*Microtus pennsylvanicus*
	Montane	*Microtus montanus*
	Wolf, gray	*Canis lupus*
	Wolverine	*Gulo gulo*

[1] Authority for scientific names is *Common and Scientific Names of Fishes from the United States and Canada,* fifth ed., American Fisheries Society, Bethesda, MD, 1990.

[2] Authority for scientific names of North American species is *Standard Common and Current Scientific Names for North American Amphibians and Reptiles* (Collins, Ed.), third ed., Univ. of Kansas, Lawrence, 1990. Authority for *Bufo bufo* is *Amphibian Species of the World: A Taxonomic and Geographic Reference* (Frost, Ed.), Allen Press, Lawrence, KS, 1985.

[3] Authority for scientific names is *The Check-List of North American Birds,* American Ornithologists' Union, Allen Press, Lawrence, KS, 1983.

[4] Authority for scientific names of North American species is *Revised Checklist of North American Mammals North of Mexico, 1991* (Jones *et al.,* Eds.), Occas. Pap. Mus. Texas Tech Univ. No. 146, Lubbock, TX, 1992. Authority for non-North American mammals is *Mammal Species of the World: A Taxonomic and Geographic Reference* (Wilson and Reeder, Eds.), second ed., Smithsonian Inst. Press, Washington, DC, 1993.

Index

A

Abundance, 2
Adaptive cluster sampling, 67–72
  incomplete detectability, 71–72
  networks, 68–69
  sample size, 71–72
  stratified, 68–71
Ad hoc methods, 77
Akaike's Information Criterion, 109–110
Allocation, *see* Stratified random sampling
Among-unit variation, 19–24, 174–176
Analysis of variance, 153

B

Badger, 312
Baseline research, 3–4
Bats, 309–311, 312
Bear, 306, 311, 313
  black, 304, 305, 311, 312
  grizzly (brown), 311, 312, 313
Bias, 38–41
Bobcat, 305, 313
Bobwhite, northern, 281, 287, 289
Bootstrapping
  defined, 34–35
  confidence intervals, 34–36
  percentile method, 34–35
  percentile-*t* method, 35
Bowden's estimator, 101, 312, 313
Breeding Bird Survey, 272
Brant, black, 274
Bunting, lark, 264

Burbot, 214

C

CAPTURE, *see also* Capture–recapture
  Model M_0, 90–91, 97
  Model M_b, 91, 98
  Model M_{bh}, 91, 99
  Model M_h, 91, 98–99
  Model M_t, 91, 97–98
  Model M_{tb}, 91, 99
  Model M_{tbh}, 91, 100
  Model M_{th}, 91, 99
  relationships among models, 91
  simulation procedure, 100
Capture–recapture
  amphibians and reptiles, 187–188,
    244–249
  assumptions, 84–88, 103–105
  birds, 277–285
  catch-per-unit-effort, 102–103, 216
  change-in-ratio, 103
  Chapman's estimator, 89–90, 280–282
  closed population, 3, 84–85
  fish, 200, 214–217
  Jolly–Seber model, 103–105, 215–216,
    248–249, 283–285
  Lincoln–Petersen estimator, 84–98,
    187–188, 215, 244–245, 278, 280–281,
    310
  mammals, 307–313
  mark-resight, 101–102, 246, 278, 280–283,
    307, 309, 311–313

Capture–recapture (continued)
  nested grid, 308
  open population, 3, 84, 85
  program CAPTURE, 90–100, 105, 220,
      246, 278, 280–281, 308, 313
  program EAGLES, 101–102
  program JOLLY, 104
  program MCAPTURE, 100–101
  program NOREMARK, 101, 246, 247,
      279, 309, 312–313
  program POPAN4, 104
  robust design, 84, 105–106
  trapping web, 245–246, 279, 308–309
Catch per effort (index), 213–214
Catch-per-unit-effort, 102–103, 216; *see also*
    Capture–recapture; Removal methods
Census
  amphibians and reptiles, 234–236
  birds, 262–265
  defined, 2, 75–76
  fish, 208, 210–212
  mammals, 301–303
Central Limit Theorem, 26
Change-in-ratio, 103; *see also*
    Capture–recapture; Removal methods
Chickadee
  black-capped, 283
  mountain, 288
Chipmunk, least, 176
CIR, *see* Change-in-ratio
Closure
  demographic, 3, 84, 85
  geographic, 3, 84, 85
Closed population, 3, 84, 85
Cluster sampling, *see* Adaptive cluster
    sampling
Coefficient of variation, 28, 112, 174,
    180–182
Community, 123
Community surveys
  abundance/density, 134–141
  objective, 123
  presence-absence, 125–139
  relative abundance/density, 139–140
  spatial distribution, 125–134
Confidence intervals
  bootstrapping, 34–36
  description, 28–37
  log-transformed, 95–97
  profile likelihood, 36, 96–97

Confidence level, 28, 29, 36
Cottontail, eastern, 305, 306, 309
Cougar, *see* Mountain lion
Counts, *see also* Survey
  lek, 262–263, 275, 280–281
  roost, 262–263, 274–275, 309–311, 312
  snorkel, 203–204, 210–211, 220
Coyote, 305, 306, 307, 311, 313
Crane, sandhill, 274
Crappie,
  black, 39
  white, 39
Cumulative sum, 156–157
CPUE, *see* Catch-per-unit-effort
CUSUM, *see* Cumulative sum
CV, *see* Coefficient of variation
Cycles, 146, 148

D

Degrees of freedom, 30
Deer, 302, 306, 314
  mule, 76, 302, 305, 306, 311, 313
  red, 313; *see also* Elk
  white-tailed, 302, 305
Density
  defined, 2
  formula, 18
  variance formula, 26
  versus number of species, 124–125
Detectability, 39–40, 45, 76–77, 268–271
Detection probability, *see* Detectability
DISTANCE, 109, 110, 111, 250, 252, 309
Distance sampling, 105–112
  amphibians and reptiles, 249–252
  birds, 285–290
  critical assumptions, 106–107, 251,
      285–287
  fish, 217–218
  heaping, 107
  line transects, 106–112, 285–287,
      289–290, 313–314
  mammals, 313–314
  point transects, 108–109, 112, 285–288
  program DISTANCE, 109, 110, 111, 250,
      252, 309
  sighting function, 106–109
  smearing, 107
  variable circular plots, 108, 287
Distribution
  binomial, 93–94

continuous, 32
discrete, 32
empirical, 34
external, 33
internal, 34
negative binomial, 37–38, 175–176, 273
normal, 26–32
Poisson, 37, 112, 175–176
reference, 33
spatial, 1–2
standard normal, 27, 28, 30, 31
Double sampling, *see* Sampling
Dove, mourning, 268

E

Eagles
bald, 262, 283
golden, 289–290
EAGLES, 102
Edge effect, 44–47, 245–246, 278–279, 308
Electrofishing, 205–207, 209–210
Electroshocking, 87, 239
Element, 6–7
Elephant, African, 313
Elk, 39, 302, 305, 313
EMAP, 5–6
Encounter sampling, 9
Enumeration variation, 20, 24, 173–174
Error
coverage, 38–39
defined, 19
nonresponse, 39–40
nonsampling, 38–41, 132–134
relative, 58, 135–136
response, 39
sampling, 38
Type I, 159–161, 180, 182
Type II, 159–161, 180, 182
Estimate, 11
Estimator, 11
Expected value, 38

F

Finite population correction, 25–26
Floaters, 273, 278
Fox
gray, 305
kit, 314

red, 305
swift, 309, 311, 314
Frog, northern leopard, 247–248

G

Gear selectivity, 87
Goshawk, northern, 274
Grouse
prairie, 275
ruffed, 268
sage, 145
Guild, 139

H

Hare, snowshoe, 306, 309
Hawk, red-tailed, 282
Heaping, 107
Hellbender, 239
Heron, great blue, 274
Horvitz-Thompson estimator, 49, 59
Hydroacoustics, 207–208, 211–212
Hypothesis testing,
alternative hypothesis, 159
defined, 159
null hypothesis, 159, 180
statistical power, 159–163, 180–186
Type I error, 159, 161, 180, 182
Type II error, 159–161, 180, 182

I

IBI, *see* Index of biotic integrity
Icthyocides, 202–203
Index
assumptions, 77–83
defined, 77
frequency, 77, 125
methods, 77–83
presence-absence, 78–80, 125–134,
212–213, 236–240, 266–267, 304
relative abundance or density, 77–78,
80–83, 213–214, 240–244, 267–277,
304–307
Index of biotic integrity, 139, 142

J

Jackrabbit, 306
JOLLY, 104
Jolly–Seber model, *see* Capture–recapture
Junco, dark-eyed, 39, 288

K

Kangaroo rat, Ord's, 176
Kinglet, ruby-crowned, 288
Kriging, 207

L

Lark, horned, 264
Lek counts, 262–263, 275, 280–281
Lincoln–Petersen estimator, *see* Capture-recapture
Line transects, *see also* Distance sampling
    Emlen, 271
    Finnish, 271
Line-intercept sampling, *see* Sampling
Lizard, side-blotched, 251–252
Lynx, 314

M

Mallard, 280
Mann–Kendall test, 165–168
Mapping
    spot, 273
    territory, 273
    total, 263–265
MAPS, 284–285
Mark-resight, *see* Capture-recapture
Maximum likelihood estimator, 93–97
MCAPTURE, 100–101
Meadowlark, western, 264
Minimum number known alive, 83, 95
MNA, *see* Minimum number known alive
Model
    exponential, 158, 160, 180–182, 187–188
    linear, 157–160, 180–182
    multiplicative, 158, 160–161, 180–182, 187–188
MONITOR, 186
Monitoring
    assessment, 3
    baseline, 3–4
    defined, 3
    effectiveness, 6
    implementation, 6
    index, 5, 201–202
    inferential, 5–6
    inventory, 3
    objectives, 172

planning guidelines, 171–188
population, 4–6
Moose, 39, 76, 302, 306, 311
Mountain lion, 186, 304, 305, 315
Mouse
    deer, 176
    Great Basin pocket, 176
    harvest, 176
    northern grasshopper, 176
    silky pocket, 176

N

Network, *see* Adaptive cluster sampling
Neyman allocation, 56–57
Nonparametric methods, 165–168
Nonresponse error, 39–40
Nonsampling error, 38–41, 132–134
NOREMARK, 101–102, 246, 247, 279, 309, 312–313
Normal deviate, 27

O

Open population, 3, 84, 85
Opossum, Virginia, 305, 309, 312
Optimum allocation, 56–58
Osprey, 281
Otter, river, 313
Owl
    boreal, 276
    great horned, 267, 274
    Mexican spotted, 149–151
    northern spotted, 284

P

Parameter, 2
Parameter estimate, 18
Partridge, gray, 82, 268
Patch sampling, 235, 240
Percentile method, *see* Bootstrapping
Percentile-*t* method, *see* Bootstrapping
Pheasant, ring-necked, 268
Pig, feral, 306
Pilot study, 111, 174–175, 198, 220–223
Plot
    design, 44–48, 128, 173–174
    permanent, 177–179, 201–202
    program QUADRAT, 48
    reselection, 177–179

selection, 51–72, 128–130, 176–177
shape, 44–46
size, 46–48
temporary, 177–179, 201
unequal sizes, 48–49
Point counts (birds), 271–273
Point estimate, 18
Point transect sampling, *see* Distance
    sampling
POPAN4, 104
Population distribution, 24–25
Population monitoring, 4–6
Population viability analysis, 152–153
Power, *see* Hypothesis testing
Prairie-chicken
    greater, 280
    lesser, 280
Prairie dog, white-tailed, 172
Precision
    defined, 19
    rigorously quantifying, 24–37
Presence-absence, *see also* Index
    amphibians and reptiles, 236–240
    birds, 266–267
    defined, 77–80, 125–126
    fish, 212–213
    mammals, 304
    nonlinearity of index, 78–79
Principle of parsimony, 109
Process variation, 20, 149
Profile likelihood, *see* Confidence intervals
Pronghorn, 7, 302–303, 313
Proportional allocation, 55–56

Q

QUADRAT, 48
Quadrat sampling, 76, 208, 210, 234–235,
    240–241, 302–303

R

Rabbit, *see* Cottontail, eastern
Raccoon, 305, 306, 309, 312, 313
Racer, 249–250
Rail, yellow, 235, 265
Random distribution, *see* Distribution,
    Poisson
Randomization methods, 163–165, 306
Ranked set sampling, 64–67

Regression
    assumptions, 159
    exponential, 158, 160, 180–182, 187–188
    linear, 157–160, 180–182
Relative abundance, *see* Index
Relative density, *see* Index
Relative error, 58, 135–136
Removal methods, *see also*
    Capture–recapture
    amphibians and reptiles, 248
    catch-per-unit-effort, 102–103, 216
    change-in-ratio, 103
    defined, 90, 102–103
    fish, 200, 216–217
Rhinoceros, black, 313
Roost counts, *see also* Counts
    bats, 309–311, 312
    birds, 262–263, 274–275, 309–311, 312
Rotenone, 202–203, 208, 209

S

Salamander, Red Hills, 250
Sample, 10
Sampled population, 10
Sampling
    adaptive cluster, 67–72
    complete remeasurement, 177
    complete replacement, 177
    convenience, 12
    design, 43–44, 49
    distribution, 22–23, 26–37
    double, 65, 83, 115–117
    encounter, 9
    error, 38
    fraction, 26
    frame, 7–10, 172–173, 192–193
    haphazard, 11
    multistage, 51, 192–194
    nonrandom, 11–13
    one-stage, 49–50
    partial replacement, 177, 178, 179
    plan, 43–44
    probability proportional to size, 13
    purposive, 11
    random, 11, 13–19
    ranked set, 64–67
    rare populations, 67–72
    simple Latin square sampling +1, 63–64
    simple random, 14, 21–23, 51–52

Sampling (continued)
　stratified random, 52–59
　systematic, 59–63
　three-stage, 51
　two-phase, 115–116
　two-stage, 50–51
　two-stage cluster, 50–51
　unit, 7
　variation, 19–20, 49, 149, 152, 153–156
Sampling with partial replacement, 177, 178, 179
Scent station index, 305–306
Selection
　bias, 12, 38
　probability, 22–23
Sheep, bighorn, 311, 312
Sightability models, 117
Simple Latin square sampling +1, 63–64
Simple random sampling
　defined, 14–15
　example, 16–19
　with replacement, 14
　without replacement, 14
Skunk, striped, 309
Smearing, 107
Snake,
　garter, 246
　northern copperbelly water, 250
Snorkel counts, 203–204, 210–211, 220
Sparrow
　Brewer's, 264
　chipping, 39
Spatial autocorrelation, 60, 63, 64
Spatial distribution, 1–2, 125–134
Spatial variation, 19, 20, 149, 156
Squirrel
　gray, 309, 314
　thirteen-lined ground, 176
SSI, *see* Scent station index
Standard deviation, 26, 27, 28
Standard error, 27, 28, 30
Stratification, 52–53, 193–195
Stratified random sampling
　computing sample size, 56–58
　cost function, 58–59
　defined, 52–53
　Neyman allocation, 56–57
　optimum allocation, 56–58
　proportional allocation, 55–56
Strip transect, 235, 241–242, 265, 314

Survey
　carcass, 314
　community, 123–142
　defined, 2
　frequency, 179
　multispecies, 123
　planning guidelines, 171–188
　single species, 123
Systematic sampling
　defined, 59–60
　multiple random starts (repeated), 60–63
　periodicity, 60, 63
　single random start, 59–60, 62
　spatial autocorrelation, 60, 63

T

t value, 30–33, 58
Target population, 7, 126, 172–173, 192–193
Taylor's power law, 176
Temporal variation, 19–20, 152, 153, 154–155
Toad
　boreal, 187–188
　Woodhouse's, 89
Tortoise
　desert, 235
　gopher, 250–251
Trap
　happiness, 87, 278
　shyness, 87, 278
Trapping
　selectivity, 87, 246
　web, 245–246, 279, 308–309
Trend
　defined, 4
　graphical methods, 156–157
　program MONITOR, 186
　program TRENDS, 181–188
　regression methods, 157–163
　tests, 159–161, 163–168
　types, 146–148
TRENDS, 181–188
Trout
　brown, 7
　Colorado River cutthroat, 192, 218–223
Turkey, wild, 266, 275, 284
Turtle, painted, 238

V

Variable circular plot, *see* Distance sampling

Variance
  components, 19–24, 149–156
  defined, 24
  population, 24
  total, 154–155
  weighted total, 154–155
Variation
  among-unit, 19–24, 174–176
  components, 19–24, 149–156
  enumeration, 20, 24, 173–174
  process, 20, 149
  sampling, 19–20, 149, 152, 153–156
  spatial, 19, 20, 149, 156
  temporal, 19–20, 152–155

Vole, montane, 176
Vulture, turkey, 262

W

Warbler, yellow-rumped, 288–289
Wolf, gray, 306–307, 315
Wolverine, 114–115, 252, 314
Wren, house, 272, 288

Z

z score, 27
z test, 149–150
z value, 27–31, 180–182